Computer-Assisted
Floodplain Hydrology
and Hydraulics

Computer-Assisted Floodplain Hydrology and Hydraulics

Daniel H. Hoggan

Professor, Department of Civil and Environmental Engineering
and the Utah Water Research Laboratory
Utah State University

McGraw-Hill Publishing Company

New York   St. Louis   San Francisco   Auckland
Bogotá   Hamburg   London   Madrid   Mexico
Milan   Montreal   New Delhi
Paris   São Paulo   Singapore
Sydney   Tokyo   Toronto

Library of Congress Cataloging-in-Publication Data

Hoggan, Daniel H.
    Computer-assisted floodplain hydrology and hydraulics / Daniel H.
Hoggan.

    p.   cm.
    Includes bibliographies and index.
    ISBN 0-07-029350-3
    1. Hydrology—Data processing.   I. Title.
GB656.2.E43H64 1989
551.48'9'0285—dc19

88-29028
CIP

1234567890 DOC/DOC 89432109

ISBN 0-07-029350-3

*The editors for this book were Joel Stein and Laura Givner, the
designer was Naomi Auerbach, and the production supervisor was
Richard A. Ausburn. It was set in Century Schoolbook and composed
by the McGraw-Hill Publishing Company Professional & Reference
Division composition unit.*

Printed and bound by R. R. Donnelley & Sons Company.

*For more information about other McGraw-Hill materials,
call 1-800-2-MCGRAW in the United States. In other
countries, call your nearest McGraw-Hill office.*

*My greatest love and appreciation are expressed
to my family—R'lene, Sharlene, Stuart, Terri, and Todd—
for their loving support and patience
during the many hours it took
to complete this work.*

Contents

Figure and Table Credits   xv
Preface   xix

Chapter 1. Analysis of Floodplain Hydrology and Hydraulics   1

   Introduction   1
      The Origin and Nature of Floods   1
      Purposes of Floodplain H&H Analysis   2
      Use of Computer Simulations   2
      Methods of Analysis   2
   Part 1: Basin Hydrology Simulation   3
      Precipitation-Runoff Process   3
      The Discharge Hydrograph   4
   Part 2: Frequency Analysis   6
      Methods for Determining Frequency Curves   6
   Part 3: Water Surface Profile Computations   9
   Application of H&H Analytical Techniques   10
      Illustration of Flood Damage Determination   10
      Guidelines for an H&H Study to Determine Discharge Frequency   11
   Simulation Programs for H&H Analysis   13
      General   13
      Hydrologic Engineering Center Computer Programs 13
   References   16

Chapter 2. Rainfall and Rainfall Loss Analysis   19

   Rainfall Analysis   19
      Basin Rainfall   19
      Spatial Distribution of Rainfall   19
      Temporal Distribution of Rainfall   22
   Rainfall Loss Analysis   24
      Introduction   24
      Loss Functions in HEC-1   26
   Workshop Problem 2.1   Basin Rainfall and Rainfall Loss Analysis for the Rahway
                          River Basin   39
   References   44

Chapter 3. Unit Hydrograph Analysis and Base Flow   45

Unit Hydrograph Analysis   45
  Unit Hydrograph Concepts   45
  Synthetic Unit Hydrographs   47
Base Flow   60
Workshop Problem 3.1   SCS Loss Rate and Clark Unit Hydrograph   62
References   67

Chapter 4. Flood Routing Analysis   69

Introduction   69
  Concept   69
  Purposes   69
  Methods   70
Hydrologic Routing Methods in HEC-1   72
  Modified Puls Method   72
  Muskingum Method   82
  Applicability of Routing Methods   89
Workshop Problem 4.1   Flood Routing in the Rahway River Basin   90
References   92

Chapter 5. The HEC-1 Flood Hydrograph Package and Parameter Estimation   95

Overview of the HEC-1 Program Capabilities   95
  Introduction   95
  Choice of Metric or English Units   95
  Data Storage System   96
  Basic Modeling Capability   96
  Major Program Options   97
  Structuring the Model   99
Data Input Requirements   100
  Model Parameters   100
  Data Input File   100
  Input File Organization   101
  Data Input Options   105
  Example Input File   106
Program Output   109
  General   109
  Input File Listing   109
  Intermediate Simulation Results   109
  Summary Output   111
  Example Output   111
Parameter Estimation   116
  Introduction   116
  Parameter Estimation in HEC-1   117
  HEC-1 Input for Parameter Estimation   121
  Output   122
  Guidelines for Estimating Parameters   125
  Estimating Routing Parameters   127
Workshop Problem 5.1   Developing a Single-Basin Model for the Rahway River
                        above the Gage near Springfield   127
Workshop Problem 5.2   Parameter Estimation, Rahway River Basin   130
References   134

Chapter 6. River Basin Modeling with HEC-1   135

Modeling Concepts   135
  Introduction   135
  Model Components   135
  Subdividing a Basin   136
  Estimation of Model Parameters   139
Developing Input for a River Basin Model   140
  Introduction   140
  Input for Hydrologic Components   140
  Computational Sequence in HEC-1 (the stack principle)   144
  Example of a River Basin Model   144
Regionalization of Unit Hydrograph and Loss Function Parameters   153
  Introduction   153
  Basic Steps in Regional Analysis   153
  Basin Characteristics Used in Regionalization   158
  Correlation Techniques   158
Workshop Problem 6.1   Regionalization of Unit Hydrograph Parameters in the
                        Region of the Rahway River Basin   164
Workshop Problem 6.2   Development of a Multiple-Subbasin Model for the Rahway
                        River Basin   166
References   171

Chapter 7. Frequency Analysis   173

Introduction   173
  Flood Risk   173
  Deterministic vs. Stochastic Models   173
  Paired Data vs. Times-Series Data   174
Fundamentals of Frequency Analysis   174
  Frequency Concepts   174
  Frequency Analysis   177
  Data Required for Frequency Analysis   182
Streamflow Frequency Analysis   186
  Annual Series vs. Partial-Duration Series   186
  Factors Affecting Homogeneity of Data   187
  Frequency Analysis Techniques   188
  Log Pearson Type III Distribution   193
  Advantages and Disadvantages of the Analytic Method   203
Frequency Analysis by Computer   203
  Computer Program HECWRC   203
  Flood Volume Frequency Analysis with the Computer Program STATS   205
Frequency Analysis of Ungaged Basins   205
  General   205
  Selection of a Technique   207
  Field Reconnaissance at Ungaged Locations   211
Workshop Problem 7.1   Frequency Analysis of Streamflow Record for the Gage
                        near Springfield (3945)   211
References   213

Chapter 8. Use of Design Storms in Rainfall-Runoff Simulation and Frequency Analysis   215

Introduction   215
  Design Storm Data   215
  Procedures for Developing a Design Storm   219
  Depth-Area Option of HEC-1   225

Development of Discharge-Frequency Curves Using Design Storms   234
 Design Storm Frequency vs. Design Flood Frequency   234
Probable Maximum Storm   235
 Concept   235
 Probable Maximum Precipitation   236
 Probable Maximum Storm Determination with Computer Program HMR52   237
Workshop Problem 8.1   Developing Design Storms and Using the HEC-1
                        Depth-Area Option   244
References   248

Chapter 9. Analysis of Urbanizing Basins   249

Effects of Urbanization on Runoff   249
 Introduction   249
 Accounting for the Effects of Urbanization in Rainfall-Runoff Analysis   249
 Effects of Urbanization on Routing   253
Kinematic-Wave Rainfall-Runoff Modeling   254
 Introduction   254
 Kinematic-Wave Equations   255
 Elements Used in Kinematic-Wave Computations in HEC-1   257
 Example Problem: Kinematic-Wave Application to Waller Creek   264
 Selection of Computational Increments Δx and Δt 268
Development of Frequency Curves in Urbanizing Areas   270
 Effects of Urbanization on Frequency Curves   270
 Methods of Developing Frequency Curves in Urbanizing Areas   270
Multiplan-Multiflood Analysis   275
 Multiflood Ratios   275
 Multiplan Analysis   277
Workshop Problem 9.1   Multiplan-Multiflood Analysis of the Rahway River
                        Basin   279
References   282

Chapter 10. Water Surface Profile Analysis   283

Introduction   283
 Purposes   283
 Computation of Profiles with HEC-2   283
Theoretical Basis and Limiting Assumptions in HEC-2   284
 General   284
 Classifications of Open Channel Flow   284
 Critical Depth and Its Significance   288
 Velocity Distribution in a Cross Section   292
 Pressure Distribution in a Channel Cross Section   295
Water Surface Profile Computations   296
 Energy Principles   296
 Determining Friction Loss in a Channel Reach   299
 Computing Water Surface Profiles   301
Workshop Problem 10.1   Standard Step Hand Calculations for the Red Fox
                         River   310
References   316

Chapter 11. Water Surface Profiles Program HEC-2   317

Program Capabilities   317
 General   317
 Basic Options   318

Data Requirements for Water Surface Profile Computations   321
  Discharge   322
  Flow Regime   322
  Starting Water Surface Elevation   322
  Roughness Coefficient and Other Energy Loss Coefficients   324
  Development of the Geometric Model   335
  Checking of Input Data Prior to the Preparation of an Input File   340
The HEC-2 Data Input File   344
  Data Organization and Format   344
  Example Problem: Development of a Data Input File for HEC-2   345
  Checking an Input File for Errors   351
Workshop Problem 11.1   Water Surface Profile Computations with HEC-2   352
References   353

Chapter 12. Review of Computed Water Surface Profiles   355

Introduction   355
Description of HEC-2 Program Output   355
  Standard Output   355
  Optional Output   360
Output Analysis   362
  Changes in Key Variables That May Reflect Errors in Input   362
  Example Problem: Analysis of Output   363
Project Review   366
  Steps   369
  Questions   369
  Verification   374
  Sensitivity of Data   375
Workshop Problem 12.1   Output Analysis—Four Case Studies   376
References   382

Chapter 13. Water Surface Profiles through Bridges   383

Introduction   383
  General   383
  Nature of Flow through Bridges   384
  Classes of Low Flow through Bridges   385
  Other Flow Conditions at Bridges   386
The Normal-Bridge Method   387
  Defining the Bridge Structure   387
  Guidelines for Locating Bridge Cross Sections   388
  Example Problem: Culvert Analysis   390
  Optional Method of Defining a Bridge with X2 and GR Lines   396
  Output for the Normal-Bridge Method   397
The Special-Bridge Method   397
  Computational Theory and Methodology   397
  Location of Cross Sections for Special-Bridge Models   406
  Example Problem: Special-Bridge Method   407
  Output for the Special-Bridge Method   411
Externally Computed Bridge Losses   412
Selection of a Method for Bridge Analysis   413
  General   413
  Bridge on Fill   414
  Low-Profile Bridge   414
  High Bridge   414
  Complex and Multiple Bridge Openings   414
  Culverts   417
  Bridges That Are a Minor Obstruction to Flow   417

Bridges under High Submergence   417
Dams and Weirs   417
Perched Bridges   418
Low-Water Bridges   418
Skewed Bridges   418
Parallel Bridges   419
Workshop Problem 13.1   Special-Bridge Method   419
Workshop Problem 13.2   Normal-Bridge Method   421
References   422

Chapter 14. Floodway and Channel Improvement Analyses   423

Floodway Determination   423
Introduction   423
Guidelines for Establishing a Floodway   423
Modeling Procedures   424
Undulating Flow Boundaries   425
Analysis of Existing Conditions   426
Determinations of a Floodway with HEC-2 Encroachment Methods   427
Example Problem: Floodway Determination   433
Problems Encountered in Floodway Determination   440

Channel Improvement   445
Introduction   445
The CI Line   445
Multiple Channel Improvements   448
Modification of Channel Reach Length   451
Example Problem: Channel Improvement   451
Volume of Excavation   453
Use of the Channel Improvement (CHIMP) Option with Other Options   454
Workshop Problem 14.1   Floodway Determination on North Buffalo Creek   455
Workshop Problem 14.2   Red Fox River Channel Improvement   458
References   461

Chapter 15. Supercritical Flow, Split Flow, and Other HEC-2 Options   463

Supercritical-Stream Analysis   463
The Nature of Supercritical Flow   463
Supercritical-Flow Analysis with HEC-2   465

Split-Flow Analysis   469
Introduction   469
Program Capabilities   471
Computational Procedure   476
Program Limitations   476
Input Requirements   477
Modeling Procedures   477
Example Problem: Red Fox River Split Flow   481

Analysis of Ice-Covered Streams   483
Introduction   483
Hydraulic Parameters   484
Ice Jams   487
Example Problem: Ice Stability Analysis   488
Water Surface Profile Analysis of Ice-Covered Streams with HEC-2   490

Other Options in HEC-2   492
Tributary Stream Profile Computations   492
Determination of Manning's n   493
Computation of Storage-Outflow Data   495

Workshop Problem 15.1   Yuba River Split-Flow Analysis   497
References   503

Appendix: Conversion Factors   505

Index   507

Figure and Table Credits

Chapter 1

Figures 1.4 and 1.7 Hydrologic Engineering Center, *Hydrologic Engineering in Planning,* Training Document 14, U.S. Army Corps of Engineers, Davis, Calif., April 1981.

Chapter 2

Figure 2.6 Hydrologic Engineering Center, *Hydrologic Engineering in Planning,* Training Document 14, U.S. Army Corps of Engineers, Davis, Calif., April 1981.

Figure 2.7 Hydrologic Engineering Center, *HEC-1 Flood Hydrograph Package, Users Manual,* U.S. Army Corps of Engineers, Davis, Calif., 1981 (rev. 1987).

Figures 2.8–2.10; Table 2.4a–d Soil Conservation Service, *Urban Hydrology for Small Watersheds,* Technical Release No. 55, U.S. Department of Agriculture, January 1975 (rev. June 1986).

Chapter 3

Figure 3.2 Hydrologic Engineering Center, *Hydrologic Engineering in Planning,* Training Document 14, U.S. Army Corps of Engineers, Davis, Calif., April 1981.

Figure 3.3 *Hydrologic Engineering Methods for Water Resources Development,* vol. 4: *Hydrograph Analysis,* U.S. Army Corps of Engineers, Davis, Calif., October 1973.

Figures 3.6 and 3.7; Table 3.2 Soil Conservation Service, *National Engineering Handbook,* Sec. 4: *Hydrology,* U.S. Department of Agriculture, March 1985.

Figure 3.8 Hydrologic Engineering Center, *HEC-1 Flood Hydrograph Package, Users Manual,* U.S. Army Corps of Engineers, Davis, Calif., 1981 (rev. 1987).

Chapter 4

Figures 4.5–4.7 Hydrologic Engineering Center, *Hydrologic Engineering in Planning,* Training Document 14, U.S. Army Corps of Engineers, Davis, Calif., April 1981.

Figure 4.11 Hydrologic Engineering Center, *HEC-1 Flood Hydrograph Package, Users Manual,* U.S. Army Corps of Engineers, Davis, Calif., 1981 (rev. 1987).

Figure 4.14 *Routing of Floods through River Channels,* U.S. Army Corps of Engineers Engineering Manual EM 1110-2-1408, March 1960.

Chapter 5

Figures 5.1 and 5.3; Tables 5.1–5.5 Hydrologic Engineering Center, *HEC-1 Flood Hydrograph Package, Users Manual,* U.S. Army Corps of Engineers, Davis, Calif., 1981 (rev. March 1987).

Chapter 6

Figures 6.9–6.19 Hydrologic Engineering Center, *HEC-1 Flood Hydrograph Package, Users Manual,* U.S. Army Corps of Engineers, Davis, Calif., 1981 (rev. March 1987).

Figure 6.20 Hydrologic Engineering Center, *Regional Frequency Study, Upper Delaware and Hudson River Basins, New York District,* U.S. Army Corps of Engineers, Davis, Calif., November 1974.

Tables 6.1 and 6.2 Hydrologic Engineering Center, *Hydrologic-Hydraulic Simulation,* Special Projects Memo 469, U.S. Army Corps of Engineers, Davis, Calif., November 1976.

Chapter 7

Figure 7.12; Tables 7.1–7.5 "Hydrologic Frequency Analysis," engineer manual, U.S. Army Corps of Engineers, Apr. 1, 1985. (Draft.)

Figure 7.14 Interagency Advisory Committee on Water Data, *Guidelines for Determining Flood Flow Frequency,* Bulletin 17B, U.S. Geological Survey, Reston, Va., 1982.

Chapter 8

Figure 8.1 National Weather Service, *Rainfall Frequency Atlas of the United States, 30-Minute to 24-Hour Durations, 1- to 100-Year Return Periods,* Technical Paper No. 40, U.S. Department of Commerce, 1961.

Figure 8.4 National Weather Service, *Five- to 60-Minute Precipitation Frequency for the Eastern and Central United States,* NOAA Technical Memorandum NWS HYDRO-35, Silver Spring, Md., 1977.

Figures 8.2, 8.3, 8.5, and 8.7 Hydrologic Engineering Center, *Hydrologic Analysis of Ungaged Watersheds, Using HEC-1,* Training Document 15, U.S. Army Corps of Engineers, Davis, Calif., 1982.

Figure 8.6 *Precipitation Frequency Atlas of the Western United States, Atlas 2,* vols. 1–13, National Oceanic and Atmospheric Administration, Washington, D.C., 1973.

Figure 8.8 Hydrologic Engineering Center, *HEC-1 Flood Hydrograph Package, Users Manual,* U.S. Army Corps of Engineers, Davis, Calif., 1981 (rev. 1987).

Figure 8.15 *Probable Maximum Precipitation Estimates, United States East of the 105th Meridian,* HMR No. 51, National Oceanic and Atmospheric Administration, Washington, D.C., 1978.

Figures 8.16–8.21 Hydrologic Engineering Center, *HMR52 Probable Maximum Storm (Eastern United States), Users Manual,* U.S. Army Corps of Engineers, Davis, Calif., 1984.

Chapter 9

Figures 9.3–9.6 Hydrologic Engineering Center, *Introduction and Application of Kinematic Wave Routing Techniques Using HEC-1,* Training Document 10, U.S. Army Corps of Engineers, Davis, Calif., May 1979.

Figure 9.8; Table 9.3 Espey, Houston and Associates, *Comprehensive Drainage Study, Waller Creek Drainage Basin,* City of Austin, Tex., 1976.

Figure 9.11 Gary W. Brunner, "Cascade vs. Single-Plane Overland Flow Modeling," master's thesis, Pennsylvania State University, University Park, May 1985.

Table 9.4 R. J. Cermak, *Application of HEC-1 Kinematic Wave,* U.S. Army Corps of Engineers, Hydrologic Engineering Center, Davis, Calif., 1981.

Chapter 10

Figures 10.3 and 10.9; Table 10.1 Ven T. Chow, *Open Channel Hydraulics,* McGraw-Hill, New York, 1959. Reproduced with permission.

Figures 10.6 and 10.7 David H. Schoellhamer, "Calculation of Critical Depth and Subdivision Froude Number in HEC-2," master's thesis, University of California, Davis, 1982.

Figure 10.8 Hydrologic Engineering Center, *HEC-2 Water Surface Profiles, Users Manual,* U.S. Army Corps of Engineers, Davis, Calif., 1982.

Figures 10.16–10.18 and 10.20 Hydrologic Engineering Center, *Water Surface Profiles,* vol. 6: *Hydrologic Engineering Methods for Water Resource Development,* U.S. Army Corps of Engineers, Davis, Calif., 1975.

Figures P10.1–P10.4 Hydrologic Engineering Center, *Water Surface Profiles,* vol. 6: *Hydrologic Engineering Methods for Water Resource Development,* U.S. Army Corps of Engineers, Davis, Calif., 1975, and *Guide for Computing Water Surface Profiles,* U.S. Bureau of Reclamation, November 1957.

Chapter 11

Figure 11.1; Table 11.1 Hydrologic Engineering Center, *HEC-2 Water Surface Profiles, Users Manual,* U.S. Army Corps of Engineers, Davis, Calif., 1982.

Figures 11.3 and 11.4 Hydrologic Engineering Center, *Accuracy of Computed Water Surface Profiles,* Research Document 26, U.S. Army Corps of Engineers, Davis, Calif., 1986.

Figure 11.5 Jacob Davidian, "Computation of Water-Surface Profiles in Open Channels," Chap. A15 in *Techniques of Water-Resources Investigations of the United States Geological Survey,* Book 3: *Applications of Hydraulics,* U.S. Department of the Interior, 1984.

Tables 11.2 and 11.3 Ven T. Chow, *Open Channel Hydraulics,* McGraw-Hill, New York, 1959. Reproduced with permission.

Chapter 12

Tables 12.1–12.3 Hydrologic Engineering Center, *HEC-2 Water Surface Profiles, Users Manual,* U.S. Army Corps of Engineers, Davis, Calif., 1982.

Chapter 13

Figure 13.1 Emmett M. Laursen, "Bridge Backwater in Wide Valleys," *Journal of the Hydraulics Division,* ASCE, vol. 96, no. HY4, April 1970.

Figures 13.2, 13.11, and 13.25 Joseph N. Bradley, "Hydraulics of Bridge Waterways," *Hydraulic Design Series No. 1,* 2d ed., Bureau of Public Roads (now the Federal Highway Administration), 1970, rev. March 1978.

Figure 13.12 Bill S. Eichert and John Peters, "Computer Determination of Flow through Bridges," *Journal of the Hydraulics Division,* ASCE, vol. 96, no. HY7, July 1970.

Table 13.1 Hydrologic Engineering Center, *HEC-2 Water Surface Profiles, Users Manual,* U.S. Army Corps of Engineers, Davis, Calif., September 1982.

Chapter 14

Figures 14.1 and 14.19 Hydrologic Engineering Center, *Floodway Determination Using Computer Program HEC-2,* Training Document 5, rev. ed., U.S. Army Corps of Engineers, Davis, Calif., January 1988.

Figures 14.4, 14.6, 14.8, 14.10, and 14.20; Table 14.1 Hydrologic Engineering Center, *HEC-2 Water Surface Profiles, Users Manual,* U.S. Army Corps of Engineers, Davis, Calif., September 1982.

Chapter 15

Figure 15.1 Ven T. Chow, *Open Channel Hydraulics,* McGraw-Hill, New York, 1959. Reproduced with permission.

Figures 15.9–15.14 Hydrologic Engineering Center, *Application of the HEC-2 Split Flow Option,* Training Document 18, U.S. Army Corps of Engineers, Davis, Calif., April 1982.

Figure 15.17 Ernest Pariset, Rene Hausser, and Andre Gagnon, "Formation of Ice Covers and Ice Jams in Rivers," *Journal of the Hydraulics Division,* ASCE, vol. 92, no. HY6, November 1966.

Tables 15.1 and 15.2 *Ice Engineering,* EM 1110-2-1612, U.S. Army Corps of Engineers, Office of the Chief of Engineers, October 1982.

Preface

Purpose

This book is intended as a comprehensive technical reference and as a text for teaching the analysis of floodplain hydrology and hydraulics (H&H) using practical, state-of-the-art computer modeling techniques developed by the Hydrologic Engineering Center (HEC) of the U.S. Army Corps of Engineers. It is designed to be used both in short courses for professionals and in university courses on surface water hydrology and open channel hydraulics. The book is intended to be used in conjunction with HEC program user's manuals. Sufficient detail and example problems are provided to enable individuals to learn, through self-study, the H&H analytic techniques and the simulation models employed.

The book focuses on H&H analysis, typified by floodplain information studies, which generally consist of three major components: basin hydrology simulation, flood-frequency analysis, and water surface profile computations. It covers a broad range of problems encountered in this type of analysis and explains the techniques used to solve them. The major topics addressed include the development of single- and multiple-basin rainfall-runoff models, the determination of discharge-frequency curves through the use of historical data and/or design storms, the analysis of ungaged basins and urbanizing basins, and the development of models for computing water surface profiles, including the effects of bridges, encroachments, and channel improvements.

Organization

The book is designed with flexibility to meet the needs of readers with a variety of interests. An engineering investigation or a course of study may focus on all or only a part of a complete floodplain information study. Chapter 1 gives an overview of a complete study, showing how the different parts fit together, and indicates computer programs that may be applied. Chapters 2 through 6 cover the basics of basin hydrology simulation, including regional analysis for estimating model parameters for ungaged basins.

Chapters 7 through 9 cover frequency analysis of historical data and methods to determine frequency curves in ungaged basins and urbanizing areas. Design storms and kinematic-wave modeling techniques are included. Chapters 10 through 15 deal with water surface profile computations, beginning with the basics and continuing through normal- and special-bridge applications, encroachments, channel improvements, and options for analyzing supercritical flow, split flow, and ice-covered flow. A reader may draw from any or all of these parts in developing a course or solving an engineering problem.

The scope and organization of the book are patterned to a large extent on training courses given by the Hydrologic Engineering Center. Basic material on basin hydrology simulation (Chaps. 2 through 6) is normally covered in a 1-week (40-hour) course. An expanded course, including the basics plus frequency analysis and other topics found in Chaps. 7 through 9, would normally be given in a 2-week period. Water surface profile analysis, the subject of Chaps. 10 through 15, may be covered in one or two courses. Basic water surface profile computations are normally taught in a 1-week course. Special topics, such as supercritical flow, split flow, and ice-covered flow, may be combined with some of the more advanced aspects of the basic topics in a separate 1-week course. Basic basin hydrology simulation combined with basic water surface profile analysis is another alternative that can be covered in a 2-week course.

There appears to be an increasing interest at universities in giving students an opportunity to learn practical simulation programs such as HEC-1 and HEC-2. Several universities are already offering courses that include these two programs. Selected topics from this book could be used as parts of general courses; also, there is sufficient material to develop a complete upper-division course. The 18 workshop problems provided are more than enough to support one laboratory session a week for a quarter or a semester.

A few of the chapters are devoted to a review of fundamental theory and methodology used in H&H analysis. Chapters 2 through 4 cover point rainfall averaging methods, rainfall loss functions, unit hydrograph theory, and flood routing. The first part of Chap. 7 discusses probability theory basic to flood-frequency analysis. Chapter 10 reviews the theory and methodology of water surface profile computations. This basic material is provided as a convenience to readers who require additional background to understand the models or need some refreshing. Others may wish to bypass these parts.

Example problems are incorporated throughout the text, and workshop problems are included at the ends of chapters. Workshop problems provide an opportunity to gain experience in the application of the techniques discussed. Hand calculations are required for some problems to teach basic principles. Computer solutions are required for most of the problems to provide experience with program input and output. The same river basin, the Rahway River basin in New Jersey, is used in the workshop problems in the first nine chapters. The Red Fox River in Colorado is used in several problems in Chaps. 10 through 15. Using a common drainage basin in a series of problems enables the reader to more easily visualize how different parts of the analysis fit together. Another advantage of this approach is that it minimizes the amount of data entry required. In several instances an input file developed for one problem can be revised for use in subsequent problems.

Acknowledgments

The author enjoyed an outstanding opportunity to work at the Hydrologic Engineering Center under temporary assignments for approximately three years—from 1985 to 1988. The work during this period included developing and applying real-time water control software and assisting with technical support of the HEC-1 and HEC-2 programs. This work experience was jointly supported by the College of Engineering at Utah State University and the Hydrologic Engineering Center of the U.S. Army Corps Engineers. This support is gratefully acknowledged.

The idea for writing this book evolved from this HEC experience. The book was perceived as being potentially useful as an additional means of transferring the analytic technology developed by the HEC to a wide spectrum of users. Appreciation is expressed to two individuals whose joint interest in the idea was essential for its realization: Bill S. Eichert, director of the Hydrologic Engineering Center, and L. Douglas James, director of the Utah Water Research Laboratory.

The staff of the Hydrologic Engineering Center, past and present, deserve credit for the continuing development of HEC-1, HEC-2, and other HEC programs described in the book. The initial formulation of HEC-1 is credited to the first director of the HEC, Leo R. Beard, and the initial formulation of HEC-2 is credited to the current director, Bill S. Eichert.

HEC training materials were the primary resources used in developing the book. These include program user's manuals, research documents, project reports, training documents, and training course lectures. Many of the figures and tables in the text were adapted from figures and tables in HEC documents. The workshop problems were adapted from workshop problems developed at the HEC. Training course lecture materials prepared by the following staff members were used as basic sources of reference.

Vern Bonner	Paul Ely*	Harold Kubik
Gary Brunner	Arlen Feldman	Al Montalvo
Mike Burnham	David Ford	Alan Oto*
J. J. DeVries*	Michael Gee	John Peters
Harry Dotson	David Goldman	Jerry Wiley
Gary Dyhouse†	Richard Hayes	

Most of these same individuals reviewed all or part of the manuscript and offered many helpful comments. Others who reviewed all or part of the manuscript and provided valuable comments include Lynne Stevenson, Bill Johnson, Daryl Davis, Randy Hills, Art Pabst, Ron Dieckmann, Bill Eichert, and L. Douglas James.

Special appreciation is expressed to John Peters, whose advice and encouragement throughout the project were extremely helpful.

The figures, in camera-ready form, and the tables for the book were skillfully prepared by Matt Lohof and Debbie Cook, respectively.

The cooperation and contribution of all the aforementioned individuals are sincerely appreciated and gratefully acknowledged. Everyone at the HEC was most cordial and cooperative during the course of the book's preparation. The acknowledgment of the contribution of these many individuals should not be interpreted as their endorsement of the final product. Any errors or deficiencies in presentation that may exist in the book are solely the author's.

Supplemental Workshop Problem Materials

Solutions to the workshop problems are presented in a separate manual, and the data input files printed in the text for the workshop problems are provided on a PC disk. Both the solutions manual and the disk are available from the author. The disk will substantially reduce the time required for data entry in working the problems.

Additional training materials are being developed for future release.

Send inquiries to *Floodplain H&H Supplemental Materials, Box 146, Providence, UT 84332.*

Daniel H. Hoggan

---

* Formerly with the HEC.

† With the Corps, St. Louis District.

Analysis of Floodplain Hydrology and Hydraulics

Introduction

The origin and nature of floods

A flood occurs when a rainstorm or melting snow generates a large amount of runoff that overtops the banks of a watercourse and flows onto the floodplain. Since the floodplain is a desirable place to locate homes and businesses, these high flows generally cause damage and, sometimes, injuries and deaths.

Floods vary greatly in intensity and duration depending on storm patterns, drainage basin characteristics, and other factors. A thunderstorm on a small drainage basin may generate a flood with a very high peak flow but of short duration. On a large basin, the peak flow from a similar storm may be significantly attenuated by storage and resistance in the catchment before it reaches the basin outlet. Storms associated with warm fronts, on the other hand, are likely to be of lesser intensity but of much greater areal extent and longer duration. Unused storage capacity in a catchment that may be sufficient to attenuate peak runoff and prevent flooding from a convective storm of short duration may be ineffective for a severe storm of this type.

The nature of floods and their impact depend on both natural and human-made conditions in the floodplain. Economic development and the installation of flood protection measures have political, economic, and social dimensions as well as engineering aspects. Hydrologic and hydraulic (H&H) analysis of floods provides a sound technical basis for facilities design as well as for management decision making that must weigh numerous other factors.

Purposes of floodplain H&H analysis

Many investigations related to the planning, design, and operation of systems within a basin or floodplain require estimates of flood flows. Typical requirements include the following:[1]

1. *Floodplain information studies.* Development of information on specific flood events such as the 10-, 100-, and 500-year frequency events.

2. *Evaluations of future land-use alternatives.* Analysis of a range of flood events (different frequencies) for existing and future land uses to determine flood hazard potential, flood damage, and environmental impact.

3. *Evaluation of flood loss reduction measures.* Analysis of a range of flood events (different frequencies) to determine flood damage reduction associated with specific design flows.

4. *Design studies.* Analysis of specific flood events for sizing facilities to assure their safety against failure.

5. *Operation studies.* Evaluation of a system to determine if the demands placed upon it by specific flood events can be met.

Definitive water surface profile information is needed for determining potential flood elevations and depths, areas of inundation, levee heights, rights-of-way limits, and the sizing of channel works.

Use of computer simulation

Computer programs used for modeling H&H systems have become highly developed over the past two decades. Advances in computer technology have made possible the simulation and analysis of the physical processes in an integrated and comprehensive framework. The high speed with which complex mathematical expressions can be solved and large amounts of data processed has expanded the realm of investigation and analysis significantly. Programs for analyzing floods have been in the forefront of this technological thrust.

Methods of analysis

The methods used in a floodplain H&H analysis are determined to a large extent by the purposes and scope of the project and the data available. Some of the techniques suitable for analyzing a large river basin may not be practical or necessary for analyzing a small urban watershed. A major floodplain information study usually consists of three parts: basin hydrology simulation, flood-frequency analysis, and

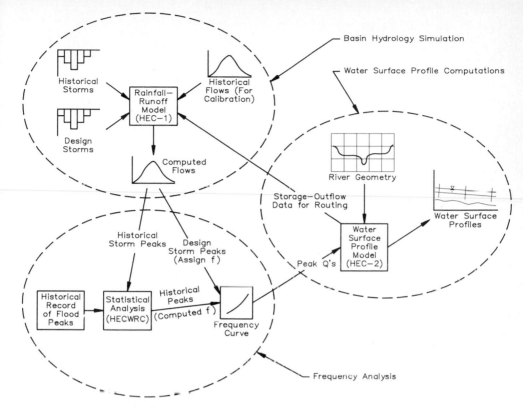

Figure 1.1   Schematic of a floodplain H&H study.

water surface profile computations.[2] The basin hydrology simulation may not be necessary if sufficient historical river discharge data is available for computing flood frequencies or if regional regression criteria are to be applied. Usually, however, the lack of gaged data at stream locations of interest and changing watershed conditions make frequency analysis of historical data alone impractical. A schematic of the H&H study process is shown in Fig 1.1.

Part 1: Basin Hydrology Simulation

The basin hydrology component of a study focuses on the simulation of the precipitation-runoff process in the drainage basin. The primary output of these computations is discharge hydrographs at locations of interest.

Precipitation-runoff process

Runoff consists of precipitation and other flow contributions collected from a drainage basin that appear in the surface stream at the outlet

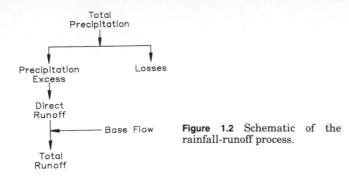

Figure 1.2 Schematic of the rainfall-runoff process.

of the basin.[3] It is generally divided into two basic parts: direct runoff and base flow. The transformation of precipitation into runoff is a complex process, influenced by numerous climatic and physiographic factors. A simplified diagram of the process, as modeled by the simulation program HEC-1, is shown in Fig. 1.2. Some of the precipitation that falls is lost as a contribution to runoff, due to infiltration and other abstractions. The precipitation excess, which remains, is transformed into direct runoff and added to base flow to produce total runoff.

In a basin that is divided into subbasins for modeling, the routing and combining of discharge hydrographs from individual subbasins are additional parts of the simulation of the rainfall-runoff process.

The discharge hydrograph

A discharge hydrograph essentially reflects the hydrologic nature of a flood. It is an integral expression of the physiographic and climatic conditions that govern the relationship between rainfall and runoff in a drainage basin. A hydrograph shows the time distribution of runoff at a point of interest, defining the complexities of the basin with a single empirical curve.[3]

To put a hydrograph in perspective, it can be compared with a flood wave as shown in Fig. 1.3. The profile of a flood wave is depicted at two different times on plots of elevation vs. distance. The elevation scale is greatly exaggerated relative to the distance scale. The plot at the left is a snapshot of the wave profile taken at time 1 when the front of the wave has just arrived at gage A. The plot at the right is a snapshot at time 2, immediately after the wave has passed gage A. In contrast, the hydrograph is a plot of discharge vs. time at a given location. It shows the change in discharge that occurs between time 1 and time 2 as the wave moves past gage A.

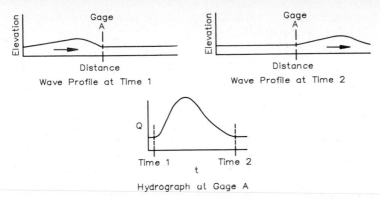

Figure 1.3   Comparison of a flood wave and a hydrograph at a gage site.

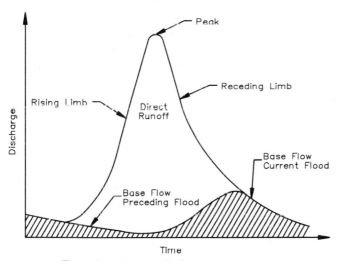

Figure 1.4   Typical single-storm hydrograph.

A typical hydrograph for a single storm (Fig. 1.4) consists of a rising limb, a peak, and a receding limb. The shape of the rising limb depends on the duration and intensity of the rainfall, antecedent moisture conditions, and drainage characteristics of the basin. The peak represents the highest concentration of runoff and usually occurs soon after the rainfall has ended, the precise time depending on the spatial distribution of rainfall. The receding limb represents withdrawal of water from storage in the form of base flow after surface inflow to the channel has ceased. The base flow during the initial period of the hydrograph is due to prior rainfall; some of the base flow during recession is due to current rainfall.

Part 2: Frequency Analysis

Methods for determining frequency curves

General. Frequency curves indicate the probability of the occurrence of floods of different severity. They have wide application in planning, design, and operations in the floodplain. Three different methods are generally used for determining frequency curves, and the choice of one or more is governed by the availability of historical data. The three methods are statistical analysis of measured peak discharges, regional frequency analysis, and precipitation-runoff analysis.

Statistical analysis of observed peak discharges. If streamflow records for several years are available at a location of interest, a discharge-frequency curve can be developed with well-defined statistical procedures.[4] The reliability of the frequency curve is dependent on the length of the period of record—curves based on 50-year records or longer records are generally considered the most reliable. With a shorter record, the frequency curve developed from the data can be used to calibrate a rainfall-runoff model. The model can then be executed using design storms to compute additional rare flood events from which a more reliable frequency curve for the rare events can be developed.

Regional frequency analysis. Discharge-frequency curves may be developed at gaged locations in a region that is meteorologically and hydrologically homogeneous. Mathematical equations relating selected parameters of these curves with watershed and climatological characteristics are determined, usually with regression analysis. Once developed, these equations are applied at ungaged locations to obtain discharge-frequency curves. However, this procedure is not suitable in basins with significant development. The existence of urbanization, channelization, reservoirs, etc., impose dissimilarities between gaged and ungaged locations that negate the relationships established between curve parameters and basin characteristics. Regional analysis is illustrated in Fig. 1.5.

Precipitation-runoff analysis. The development and use of a mathematical simulation model of the precipitation-runoff process is an alternative for determining discharges at points of interest. Historical or design storms are used as input for the model, and a discharge-frequency

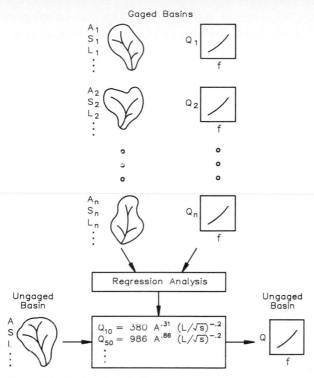

Figure 1.5   Regional analysis.

curve is developed from the peak discharges computed. This approach is adaptable to a wide range of conditions. The effects of storage reservoirs, channel modifications, urbanization, and other forms of development can be accommodated in the model, and peak discharges can be determined at virtually any point of interest. On the other hand, modeling can be time-consuming and expensive, and the association of frequencies with calculated peak discharges can be difficult, especially for design storms.

Three of the approaches that are used in precipitation-runoff modeling are (1) analysis of period-of-record data with a "continuous" simulation model, (2) analysis of period-of-record single-event data with a "single-event" simulation model, and (3) analysis of design storms with a "single-event" simulation model.

Analysis of period-of-record data with a "continuous" simulation model.   Precipitation data for an entire period of record is input into a continuous simulation model so that continuous discharge hydrographs at points of interest on a stream can be computed. The record must include sufficiently large storms to define the high ends of the frequency curves.

Conventional statistical analyses of the resulting peak discharges are performed to obtain the discharge-frequency curves. Continuous simulation models are relatively sophisticated, incorporating continuous soil moisture accounting and requiring substantial amounts of data and extensive parameter estimation.

Analysis of period-of-record single-event data with a "single-event" simulation model. Single-event data selected from an entire period of record is input into a single-event simulation model so that annual or partial duration series peak discharges can be computed. Conventional statistical analysis of the peak discharges is performed to obtain the discharge-frequency curves. This can be laborious and time-consuming, considering that all the events that could conceivably produce the peak discharges are modeled. This approach can be simpler and less costly than continuous simulation, but it presents difficulties in selecting the proper storm events and in estimating antecedent moisture conditions to associate with each storm.

Analysis of design storms with a "single-event" simulation model. Design storms associated with specific frequencies are input into a single-event simulation model so that discharge hydrographs at points of interest can be obtained. Each storm results in a peak discharge and a corresponding frequency value that can be used to define a single point on a discharge-frequency curve. Analyzing storms with a range of frequencies will produce a sufficient number of points to define the discharge-frequency curve. It is desirable, if not essential, to verify or adjust these assigned frequencies with known frequencies derived from long-term gaged data. This is the simplest and least costly of the three methods, but it is not entirely without problems. Estimating antecedent moisture conditions and assigning appropriate frequencies to the runoff can be difficult. This method is depicted in Fig. 1.6.

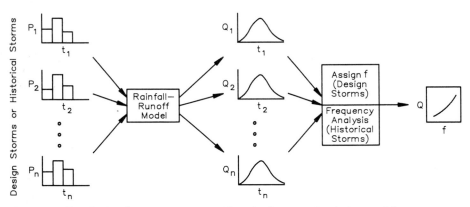

Figure 1.6 Developing frequency curves with a single-event simulation model.

Part 3: Water Surface Profile Computations

Just as the discharge hydrograph is a reflection of the hydrologic nature of a flood, the water surface profile is a reflection of the hydraulics of flood flow in a channel and floodplain. The water surface profile changes along a watercourse due to changes in natural channel geometry, surface roughness, and developments such as levees, bridges, and channel improvements. The specific effect of these changes on the water surface profile depends on the type and class of flow. Water surface profile changes associated with a variety of channel conditions are depicted in Fig. 1.7.

The flow in natural streams varies in time and space, and channel boundaries are not static during flood events, but simplifying assumptions are made in computer programs, such as HEC-2, to make systematic analysis practical. The simplifying assumptions in HEC-2 include "rigid boundary conditions," "steady, gradually varied flow," "one-dimensional flow," and "constant fluid properties." These assumptions have been found to be satisfactory for many applications.[1]

The conservation of energy is the dominant principle used in the computation of steady-flow profiles in the HEC-2 program, described later. System energy is continuously accounted for in proceeding from one point in a stream system to another. The energy equation and en-

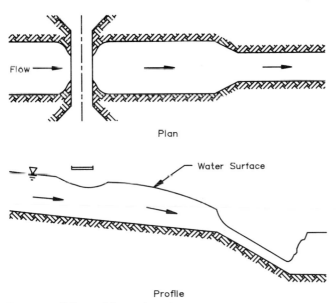

Plan

Water Surface

Profile

Figure 1.7   Effects of channel conditions on the water surface profile.

ergy loss equations are solved between adjacent streamflow cross sections in a numerical computational procedure called the "standard step method."

Application of H&H Analytical Techniques

Illustration of flood damage determination

The integrated application of floodplain H&H analytical techniques can be illustrated in the determination of flood damage. Key functional relationships used to define flood damage are discharge-frequency, stage-discharge, and stage-damage. These can be combined to determine a damage-frequency curve for use in estimating expected annual damage (Fig. 1.8). The discharge for a specific frequency is obtained from Fig. 1.8a. The stage and cost of damage for this discharge are obtained from curves b and c, respectively. The cost of damage and the corresponding frequency define a point on the damage-frequency curve, d. Several damage-frequency points are determined to define the damage-frequency curve. The expected annual damage, which is represented by the area under this curve, is compared with annualized project costs in economic evaluations.

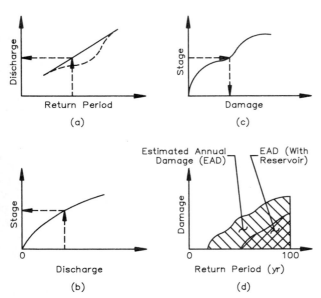

Figure 1.8 Key functional relationships in estimating expected annual damage. (a) Discharge-frequency; (b) stage-discharge; (c) stage-damage; (d) damage-frequency.

Benefits from a project may accrue at several locations in a basin. Typically, the relationships shown in Fig. 1.8 are required for an index station in each "damage reach." Land-use changes and alternative project configurations are reflected in one or more of these relationships. For example, the effect of adding a reservoir is shown with dashed lines in Fig. 1.8a. In a, the reservoir tends to reduce the peak discharge for a range of frequencies except at the high end of the frequency curve—if a flood is large enough to fill the design storage, the reservoir is ineffective in reducing the peak associated with further inflow.

Precipitation-runoff modeling and/or frequency analysis is required for the development of the discharge-frequency relationships in Fig. 1.8a; water surface profile computations are required for the development of the stage-discharge relationships in Fig. 1.8b; and stage-damage relationships in Fig. 1.8c are obtained from economic analysis, including field surveys and inventories of structures in the floodplain.

Guidelines for an H&H study to determine discharge frequency

The desired end products of a floodplain H&H analysis often are discharge-frequency curves and stage-discharge rating curves at key locations throughout a basin. The following steps are suggested for a basin with streamflow data available at some locations, but with most of the basin ungaged. Details of these procedures are covered in subsequent chapters.

1. On a map of the basin, locate watershed boundaries, existing and planned project sites, index points where discharge-frequency curves are required, and locations of gages. The location of project sites and index points should be based on consultations with planners and economists who are part of the study team.

2. Delineate subbasin boundaries consistent with step 1, taking into account physiographic and climatic variability in the basin.

3. Obtain streamflow records for gages within the basin and for gages, in close proximity, outside of the basin. Develop discharge-frequency curves in accordance with appropriate criteria[4] for all sites which have 10 or more years of recorded data. For relatively short records, say those covering less than 25 years, it may be desirable to calibrate a rainfall-runoff model to the frequency curve.

Then generate a more reliable curve for rare events using design storm data. See steps 9 and 10.

4. Obtain rainfall data for stations in and near the basin for five or more major storm events of record within the range of storm frequencies being analyzed, if possible.

5. Estimate unit hydrograph parameters and loss-rate parameters for any gaged subbasins.

6. Employ regional analysis to determine parameter estimates for ungaged subbasins.

7. Develop estimates of routing parameters for stream reaches.

8. Develop a multisubbasin model and adjust parameters using parameter estimation procedures and data from "observed" events.

9. Develop hyetographs for design storms with selected exceedance frequencies.

10. Estimate rainfall loss function parameters for the design storms to produce peak discharges consistent with discharge-frequency estimates developed from gaged data.

11. Compute a discharge hydrograph at each index location for each design event.

12. If regional frequency criteria are available for the basin, develop discharge-frequency curves at index locations using the regional frequency information.

13. Plot the regional frequency curves and the frequency curves developed with precipitation-runoff analysis on the same graph.

14. On the basis of the results of the preceding step and any other pertinent information available, adopt consistent frequency curves for the index locations.[5]

15. Develop steady-flow water surface profiles to enable the development of stage-discharge rating curves for the index locations. It may be necessary to develop water surface profiles prior to step 7 to determine storage-outflow data if a storage routing method is being used. In this case, the final stage-discharge rating curves are determined from a second computation of water surface profiles, made after the routed hydrograph is computed.

16. For future land use and future project conditions, modify the calibrated precipitation-runoff model to reflect the projected future conditions. For example, unit hydrograph and loss-rate parameters might be modified to reflect urbanization; routing parameters might be modified to reflect channel modifications; and reservoir

routing parameters might be added to evaluate a proposed reservoir. If channel modifications are being simulated, it may be necessary to compute a new set of water surface profiles to reflect the modified channel hydraulics.

17. Run the modified model with the design storms in step 9 to develop future-condition frequency curves. An assumption frequently made in this situation is that the frequency of runoff for a given design storm is the same for both existing conditions and modified conditions.

Simulation Programs for H&H Analysis

General

Computer programs developed by the U.S. Army Corps of Engineers, Hydrologic Engineering Center (HEC), for floodplain analysis have been used widely, not only by the Corps but also by other government agencies and consulting engineering firms. These programs constitute one set of analytical tools that can be used effectively. No attempt has been made in this book to compare them with similar programs that are available. The basic methodology and modeling techniques discussed regarding floodplain H&H should be helpful to those employing other programs.

Hydrologic Engineering Center
computer programs

General. The HEC maintains and distributes an extensive library of computer programs for hydrologic engineering and water planning. These programs have been developed over the years to provide effective analytical tools incorporating state-of-the-art technology. The HEC-1 Flood Hydrograph Package and the HEC-2 Water Surface Profiles Program are discussed in detail in subsequent chapters. Other HEC programs discussed include MLRP, Multiple Linear Regression Program; HECWRC, Flood Flow Frequency Analysis Program; STATS, Statistical Analysis of Time Series Data; and HMR52, Probable Maximum Storm (Eastern United States).

Personal computer versions. All the programs mentioned above are available in personal computer (PC) versions, MS DOS–compatible. Since the programs are batch-oriented, using one of them typically requires creating and checking a data input file, executing the program, and summarizing and displaying the output. To facilitate the use of

HEC-1 and HEC-2 on the PC, menu and utility programs have been developed to integrate and execute these functions. A full-screen text-editing program named COED provides on-line help screens and input variable documentation for preparing data input files.[6] Utility programs are available with HEC-2 for error checking of input files, for creating predefined and user-defined tables of output, and for generating plots of cross sections and profiles.[7] These application and utility programs can be selected by the user from a menu and executed.

Procurement. The Hydrologic Engineering Center computer programs are available to the public and to other government agencies. Since HEC-1 and HEC-2 are popular in the private sector, several private companies and organizations market these programs or similar versions of their own. Ads for marketing their software appear frequently in civil engineering magazines. Some suppliers offer telephone hot line support, engineering assistance, and auxiliary programs.

Software suppliers. For the HEC software suppliers list, write: *The Hydrologic Engineering Center, U.S. Army Corps of Engineers, Attention: Training Division, 609 Second Street, Davis, CA 95616.*

HEC-1 Flood Hydrograph Package. The HEC-1 Flood Hydrograph Package[8] calculates discharge hydrographs from single-storm events for basins of virtually any degree of complexity. The program has several optional capabilities, including parameter estimation, multiplan-multiflood analysis, precipitation depth–area computations, and flood damage analysis.

Use of HEC-1 for precipitation-runoff modeling requires subbasin boundary delineation, precipitation data, and runoff and routing parameters. Parameter estimation is an essential part of precipitation-runoff modeling, and HEC-1 has the capability to automatically estimate parameters for single basins that have observed precipitation and streamflow data available. Estimated parameters for gaged basins can be used in regional analysis (see chap. 6) to develop criteria for estimating parameters for ungaged basins.

HEC-2 Water Surface Profiles Program. The HEC-2 Water Surface Profiles Program[9] computes one-dimensional steady, gradually varied flow water surface profiles in streams with virtually any geometry. The program uses an iteration procedure called the "standard step method" to solve energy and energy loss equations between adjacent flow cross sections. Required data for modeling includes starting water surface elevation, discharge, cross-sectional geometry and reach lengths, Manning's n values and other energy loss coefficients, and characteristics of bridges, weirs, and other control features.

MLRP—Multiple Linear Regression Program. The Multiple Linear Regression Program[10] is used in regional analysis to analyze relationships between dependent variables (rainfall-runoff model parameters) and independent variables (basin characteristics). In addition to performing a typical multiple linear regression analysis, the program automatically deletes the least significant variable after each iteration of the analysis. All independent and dependent variables can be stored and then selected for analysis as desired. The program can combine input variables to form new variables, and it can make logarithmic, square root, and reciprocal transformations. It also computes and tabulates residuals from the prediction equation.

HECWRC—Flood Flow Frequency Analysis. The HECWRC flood flow frequency program[11] performs frequency computations of flood peaks according to Bulletin 17B guidelines.[4] The program performs the following functions:

1. Arrays data and computes plotting positions for graphical analysis by the Weibull, median, or Hazen formulas.
2. Computes an analytical frequency curve using the Log Pearson Type III theoretical distribution.
3. Weights the computed skew coefficient with a generalized skew coefficient if one is input into the program.
4. Automatically analyzes a broken record as a continuous record.
5. Applies the conditional probability adjustment for missing data at the low end of the frequency curve.
6. Automatically deletes zero flood events and applies conditional probability adjustment to determine the frequency curve.
7. Tests for high and low outliers and makes appropriate adjustments if indicated.
8. Weights plotting positions and statistics according to historical events that are input.
9. Computes confidence limit curves for 0.05 and 0.95 levels or other levels as specified.
10. Computes frequency curve ordinates with and without the expected probability adjustment.

HMR52—Probable Maximum Storm (Eastern United States). The HMR52 computer program[12] computes basin average precipitation for probable maximum storms (PMSs) according to criteria in Hydrometeorological Report (HMR) No. 52 for watersheds east of the 103d meridian in the United States. The program computes the spatial-average probable maximum precipitation (PMP) for any subbasins or combination of subbasins defined in input. It optimizes the storm-area size and orientation to produce maximum basin average precipitation. The user must specify the desired centering and time distribution for the storm. A precipitation data file is produced that can subsequently be input into a rainfall-runoff model such as HEC-1.

STATS—Statistical Analysis of Time Series Data. The computer program STATS[13] reduces large volumes of time-series data, such as daily or monthly data, to a few meaningful statistics or curves. It computes flow-duration curves, annual maximums, annual minimums, departures of annual and monthly values from their respective means, and annual volume-duration exchange of high and low events. Analytical frequency analysis can be applied to data that follows a theoretical distribution, or graphical methods can be applied to data, such as regulated flows, that does not. Input data can be transformed into logarithms for analysis if desired.

References

1. Hydrologic Engineering Center, *Hydrologic Engineering in Planning,* Training Document 14, U.S. Army Corps of Engineers, Davis, Calif., April 1981.
2. Arlen D. Feldman, "HEC Models for Water Resources System Simulation: Theory and Experience," *Advances in Hydroscience,* vol. 12, Academic Press, Inc., New York, 1981.
3. Ven T. Chow, "Runoff," Sec. 14 in Ven T. Chow (ed.), *Handbook of Applied Hydrology,* McGraw-Hill, New York, 1964.
4. Interagency Committee on Water Data, *Guidelines for Determining Flood Flow Frequency,* Bulletin 17B, U.S. Geological Survey, Office of Water Data Coordination, Reston, Va., 1982.
5. Hydrologic Engineering Center, *Adoption of Flood Flow Frequency Estimates at Ungaged Locations,* Training Document 11, U.S. Army Corps of Engineers, Davis, Calif., 1980.
6. Hydrologic Engineering Center, "COED, Corps of Engineers Editor," U.S. Army Corps of Engineers, Davis, Calif., 1987. (Draft.)
7. Hydrologic Engineering Center, "Computing Water Surface Profiles with HEC-2 on a Personal Computer," Training Document 26, U.S. Army Corps of Engineers, Davis, Calif., 1987. (Draft.)
8. Hydrologic Engineering Center, *HEC-1 Flood Hydrograph Package,* user's manual, U.S. Army Corps of Engineers, Davis, Calif., 1981 (rev. 1987).

 9. Hydrologic Engineering Center, *HEC-2 Water Surface Profiles,* user's manual, U.S. Army Corps of Engineers, Davis, Calif., 1982 (rev. 1985).
10. Hydrologic Engineering Center, *Multiple Linear Regression,* generalized computer program, U.S. Army Corps of Engineers, Davis, Calif., 1970.
11. Hydrologic Engineering Center, *Flood Flow Frequency Analysis,* user's manual, U.S. Army Corps of Engineers, Davis, Calif. 1982.
12. Hydrologic Engineering Center, *Probable Maximum Storm (Eastern United States), HMR52,* user's manual, U.S. Army Corps of Engineers, Davis, Calif., 1984.
13. Hydrologic Engineering Center, "STATS, Statistical Analysis of Time Series Data," U.S. Army Corps of Engineers, Davis, Calif., April 1985. (Draft input description.)

2

Rainfall and Rainfall Loss Analysis

Rainfall Analysis

Basin rainfall

Rainfall is a fundamental part of the hydrologic system, and representing it adequately in a rainfall-runoff model is very important but often difficult. The spatial and temporal variability of rainfall intensity greatly affects runoff, and simulation of rainfall is difficult in many cases because of the sparseness of data. Storm events are measured by gages at specific locations, and point rainfall data is used collectively to estimate spatial and temporal distributions. For a model to accurately simulate the rainfall-runoff process, it must be able to deal with the dynamics of location, time, and amount of rainfall in the basin during a storm event. Methods to determine this information range from simple arithmetic averaging of gaged data to advanced programmed techniques applied to real-time data and radar images.

Spatial distribution of rainfall

General. It is customary in models to use average or lumped data to represent rainfall over a basin area. This is done to simplify computations and to be consistent with the availability of measured data, which is frequently very sparse. Three methods have been used extensively for spatial averaging of rainfall: the computation of the arithmetic mean of point rainfall amounts, the Thiessen polygon method, and the isohyetal method. Whether any of these methods will give reliable results for a particular event depends on gage spacing and type of storm. It is unlikely that gage spacing will be found sufficiently dense to accurately measure rainfall from convective storms. These dynamic storms exhibit very high spatial and temporal variability. On the other hand, cyclonic rainfall is relatively uniform and generally can be measured adequately with the gages available.

Arithmetic average of gaged data.   Usually, average depths of rainfall for representative parts of a basin are used to define spatial distribution. The most direct approach for determining average rainfall depth for a catchment is to calculate a simple arithmetic mean of gaged values. This is satisfactory if the gages are uniformly distributed and the topography is flat so that variation in rainfall is small—conditions that are not too common.

Thiessen polygon method.   If the rainfall is nonuniform or the gages are unevenly distributed, the rainfall at each station may be weighted in proportion to the area the gage represents. A common method of determining weighting factors is the Thiessen polygon method (Fig. 2.1). Polygons are constructed by connecting adjacent stations on a map with straight lines and drawing perpendicular bisectors to each connecting line. The perpendicular bisectors around a station enclose an area that is everywhere closer to that station than to any other station. Not all bisectors meet this criterion, so sometimes a "cut-and-try" procedure is needed to complete the polygons.

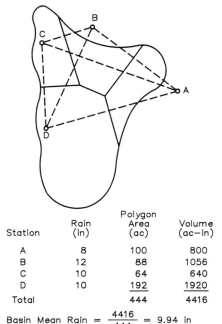

Station	Rain (in)	Polygon Area (ac)	Volume (ac–in)
A	8	100	800
B	12	88	1056
C	10	64	640
D	10	192	1920
Total		444	4416

Figure 2.1   Thiessen polygon method.

$$\text{Basin Mean Rain} = \frac{4416}{444} = 9.94 \text{ in}$$

To compute the average rainfall, the area represented by each gage is multiplied by the gaged rainfall to obtain a volume for the area. The volumes are summed, and the sum is divided by the total area to obtain the basin average rainfall. The polygons must be reconstructed each time a gage is added to or removed from the network, or each time the amount for any station is missing. The effect of this method is to represent the area closest to a gage by the measurements of the gage.

Isohyetal method.   The assumption in the Thiessen method that a gage best represents the area closest to it may be invalid in areas where topography has a significant effect on rainfall. An isohyetal map, consisting of contours of equal rainfall, can be drawn to conform to elevation contours and other factors affecting rainfall and thus may provide a more accurate determination of rainfall distribution. The improved accuracy depends on how skillfully these other factors are incorporated. Reference to maps of normal annual precipitation for the area, if available, may facilitate drawing the isohyets. If the isohyets are merely interpolated linearly between gages, the computed average rainfall will not differ significantly from the average rainfall computed with the Thiessen method.[1]

After the isohyetal map is completed, the mean depth for each subarea of the basin enclosed between a pair of the isohyets is determined. This is done by locating the centroid of each area and interpolating between the centroid and the two adjacent isohyets. The subarea between a pair of isohyets is measured with a planimeter or by some other method and is multiplied by the mean depth to compute an incremental volume. The average depth of rainfall is determined by summing these incremental volumes and dividing the sum by the total area.

Combination of isohyetal and Thiessen methods.   A disadvantage of the isohyetal method is that subareas must be remeasured for each storm event with different isohyets. This disadvantage can be overcome with a method combining Thiessen polygons with isohyets (Fig. 2.2). The mean precipitation of each polygon is computed or estimated on the basis of the isohyets. The mean precipitation of the basin is then computed as in the Thiessen method. In many storms it is possible to estimate the mean rainfall of each polygon by inspection. Since the isohyets reflect the influence of topography, more accuracy may be attained by their use. Using the Thiessen polygon method permits the use of a tabular form for computing basin precipitation, thus saving time.[2]

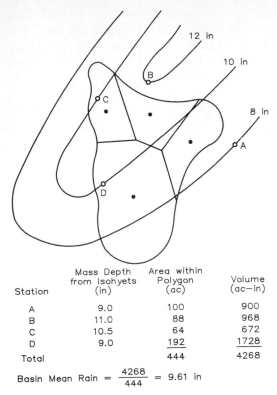

Station	Mass Depth from Isohyets (in)	Area within Polygon (ac)	Volume (ac–in)
A	9.0	100	900
B	11.0	88	968
C	10.5	64	672
D	9.0	192	1728
Total		444	4268

$$\text{Basin Mean Rain} = \frac{4268}{444} = 9.61 \text{ in}$$

Figure 2.2 Combination of isohyetal and Thiessen methods.

Temporal distribution of rainfall

Temporal distribution of rainfall is determined with recording precipitation gages, which measure rainfall amounts according to time. Recording gages provide a continuous record of rainfall amounts that can be translated into a mass curve. For example, the mass curve in Fig. 2.3 labeled "Station A" represents the temporal distribution of rainfall for Gage A in Fig. 2.1. Variations in the slope of this curve, which portrays accumulations of rainfall with time, reflect changes in intensity. Steep slopes indicate high intensity.

Determining the time distribution of basin average rainfall. The determination of the temporal distribution of average rainfall for a basin, such as the one in Fig. 2.1, generally requires the use of an index gage, because not all gages are recording gages. For example, assume that Gage A is the only recording gage in the basin and that its mass curve adequately represents the temporal distribution of rainfall at

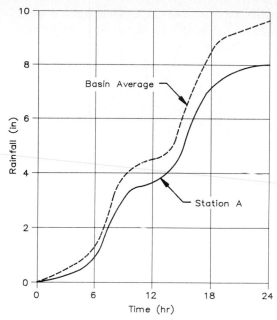

Figure 2.3 Mass curves.

the other gages. In Table 2.1, the 2-h ordinates of the mass curve for Gage A are entered in column (2). The percentage of total rainfall occurring in each interval is computed and entered in column (3). The basin average rainfall of 9.61 in from Fig. 2.2 is multiplied by these percentages to obtain ordinates of a mass curve for the basin [column

TABLE 2.1   Time Distribution of Basin Average Rainfall Determined from an Index Gage

Time, h	Gage A rain, in		Basin average rain, in	
	Cumulative	Ratio	Cumulative	Incremental
0	0.0	0.000	0.00	0.00
2	0.2	0.025	0.24	0.24
4	0.4	0.050	0.48	0.24
6	0.6	0.100	0.96	0.48
8	2.4	0.300	2.88	1.92
10	3.4	0.425	4.08	1.20
12	3.6	0.450	4.32	0.24
14	4.0	0.500	4.80	0.48
16	5.6	0.700	6.72	1.92
18	7.0	0.875	8.40	1.68
20	7.6	0.950	9.12	0.72
22	7.8	0.975	9.36	0.24
24	8.0	1.000	9.60	0.24

(4)]. The incremental amounts per 2-h period needed for the basin hyetograph [column (5)] are obtained by taking the difference between adjacent ordinates of the basin mass curve.

If more than one of the gages is a recording gage, this same procedure applies except that a simple or weighted average of the mass curves for the recording gages is used in lieu of the mass curve for the single Gage A. However, caution should be exercised when recording gage data is averaged. Temporal information can be lost in the averaging process. This is illustrated in Fig. 2.4, in which the averaging of

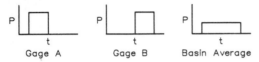

Figure 2.4   Loss of temporal information by averaging data from two recording gages.

storm data from Gages A and B does not produce a representative time distribution of rainfall. The gage data probably represents a fast-moving storm of short duration rather than the storm of longer duration obtained by averaging.

Rainfall data input into HEC-1.   A rainfall hyetograph is used for runoff calculations in HEC-1. Data input can be in the form of either basin average data or gage data. Basin average rainfall is specified as a storm total and incremental or cumulative amounts according to a time interval. Gage data is input as storm totals for selected gages and relative weights assigned to these gages. The total basin rainfall is computed as a weighted average of the gage data. Time interval increments of rainfall entered for recording gages are used as an index for temporal distribution of the total basin average.

Synthetic rainfall distributions developed for design storms also can be used in HEC-1. The use of triangular rainfall distributions associated with standard project storms, probable maximum precipitation, and storms of various durations developed from depth-area data are discussed in Chap. 8.

Rainfall Loss Analysis

Introduction

Definition of loss.   Not all the rainfall from a storm contributes to a flood; some is detained within the basin. Rainfall which does not contribute to direct runoff is considered to be a "rainfall loss." Rainfall losses that are significant in analyzing flood events include land-cover interception, depression storage, and infiltration. Evapotranspiration, important in long-term continuous modeling of runoff, has a negligible effect in most flood events. Interception and depression storage result from the retention of water on the surfaces of vegetation, in local depressions in the ground surface, in pavement cracks, etc. Infiltration, generally the most significant loss during flood events, is the movement of water into the ground.

Infiltration. Factors that affect the rate of infiltration include the type, extent, and conditions of vegetal cover; temperature; physical properties of the soil, including moisture content; water quality; land use; and rainfall intensity and duration. The rate at which water can penetrate the ground surface is highly dependent on surface conditions. Sealing of the surface by the washing in of fine materials may sometimes result in low infiltration rates, even when underlying soils are highly permeable. Water cannot continue to enter the soil more rapidly than it can be transmitted downward, so transmissibility of the soil is an important factor. The storage capacity of the soil, which depends upon porosity, thickness of the soil layer, and the amount of moisture present, also is important. The infiltration that occurs early in a storm is largely controlled by the available storage volume. After the soil is saturated, the infiltration rate becomes equal to the transmission rate.

A relationship between infiltration capacity and time was identified in early studies by Horton.[3] If rainfall intensity exceeds the infiltration capacity (i.e., the maximum rate at which water will infiltrate under existing conditions), the capacity tends to decrease exponentially with time and approaches an equilibrium value (Fig. 2.5).

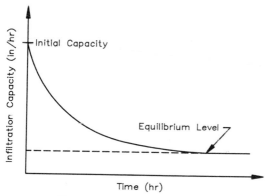

Figure 2.5 Variation of infiltration with time.

Loss functions in HEC-1

Formulations of rainfall loss functions have been developed, ranging from very simple lumped-parameter methods to complex systems that continuously account for soil moisture and evapotranspiration. Characteristics of several methods are shown in Table 2.2. Four of these are included in HEC-1: the initial and uniform, the exponential, the SCS, and the Holtan. The Green-Ampt method is being added in a forthcoming version of the program.

Initial and uniform loss rates. In this conceptualization of the rainfall loss process, all rainfall is lost until an initial loss is satisfied; then the remainder is lost at a constant rate. Because of its simplicity, this loss-rate function is used widely, particularly for design storms. It is best suited for large areas and studies where lumped estimates of losses are acceptable.

Example computation of initial and uniform loss rates. When observed runoff data is available, one of the best ways to determine loss rates is to analyze the runoff hydrograph relative to the rainfall event. Average rainfall losses for a basin can be estimated from hydrograph analysis, and these can be distributed in the form of initial and uniform losses. First, the volume of rainfall excess from a storm is determined by separating the base flow from the observed hydrograph, and then the remaining area under the hydrograph is measured. Without showing the hydrograph, it is assumed that the total rainfall excess of 2.50 in shown in column (5) of Table 2.3 was determined this way. The time intervals and corresponding rainfall amounts from the observed storm hyetograph are shown in columns (1) and (2). The initial loss of 0.50 in, shown in column (3), is estimated as the total amount of rainfall that occurred between the start of rainfall and the initial rise of the hydrograph.

What remains to be done is to determine the total uniform loss and distribute it with time. The basin average rainfall minus the rainfall excess gives the total loss, and the initial loss can be subtracted from this total to obtain the uniform loss of 1.62 in to be distributed.

$$\text{Total uniform loss} = 4.62 \text{ in} - 2.50 \text{ in} - 0.50 \text{ in} = 1.62 \text{ in}$$

Since all the rainfall during the first 2-h time interval is accounted for by the initial loss, the uniform loss must be distributed over the remaining time intervals. This is accomplished with the following trial-and-error procedure: (1) On the rainfall hyetograph (Fig. 2.6) a

TABLE 2.2 Characteristics of Rainfall Loss Methods

Loss technique	Units of computation: rate or percentage of rain	Nonlinear function of time or moisture	Evapotranspiration recovery during dry periods	Function of rainfall intensity	Direct accounting of detention storage	Estimation from geographic characteristics
Coefficient	%	Linear constant			X	Poor
Initial and uniform	in/h				X	Poor
Horton	in/h	Time				Fair
Holtan	in/h	Moisture	X			Fair
Exponential (HEC-1)	in/h	Moisture		X	X	Poor
SCS	%	Moisture			X	Good
Soil moisture accounting	in/h	Moisture	X	X	X	Fair–good
API	%	Moisture	X			Poor
Green-Ampt	in/h	Moisture		X		Fair

TABLE 2.3   Initial and Uniform Losses Derived from Rainfall Excess

Time, h (1)	Rainfall, in (2)	Initial loss, in (3)	Uniform loss, in (4)	Rainfall Excess, in (5)
	0.00	0.00	0.00	0.00
2	0.50	0.50	0.00	0.00
4	1.00		0.25	0.75
6	1.25		0.25	1.00
8	0.75		0.25	0.50
10	0.50		0.25	0.25
12	0.25		0.25	0.00
14	0.25		0.25	0.00
16	0.12		0.12	0.00
Sum	4.62	0.50	1.62	2.50

horizontal dashed line is drawn representing a trial estimate of the uniform loss rate f_u. The location of this line will determine how many time intervals will have rainfall excess. In locating the line between 0.25 in and 0.50 in, in this case, it is assumed that there are four intervals with rainfall excess, but this may have to be adjusted in a subsequent trial. (2) An equation relating the sum of time interval rainfall excess amounts to total rainfall excess is written and solved.

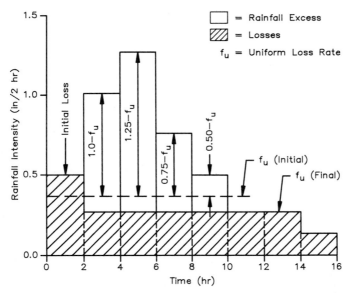

Figure 2.6 Determining a uniform loss rate on a rainfall hyetograph.

$$(1.00 \text{ in} - f_u) + (1.25 \text{ in} - f_u) + (0.75 \text{ in} - f_u) + (0.50 \text{ in} - f_u) = 2.50 \text{ in}$$
$$f_u = 0.25 \text{ in}$$

(3) This solution for the uniform loss rate is compared with the trial assumption represented by the horizontal line. In this case, f_u is between 0.25 and 0.50, so the solution is complete. If f_u had been outside of this range, more or fewer ordinates of rainfall excess would exist, and a second trial computation with another horizontal line representing f_u would be required.

Three data inputs are required by HEC-1 when this method is used: the initial loss, the constant loss rate, and the percentage of imperviousness. No losses are computed for the part of the basin area that is specified as impervious.

Nonuniform loss rates

Exponential loss function.   Although this is an empirical method developed for the HEC-1 program, its form reflects the physical processes involved in rainfall loss. The loss rate is a function of rainfall intensity and accumulated loss, which is related to soil moisture storage. A graph of the function on semilog paper (Fig. 2.7) shows an initial infiltration rate STRKR, which decreases at an exponential rate AK, related to cumulative loss CUML. During the initial period, when there are additional losses due to deficiencies in soil moisture and factors other than infiltration, another loss function DLTK, representing initial loss, is superimposed on the function AK. The cumulative rainfall loss during this initial period is DLTKR.

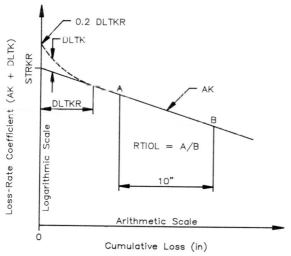

Figure 2.7   Exponential loss function.

RTIOL is the ratio of the loss-rate coefficient at one point on the exponential-loss curve to that corresponding to another point representing 10 in (or 10 mm) more of accumulated loss. This variable is considered to be a function of the ability of the surface of a basin to absorb precipitation and should be reasonably constant for large, relatively homogeneous areas. ERAIN is the exponent of precipitation for the rain loss function and reflects the influence of precipitation rate on basin average loss characteristics.

The loss rate is computed with the following equations:

$$DLTK = 0.2DLTKR[1 - (CUML/DLTKR)]^2 \qquad (2.1)$$

$$AK = STRKR/(RTIOL^{0.1CUML})$$

A serious drawback to the method is the lack of relationships between its parameters and basin characteristics, making it difficult to apply. It is virtually essential to use the HEC-1 parameter estimation option to define the parameters. A set of guidelines for doing this is presented in Chap. 5. Generally, it is more convenient to use one of the two-parameter versions of the function that can be formulated. A version with an initial and constant loss rate and a version with a loss rate that decays exponentially with no initial loss can easily be formulated.[4]

Input required by HEC-1 for this method consists of the variables STRKR, DLTKR, RTIOL, ERAIN, and RTIMP (percentage of subbasin that is impervious).

SCS loss method. The SCS Runoff Curve Number method, developed by the Soil Conservation Service (SCS), relates accumulated rainfall excess or runoff to accumulated rainfall with an empirical curve number in Eqs. (2.2) and (2.3). The curve number (CN) is a function of land use, cover, soil classification, hydrologic conditions, and antecedent runoff conditions.

$$Q = \frac{(P - I_a)^2}{P - I_a + S} \qquad (2.2)$$

$$S = \frac{1000}{CN} - 10 \qquad (2.3)$$

where Q = runoff, in
P = rainfall, in
I_a = initial abstraction
S = potential maximum retention after runoff begins, in
CN = curve number, percent of runoff

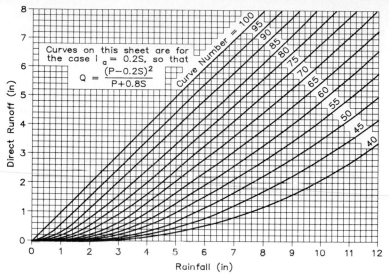

Figure 2.8 Graphical solution of rainfall-runoff equations.

Through studies of several small watersheds, I_a was found to be approximated by the empirical equation

$$I_a = 0.2S \qquad (2.4)$$

Substituting this approximation in Eq. (2.2) gives

$$Q = \frac{(P - 0.2S)^2}{P + 0.8S} \qquad (2.5)$$

Fig. 2.8 is used to obtain solutions to Eqs. (2.3) and (2.5) for a range of curve numbers and rainfall.

The variation in infiltration rates of different soils is incorporated in curve number selection through the classification of soils into four hydrologic soil groups: A, B, C, and D. These groups, representing soils having high, moderate, low, and very low infiltration rates, are as follows:

Group A soils have low runoff potential and high infiltration rates even when thoroughly wetted. They consist chiefly of deep, well- to excessively drained sands or gravels and have a high rate of water transmission (greater than 0.30 in/h).

Group B soils have moderate infiltration rates when thoroughly wetted and consist chiefly of moderately deep to deep, moderately well drained to well-drained soils with moderately fine to moderately coarse textures. These soils have a moderate rate of water transmission (0.15–0.30 in/h).

Group C soils have low infiltration rates when thoroughly wetted and consist chiefly of soils with a layer that impedes downward movement of water and soils with moderately fine to fine texture. These soils have a low rate of water transmission (0.05–0.15 in/h).

Group D soils have high runoff potential. They have very low infiltration rates when thoroughly wetted and consist chiefly of clay soils with a high swelling potential, soils with a permanent high water table, soils with a claypan or clay layer at or near the surface, and shallow soils over nearly impervious material. These soils have a very low rate of water transmission (0–0.05 in/h).[5]

A list of most of the soils in the United States with the group classification for each soil is provided in the Soil Conservation Service publication TR 55.[5] The soils in particular areas of the country can be identified from soil survey reports available from local SCS offices.

Runoff curve numbers for urban areas, cultivated and other agricultural lands, and arid and semiarid rangelands are shown in Table 2.4.[5] In determining CNs for urban areas, the percentage of impervious area and the means of conveying runoff from impervious areas to the drainage system should be considered.[6] The curve numbers in Table 2.4a are based on specific assumed percentages of impervious area. Pervious urban areas are assumed to be equivalent to pasture in good hydrologic condition, and impervious areas are assumed to have a CN of 98 and to be directly connected to the drainage system. An impervious area is considered connected if its runoff flows directly into the drainage system or occurs as concentrated shallow flow over a pervious area into the drainage system.

If the impervious-area percentages or the pervious area land-use assumptions are not applicable, an adjusted CN can be estimated from Fig. 2.9. For example, Table 2.4a gives a CN of 70 for a ½-acre lot in hydrologic soil group B, with an assumed impervious area of 25 percent. However, if the actual lot has a 20 percent impervious area and a pervious CN of 61, an adjusted (composite) CN of 68 can be read from Fig. 2.9, using these values.

When unconnected impervious areas are involved and the total impervious area is less than 30 percent, a composite CN can be obtained from Fig. 2.10. When the total impervious area is 30 percent or more, a composite value is not provided, because it is assumed that the remaining pervious area does not significantly affect runoff. A composite CN is obtained by first entering the right half of Fig. 2.10 with the

TABLE 2.4a   Runoff Curve Numbers for Urban Areas*

Cover description		Curve numbers for hydrologic soil group			
Cover type and hydrologic condition	Average percentage of impervious area†	A	B	C	D
Fully developed urban areas (vegetation established)					
Open space (lawns, parks, golf courses, cemeteries, etc.)‡					
Poor condition (grass cover < 50%)		68	79	86	89
Fair condition (grass cover 50% to 75%)		49	69	79	84
Good condition (grass cover > 75%)		39	61	74	80
Impervious areas:					
Paved parking lots, roofs, driveways, etc. (excluding right-of-way)		98	98	98	98
Streets and roads:					
Paved; curbs and storm sewers (excluding right-of-way)		98	98	98	98
Paved; open ditches (including right-of-way)		83	89	92	93
Gravel (including right-of-way)		76	85	89	91
Dirt (including right-of-way)		72	82	87	89
Western desert urban areas:					
Natural desert landscaping (pervious areas only)§		63	77	85	88
Artificial desert landscaping (impervious weed barrier, desert shrub with 1- to 2-inch sand or gravel mulch, and basin borders)		96	96	96	96
Urban districts:					
Commercial and business	85	89	92	94	95
Industrial	72	81	88	91	93
Residential districts by average lot size:					
1/8 acre or less (town houses)	65	77	85	90	92
1/4 acre	38	61	75	83	87
1/3 acre	30	57	72	81	86
1/2 acre	25	54	70	80	85
1 acre	20	51	68	79	84
2 acres	12	46	65	77	82
Developing urban areas:					
Newly graded areas (pervious areas only, no vegetation)¶		77	86	91	94
Idle lands (CNs are determined through the use of cover types similar to those for other agricultural lands.)					

*Average runoff condition, and $I_a = 0.2S$.

†The average percentage of impervious area shown was used to develop the composite CNs. Other assumptions are as follows: Impervious areas are directly connected to the drainage system; impervious areas have a CN of 98; and pervious areas are considered equivalent to open space in good hydrologic condition. CNs for other combinations of conditions may be computed through the use of Fig. 2.9 or 2.10.

‡CNs shown are equivalent to those of pasture. Composite CNs may be computed for other combinations of open space cover type.

§Composite CNs for natural desert landscaping should be computed through the use of Fig. 2.9 or 2.10 on the basis of impervious area percentage (CN = 98) and the pervious area CN. The pervious area CNs are assumed equivalent to desert shrub in poor hydrologic condition.

¶Composite CNs to use for the design of temporary measures during grading and construction should be computed through the use of Fig. 2.9 or 2.10 on the basis of the degree of development (impervious area percentage) and the CNs for the newly graded pervious areas.

TABLE 2.4b   Runoff Curve Numbers for Cultivated Agricultural Lands*

Cover description			Curve numbers for hydrologic soil group			
Cover type	Treatment†	Hydrologic condition‡	A	B	C	D
Fallow	Bare soil		77	86	91	94
	Crop residue cover	Poor	76	85	90	93
	(CR)	Good	74	83	88	90
Row crops	Straight row (SR)	Poor	72	81	88	91
		Good	67	78	85	89
	SR + CR	Poor	71	80	87	90
		Good	64	75	82	85
	Contoured (C)	Poor	70	79	84	88
		Good	65	75	82	86
	C + CR	Poor	69	78	83	87
		Good	64	74	81	85
	Contoured and ter-	Poor	66	74	80	82
	raced (C&T)	Good	62	71	78	81
	C&T + CR	Poor	65	73	79	81
		Good	61	70	77	80
Small grain	SR	Poor	65	76	84	88
		Good	63	75	83	87
	SR + CR	Poor	64	75	83	86
		Good	60	72	80	84
	C	Poor	63	74	82	85
		Good	61	73	81	84
	C + CR	Poor	62	73	81	84
		Good	60	72	80	83
	C&T	Poor	61	72	79	82
		Good	59	70	78	81
	C&T + CR	Poor	60	71	78	81
		Good	58	69	77	80
Close-seeded or broadcast legumes or rotation meadow	SR	Poor	66	77	85	89
		Good	58	72	81	85
	C	Poor	64	75	83	85
		Good	55	69	78	83
	C&T	Poor	63	73	80	83
		Good	51	67	76	80

*Average runoff condition, and $I_a = 0.2S$.

†Crop residue cover applies only if residue is on at least 5% of the surface throughout the year.

‡Hydrologic condition is based on a combination of factors that affect infiltration and runoff, including (a) density and canopy of vegetative areas, (b) amount of year-round cover, (c) amount of grass or close-seeded legumes in rotations, (d) percentage of residue cover on the land surface (good ≥ 20%), and (e) degree of surface roughness.

Poor: Factors impair infiltration and tend to increase runoff.

Good: Factors encourage average and better-than-average infiltration and tend to decrease runoff.

TABLE 2.4c   Runoff Curve Numbers for Other Agricultural Lands[a]

Cover description		Curve numbers for hydrologic soil group			
Cover type	Hydrologic condition	A	B	C	D
Pasture, grassland, or	Poor	68	79	86	89
range—continuous forage	Fair	49	69	79	84
for grazing[b]	Good	39	61	74	80
Meadow—continuous grass, protected from grazing and generally mowed for hay		30	58	71	78
Brush—brush-weed-grass	Poor	48	67	77	83
mixture with brush the ma-	Fair	35	56	70	77
jor element[c]	Good	30[d]	48	65	73
Woods—grass combination	Poor	57	73	82	86
(orchard or tree farm)[e]	Fair	43	65	76	82
	Good	32	58	72	79
Woods[f]	Poor	45	66	77	83
	Fair	36	60	73	79
	Good	30[d]	55	70	77
Farmsteads—buildings, lanes, driveways, and surrounding lots		59	74	82	86

[a]Average runoff condition, and $I_a = 0.2S$.

[b]Poor: < 50% ground cover or heavily grazed with no mulch.
 Fair: 50% to 75% ground cover and not heavily grazed.
 Good: > 75% ground cover and lightly or only occasionally grazed.

[c]Poor: < 50% ground cover.
 Fair: 50% to 75% ground cover.
 Good: > 75% ground cover.

[d]Actual curve number is less than 30; use CN = 30 for runoff computations.

[e]CNs shown were computed for areas with 50% woods and 50% grass (pasture) cover. Other combinations of conditions may be computed from the CNs for woods and pasture.

[f]Poor: Forest litter, small trees, and brush are destroyed by heavy grazing or regular burning.
 Fair: Woods are grazed but not burned, and some forest litter covers the soil.
 Good: Woods are protected from grazing, and litter and brush adequately cover the soil.

TABLE 2.4d   Runoff Curve Numbers for Arid and Semiarid Rangelands*

Cover description		Curve numbers for hydrologic soil group			
Cover type	Hydrologic condition†	A‡	B	C	D
Herbaceous—mixture of grass, weeds, and low-growing brush, with brush the minor element	Poor		80	87	93
	Fair		71	81	89
	Good		62	74	85
Oak-aspen—mountain brush mixture of oak brush, aspen, mountain mahogany, bitter brush, maple, and other brush	Poor		66	74	79
	Fair		48	57	63
	Good		30	41	48
Pinyon-juniper—pinyon, juniper, or both; grass understory	Poor		75	85	89
	Fair		58	73	80
	Good		41	61	71
Sagebrush with grass understory	Poor		67	80	85
	Fair		51	63	70
	Good		35	47	55
Desert shrub—major plants include saltbush, greasewood, creosote bush, blackbrush, bur sage, paloverde, mesquite, and cactus	Poor	63	77	85	88
	Fair	55	72	81	86
	Good	49	68	79	84

*Average runoff condition, and $I_a = 0.2S$. For range in humid regions, use the table for other agricultural lands.
†Poor: < 30% ground cover (litter, grass, and brush overstory).
  Fair: 30% to 70% ground cover.
  Good: > 70% ground cover.
‡Curve numbers for group A have been developed for desert shrub only.

percentage of total impervious area and the ratio of total unconnected impervious area to total impervious area. Then move horizontally to the left to the appropriate pervious CN, and read down to find the composite CN.

The SCS method is easy to use. First, soil types and soil group classifications for a basin are determined from SCS soil maps and publications. Next, the CNs corresponding to the appropriate land uses and soil groups are obtained from Table 2.4 and adjusted for impervious conditions if appropriate. Then the estimated CNs are entered as loss function input into a rainfall-runoff program such as HEC-1, or they are used to determine runoff directly from Fig. 2.8.

Input required for HEC-1 to use the SCS method includes, in addition to the CN, initial rainfall abstraction and percentage of the drain-

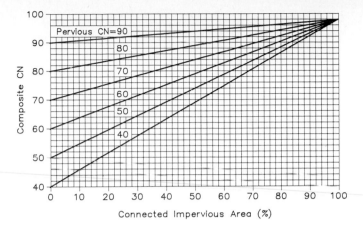

Figure 2.9 Composite CN with connected impervious area.

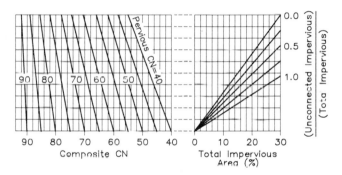

Figure 2.10 Composite CN with unconnected impervious areas and total impervious area less than 30 percent.

age basin that is impervious. The input of percentage that is impervious should be used only for directly connected impervious areas that are not already accounted for in the CN.

Holtan loss function. This function [Eq. (2.6)] uses soil moisture as a measure of infiltration rate.[7] As the moisture content of the soil increases, the infiltration rate decreases (Fig. 2.11a).

$$f = \mathrm{GI}\, a\, \mathrm{SA}^b + f_c \tag{2.6}$$

where f = infiltration rate, in/h
  GI = growth index representing plant maturity; $0 < \mathrm{GI} \le 1$
  a = infiltration rate, in/h per inch^b of available moisture storage (index of surface-connected porosity)

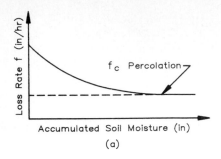

(a)

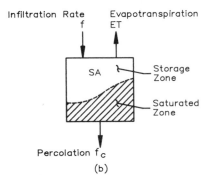

(b)

Figure 2.11 Holtan rainfall loss function. (a) Variation of loss rate with accumulated soil moisture; (b) soil moisture storage relationships.

SA = volume of unused soil moisture storage, in, in controlling zone of soil profile

b = exponent, usually 1.4

f_c = limiting rate of infiltration from controlling zone of soil profile

Soil moisture storage relationships are depicted in Fig. 2.11b. The incremental surface layer of soil is divided into two zones, a saturated lower zone (shaded) and an upper zone, SA, available for storage. The available storage is affected by three factors: (1) It is increased by percolation of water out of the element, which is governed by hydraulic conductivity f_c. (2) It is reduced by the amount of infiltration into the space. (3) It is increased as water leaves the surface through evapotranspiration (ET). The soil moisture storage availability is accounted for continuously by

$$SA_t = SA_{t-1} - f\,\Delta t + f_c\,\Delta t + ET\,\Delta t$$

The infiltration in a single period is computed as the average of the infiltration at the beginning of the time interval Δt and the infiltration at the end:

$$f = \frac{f_1 + f_2}{2}$$

The Holtan infiltration equations represent the physical processes occurring in a watershed, although some of the equations have empirical coefficients. An advantage of the Holtan method is that the parameters can be estimated from a rather general description of soil type and crop conditions. A major difficulty with the method is in the determination of the control depth on which to base the available storage SA in the surface layer.[8]

Required input into HEC-1 for the Holton loss function includes Holtan's long-term equilibrium loss rate f_c; the infiltration rate for available moisture storage a; the initial value of soil moisture storage capacity SA; the exponent of available soil moisture storage b; and the percentage of the basin that is impervious, represented as RTIMP.

Workshop Problem 2.1: Basin Rainfall and Rainfall Loss Analysis for the Rahway River Basin

Purpose

The purpose of this workshop problem is to provide an understanding of basic techniques used to compute basin average rainfall and estimate rainfall losses.

Problem statement

A floodplain study for the 25.5-mi^2 area upstream from the stream gaging station near Springfield (Fig. P2.1) requires the derivation of a unit hydrograph based on gaged data from a storm of August 27 and 28, 1971. Total storm rainfall is shown in Table P2.1. Hourly rainfall at the Springfield gage is shown in Table P2.2. The discharge hydrograph for this flood event, consisting of two peaks occurring about 14 h apart, is shown in Fig. P2.2. Separation of the two floods is shown in Fig. P2.3, and average hourly rainfall and direct surface runoff for the August 28 (second peak) part of the flood event are shown in Table P2.3.

Tasks

1. Construct a network of Thiessen polygons for the basin above the Springfield stream gage, and estimate the weighting factors for the precipitation gages. A trace of the basin map and rain gage locations should be made on a clean sheet of paper to serve as a work-

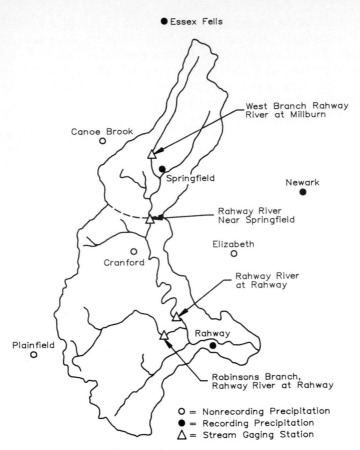

Figure P2.1 Rahway River basin.

TABLE P2.1 Total Storm Rainfall, August 27–28, 1971

Gage no.	Name	Rainfall, in
5	Canoe Brook	9.20
7	Newark	8.01
10	Plainfield	8.78
11	Rahway	8.15
1	Springfield	9.45
13	Essex Fells	8.30
20	Elizabeth	9.10
23	Cranford	8.58

TABLE P2.2   Hourly Rainfall at the Springfield Gage

Date/time		Incremental rainfalls, in	Ratio	Distributed average
8/27	1	0.00		
	2	0.00		
	3	0.00		
	4	0.04		
	5	0.12		
	6	0.06		
	7	0.10		
	8	0.08		
	9	0.43		
	10	0.76		
	11	0.37		
	12	0.70		
	13	1.15		
	14	1.36		
	15	0.18		
	16	0.06		
	17	0.04		
	18	0.09		
	19			
	20			
	21			
	22			
	23			
	24			
8/28	1			
	2	0.10		
	3	0.76		
	4	1.53		
	5	1.12		
	6	0.37		
	7	0.03		
Sum		9.45		

ing drawing. The tracing may be enlarged on a copy machine to obtain a more convenient size to work with.

2. Compute the basin average rainfall using the weighting factors for the gages, and distribute the average rainfall using the hourly data from the Springfield gage (Table P2.2) as an index.

3. Using the total storm depths for the gages, draw the isohyetal pattern with 0.5-in intervals on the basin map. Make an estimate of the basin average rainfall using the isohyetal pattern, disregarding the Thiessen polygons.

4. Compute the basin average rainfall using the combination method of isohyetal lines and Thiessen polygons described in this chapter.

5. Compare the results of the methods used in steps 1, 3, and 4.

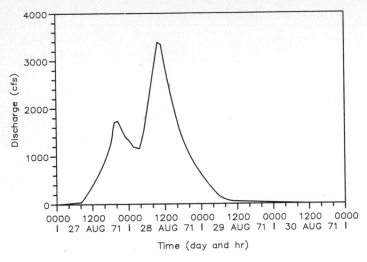

Figure P2.2 Discharge hydrograph for storm of August 27 and 28, 1971.

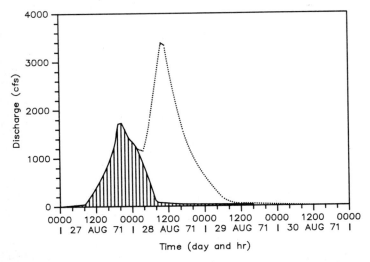

Figure P2.3 Separation of flood events in storm of August 27 and 28, 1971.

TABLE P2.3   Rahway River at Springfield, Flood of August 28, 1971

Date/time		Average precipitation, in	Loss, in	Excess, in	Observed flow, ft^3/s*	Flood 1 recession, ft^3/s*	Direct runoff, ft^3/s*
8/28	1	0.00			1,200	1,200	0
	2	0.10			1,085	1,085	0
	3	0.74			1,054	850	204
	4	1.48			1,300	700	600
	5	1.09			1,675	500	1,175
	6	0.36			2,160	320	1,840
	7	0.03			2,620	200	2,420
	8				3,080	100	2,980
	9				3,400	90	3,310
	10				3,350	80	3,270
	11				3,060	75	2,985
	12				2,820	70	2,750
	13				2,450	65	2,385
	14				2,160	60	2,100
	15				1,975	55	1,920
	16				1,760	50	1,710
	17				1,525	50	1,475
	18				1,300	50	1,250
	19				1,160	50	1,110
	20				1,010	50	960
	21				920	50	870
	22				820	50	770
	23				720	50	670
	24				609	50	559
8/29	1				500	50	450
	2				410	50	360
	3				320	50	270
	4				225	50	175
	5				177	50	127
	6				140	50	90
	7				115	50	65
	8				90	50	40
	9				80	50	30
	10				75	50	25
	11				74	50	24
	12				70	50	20
	13				60	50	10
	14				55	50	5
	15				50	50	0
	16				50	50	0
Sum					45,704	6,700	39,004

*The abbreviation "cfs" is also used for "cubic feet per second."

6. Determine the rainfall excess and the uniform loss rate for the storm on the basis of the data in Table P2.3. Assume that the initial loss is satisfied in the first time interval of rainfall.

References

1. Ray K. Linsley and Joseph B. Franzini, *Water Resources Engineering*, 3d ed., McGraw-Hill, New York, 1979.
2. *Flood Prediction Techniques,* U.S. Department of the Army Technical Bulletin 5-550-3, February 1957.
3. R. E. Horton, "The Role of Infiltration in the Hydrologic Cycle," *Transactions of the American Geophysical Union,* vol. 14, 1933, pp. 446–460.
4. Hydrologic Engineering Center, *HEC-1 Flood Hydrograph Package, Users Manual,* U.S. Army Corps of Engineers, Davis, Calif., 1981 (rev. 1987).
5. Soil Conservation Service, *Urban Hydrology for Small Watersheds,* Technical Release No. 55, U.S. Department of Agriculture, January 1975 (rev. June 1986).
6. W. J. Rawls, A. Shalaby, and R. H. McCuen, "Evaluation of Methods for Determining Urban Runoff Curve Numbers," *Transactions of the American Society of Agriculture Engineers,* vol. 24, no. 6, 1981, pp. 1562–1566.
7. H. N. Holtan and N. C. Lopez, *USDAHL-70 Model of Watershed Hydrology,* USDA Agriculture Research Service Technical Bulletin 145, 1971.
8. R. W. Skaggs and R. Khaleel, "Infiltration," Chap. 4 in *Hydrologic Modeling of Small Watersheds,* American Society of Agricultural Engineers, St. Joseph, Mich., 1982.

Unit Hydrograph Analysis and Base Flow

Unit Hydrograph Analysis

Unit hydrograph concepts.

Introduction. After rainfall and rainfall losses in a basin are determined, the next step in the analysis is to transform the rainfall excess into surface runoff. Unit hydrograph theory, introduced by Sherman,[1] has become widely used for estimating direct runoff from rainfall excess. It is the basis for one of two hydrograph generation approaches in HEC-1; the kinematic-wave method is the other. The unit hydrograph of a drainage basin represents the direct runoff resulting from a unit (e.g., 1 in) of rainfall excess distributed uniformly over the basin area at a uniform rate during a specified time interval.

The duration associated with a unit hydrograph must be the same as the time interval used to define the rainfall excess hyetograph. For example, a 1-h unit hydrograph is required for a hyetograph defined in 1-h intervals, and a 3-h unit hydrograph is required for a hyetograph defined in 3-h intervals. In HEC-1, the unit hydrograph duration is determined by the computational time interval specified in input. The program interpolates the rainfall data to this same time step irrespective of the time interval of the data entered.

Conversion of a unit hydrograph of one duration to one with another duration can be performed by lagging and combining procedures, such as the S-curve method.[2] For example, if the ordinates of a 1-h unit hydrograph are added to the ordinates of another 1-h unit hydrograph lagged 3 h, and the new ordinates are divided by 2, the result is a 3-h unit hydrograph.

Principles of linearity and time invariance. On the basis of the principles of "linearity" and "time invariance," the unit hydrograph is used to derive a total direct runoff hydrograph for a storm with any amount of effective rainfall. Ordinates of unit hydrographs of common time duration are added or superimposed numerically in proportion to the fraction of total runoff each represents to obtain the total hydrograph. The unit hydrograph of runoff for a drainage basin reflects all the combined physical characteristics of the basin and is the same for a given pattern of effective rainfall no matter when it occurs. Of course, this principle holds only if the conditions of the basin are fixed.

Selection of a computational time interval. The discharges computed with a unit hydrograph method are affected by the computational time interval Δt selected. A small time interval can produce significantly different results than a large one. As a rule of thumb, it is recommended that a Δt be used that will compute at least three or four points on the rising limb of the hydrograph. Since basin characteristics largely determine the length of the computational time interval that will produce this result, they should be considered in selecting a value. The initial value may have to be adjusted if the computed hydrograph does not have a sufficient number of ordinates on the rising limb.

Observing basic assumptions. Under natural conditions, the assumptions basic to unit hydrograph theory cannot be satisfied perfectly. However, when care is exercised in developing the model to meet the assumptions, results can be obtained that are satisfactory for practical purposes. The assumption of uniform distribution of rainfall and losses over a basin, which permits rainfall excess and losses to be treated as basin average quantities, is one that may affect the way the technique is applied. If the drainage area is too large to be represented reasonably well by uniform distribution of rainfall or losses, it may be necessary to subdivide the basin into smaller subbasins for accurate analysis.

The assumptions of linearity and time invariance also may affect the application of the technique. These require that the unit hydrograph used with a particular storm hyetograph be appropriate for the magnitude of the storm. A large storm producing considerable overbank flooding can be expected to have a quite different runoff response than a smaller storm which produces flows completely contained within the channel. Thus, unit hydrographs to be used with large design storms should, if possible, be derived from data for large historical flood events.

Derivation of a unit hydrograph. A unit hydrograph can be derived from observed hydrographs for simple or complex storms. In the "isolated-storm" approach, a complex storm is separated into single-storm segments, usually associated with the different hydrograph peaks. For a single-storm segment of fairly uniform rainfall distribution in space and time, it is assumed that the ordinates of the unit hydrograph are equal to the corresponding ordinates of the direct-runoff hydrograph divided by the total amount of direct runoff in inches. Calculations for a unit hydrograph based on data for the storm of August 28, 1971, given in Table P2.3, are tabulated in Table 3.1. Other methods for deriving unit hydrographs from complex storms are presented in basic hydrology texts.[3]

Parameter estimation techniques available in rainfall-runoff models, such as HEC-1, can be used to define parameters for synthetic unit hydrographs. Synthetic unit hydrographs are discussed later in this chapter, and parameter estimation is discussed in Chap. 5.

Data collection for deriving a unit hydrograph for a gaged watershed can be very time-consuming. It is desirable to obtain as many rainfall records for as many locations as possible within the study area to ensure that the amount and distribution of rainfall are accurately known.

Unit hydrograph application. The application of the unit hydrograph method consists of four steps: (1) determine the rainfall hyetograph for the basin; (2) estimate losses, and subtract these from the rainfall hyetograph to obtain rainfall excess; (3) transform rainfall excess into direct runoff with a unit hydrograph; and (4) add base flow to direct runoff to obtain the total-runoff hydrograph. These steps are depicted in Fig. 3.1.

In HEC-1, the rainfall-excess hyetograph is transformed into a discharge hydrograph through the use of Eq. (3.1). The program automatically sets the duration of rainfall excess equal to the computational time interval.

$$Q(i) = \sum_{j=1}^{i} P(j)U(n) \qquad (3.1)$$

where $Q(i) = i$th discharge hydrograph ordinate
$P(j) = j$th rainfall excess hyetograph ordinate
$U(n) = n$th unit hydrograph ordinate, with $n = i - j + 1$

Synthetic unit hydrographs

A number of methods have been developed for synthesizing a unit hydrograph with one or two parameters. If relationships can be defined by regression analysis between these parameters and measurable physical characteristics of basins, unit hydrographs can be syn-

TABLE 3.1 Calculation of UH Ordinates from Storm of August 28, 1971

Date/time	Direct runoff, ft³/s	UH ordinates, ft³/s	Date/time	Direct runoff, ft³/s	UH ordinates, ft³/s	Date/time	Direct runoff, ft³/s	UH ordinates, ft³/s
8/28 2	0	0	15	1920	810	4	175	74
3	204	86	16	1710	722	5	127	54
4	600	253	17	1475	622	6	90	38
5	1175	496	18	1250	527	7	65	27
6	1840	776	19	1110	468	8	40	17
7	2420	1021	20	960	405	9	30	13
8	2980	1257	21	870	367	10	25	11
9	3310	1397	22	770	325	11	24	10
10	3270	1380	23	670	283	12	20	8
11	2985	1259	24	559	236	13	10	4
12	2750	1160	8/29 1	450	190	14	5	2
13	2385	1006	2	360	152	15	0	0
14	2100	886	3	270	114	16	0	0
Sum							39,004	16,456

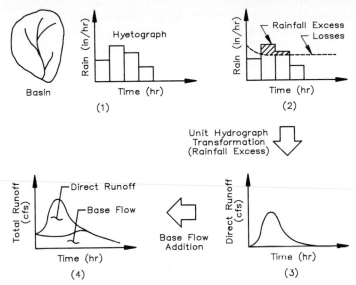

Figure 3.1 Unit hydrograph method. (1) Determine rainfall hyeto-graph; (2) subtract losses to obtain rainfall excess; (3) transform rainfall excess into direct runoff; (4) add base flow to obtain total runoff.

thesized for ungaged basins. Three commonly used synthetic unit hydrograph methods incorporated in the HEC-1 program are the Clark, Snyder, and SCS methods.

Clark method. The Clark method uses the concept of the instanta-neous unit hydrograph (IUH) to define a unique unit hydrograph for a basin.

Instantaneous unit hydrograph. The IUH conceptually is the hydrograph that would result if 1 unit of rainfall excess were spread uniformly over the basin instantaneously and allowed to run off. The rate of run-off is computed with the convolution integral:

$$Q(t) = \int_0^t u(t,\tau)I(\tau)\,d\tau \qquad (3.2)$$

where $Q(t)$ = direct runoff at time t
 $u(t,\tau)$ = the appropriate IUH ordinate
 $I(\tau)$ = the excess precipitation ordinate

In practice, the IUH is converted into a unit hydrograph of conve-nient duration through the use of techniques that are similar to those used to change the duration of any other unit hydrograph. For example, a unit hydrograph of 2-h duration can be obtained by summing the ordinates of one IUH with those of another IUH lagged 2 h, and then dividing the ordinates by 2. The S-curve method described in most hydrology texts[3,4] is a convenient method to use in changing the durations of unit hydrographs.

The Clark conceptual model.   The Clark method uses a conceptual model for an IUH consisting of a linear channel in series with a linear reservoir. These two components are modeled separately to account for translation and attenuation, respectively. The outflow from the linear channel is inflow to the linear reservoir, and the outflow from the linear reservoir is the IUH.

The linear channel component employs an area-time relationship based on the assumption that the velocity of flow over the entire area is uniform and the time required for runoff to reach the outlet is directly proportional to the distance. This relationship is defined by dividing the area of a basin into subareas with distinct runoff travel times to the outlet. These areas are delineated with isochrones of equal travel time, spaced in equal travel-time increments numbered sequentially upstream from the outlet.

The area-time relationship is used to estimate the time distribution of runoff from the basin, as illustrated in Fig. 3.2a and b. Isochrones of equal travel time for runoff are defined on the map (a), and amounts of runoff from the areas between isochrones are calculated so that a translation hydrograph for a unit of rainfall excess can be constructed (b). An ordinate in the translation hydrograph represents average flow passing the outlet during the interval $i - 1$ to i, where i represents consecutive isochrones. The five ordinates in Fig. 3.2b correspond to the five isochrones in Fig. 3.2a, each associated with a different travel time. For example, note the corresponding shaded areas between numbers 3 and 4. The parameter T_c is the travel time of runoff from the most remote point in the basin and is represented in Fig. 3.2b by the distance between the extremities of the histogram.

The linear reservoir component of Clark's model (Fig. 3.2c) represents the lumped effects of storage and resistance in the basin. Reservoir outflow and storage are related as

$$S_i = RO_i \tag{3.3}$$

where S_i = storage at end of period i
O_i = outflow during period i
R = storage coefficient

Outflow from the linear reservoir is computed with a simplified form

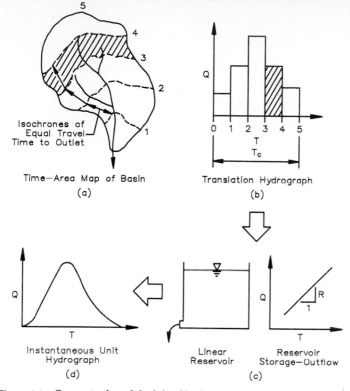

Figure 3.2 Conceptual model of the Clark unit hydrograph.

of the continuity equation:

$$\bar{I}_i - \frac{O_{i-1} + O_i}{2} = \frac{S_i - S_{i-1}}{\Delta t} \tag{3.4}$$

where $\bar{I}_i$ = average inflow in period i, determined from the area-time
relationship
O_i = outflow during period i
S_i = storage at the end of period i

Combining Eq. (3.3) with Eq. (3.4) and manipulating the terms yields

$$\bar{I}_i - \frac{O_{i-1} + O_i}{2} = \frac{RO_i - RO_{i-1}}{\Delta t} \tag{3.5}$$

Substituting

$$c = \frac{2\,\Delta t}{2R + \Delta t} \tag{3.6}$$

results in the following:

$$O_i = c\bar{I}_i + (1 - c)O_{i-1} \qquad (3.7)$$

Numerical solution of Eq. (3.7) yields the IUH shown in Fig. 3.2d. A unit hydrograph of duration Δt is computed by averaging two instantaneous unit hydrographs spaced at an interval Δt apart with

$$U_i = 0.5(O_i + O_{i-1}) \qquad (3.8)$$

Clark parameters. The parameters of the Clark unit hydrograph are the time of concentration, T_c, and the storage coefficient, R, which represents the slope of the storage-outflow curve for the linear reservoir. For gaged basins, the parameters T_c and R can be obtained from hydrographs of observed events, as illustrated in Fig. 3.3. T_c is estimated as the time from the end of a burst of rainfall excess to the inflection point on the receding limb. R is estimated by dividing the direct-runoff discharge at the inflection point by the slope of the curve

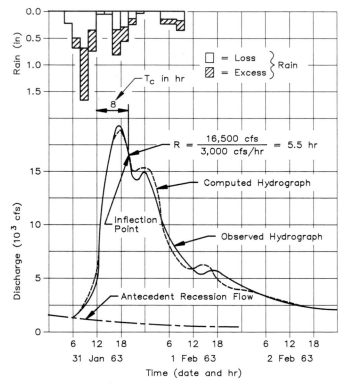

Figure 3.3 Estimating T_c and R from an observed hydrograph.

at that point. At the inflection point, inflow to the linear reservoir is zero, so the term representing inflow can be dropped, and Eq. (3.4) can be written as

$$-\frac{O_{i-1} + O_i}{2} = \frac{S_i - S_{i-1}}{\Delta t} \tag{3.9}$$

Substituting the linear relationship from Eq. (3.3) for S_i and S_{i-1} and solving for R yields

$$R = \frac{(O_{i-1} + O_i)/2}{-\left[(O_i - O_{i-1})/\Delta t\right]} \tag{3.10}$$

This indicates that R equals the average discharge at the inflection point divided by the negative slope of the hydrograph at that point (which is negative).

Another way to estimate the parameter R is by determining the volume remaining under the recession limb following the inflection point and dividing that volume by the flow at that point. In either approach, R should be determined by averaging the results of the analysis of several different hydrographs. The preferred method of determining T_c and R for gaged basins is to use the parameter estimation option in HEC-1, discussed in Chap. 5.

For ungaged basins, the time of concentration T_c can be determined from an analysis of travel times in the basin or from regression relationships that have been developed for the region. The storage coefficient R cannot be obtained from measurable watershed characteristics directly, but regression equations for R have been developed for some regions. For example, regression equations determined for the Rahway River basin in New Jersey[5] are as follows:

$$T_c = 8.29(1.0 + 0.03I)^{-1.28}(DA/S)^{0.28}$$

$$\frac{R}{T_c + R} = 0.65$$

Because of the interdependence of T_c and R, variable combinations, $T_c + R$ and $R/(T_c + R)$, have been used in defining regression relationships. The development of regional regression equations is discussed in Chap. 6.

Clark unit hydrograph computations in HEC-1.   The HEC-1 program utilizes a synthetic time-area curve derived from a generalized basin shape to obtain the translation hydrograph for a basin (Fig. 3.4). The time-area relationships for this basin are

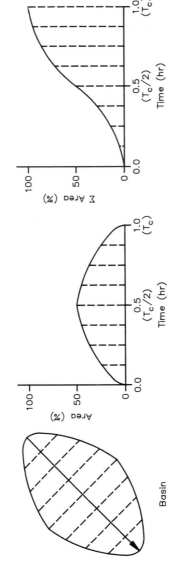

Figure 3.4 Generalized basin shape and synthetic time-area curve used in HEC-1.

$$AI = 1.414T^{1.5} \qquad 0 \le T < 0.5 \qquad\qquad (3.11)$$

$$1 - AI = 1.414(1 - T)^{1.5} \qquad 0.5 < T < 1 \qquad (3.12)$$

where AI is the cumulative area as a fraction of subbasin area and T is a fraction of the time of concentration. This curve is applicable to most basins; however, for a basin that deviates substantially from this generalized shape, a time-area curve developed specifically for the basin or from some other generalized basin shape can be used as input.

After the translation hydrograph is computed, it is routed through a linear reservoir by the program through the use of a recursive solution of Eq. (3.7). A unit hydrograph of a given duration is computed with Eq. (3.8).

A family of 2-h unit hydrographs generated by HEC-1, using a synthetic time-area curve, is shown in Fig. 3.5. These curves, computed

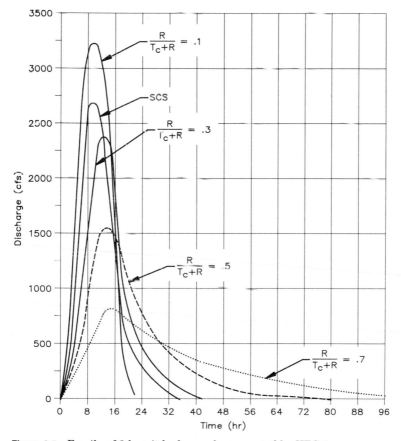

Figure 3.5   Family of 2-h unit hydrographs generated by HEC-1.

for a 50-mi^2 basin with a time of concentration of 13.3 h, demonstrate the flexibility of a two-parameter unit hydrograph in representing a wide range of runoff responses. The unit hydrographs generated with the Clark method are compared with a unit hydrograph computed with the SCS method, a one-parameter method. It would be necessary to change the basic time-area relationship in the SCS method to represent a set of conditions that is different from the average conditions for which it was derived. The steepest hydrograph in Fig. 3.5, with $R/(T_c + R) = .1$, would ordinarily represent a basin with steep slopes and a small amount of storage capacity. The flattest hydrograph in the group, at the other end of the spectrum, might be for a relatively flat basin with a large amount of storage capacity.

Snyder method. This method defines the unit hydrograph with two basic parameters: standard lag t_p, and a storage coefficient, C_p.

Concept. The Snyder method relies upon relationships of lag time and peak discharge with various physiographic basin characteristics. Originally, Snyder derived a lag-time relationship [Eq. (3.13)] for watersheds in the Appalachian mountain region of the United States.

$$t_p = C_t(LL_{ca})^{0.3} \qquad (3.13)$$

where t_p = lag time, hr
  C_t = coefficient representing basis slopes and storage
  L = length of the main channel from the basin outlet to the divide, mi
  L_{ca} = length of the main channel to a point opposite the centroid of the basin area, mi

The product, LL_{ca}, accounts for watershed shape, and C_t accounts for variations in slopes and storage. Steep slopes tend to generate low values of C_t, as low as 0.4 in Southern California, and at the other extreme, flat areas along the Gulf of Mexico have produced values as high as 8.0.[2]
  The peak discharge per unit drainage area for rainfall excess of standard duration is computed with Eq. (3.14):

$$Q_p = \frac{640C_pA}{t_p} \qquad (3.14)$$

where Q_p = Peak discharge, ft^3/s
  C_p = coefficient accounting for storage and other runoff conditions
  A = basin area, mi^2

Values of C_p range from 0.4 to 0.8, depending upon the retention or storage capacity of the basin.[2]

The standard duration of rainfall excess is a function of standard lag:

$$\Delta t = \frac{t_p}{5.5} \qquad (3.15)$$

where Δt is the duration of rainfall excess in hours. To permit lag time and peak discharge to be computed for other unit hydrograph durations, the following equation was developed:

$$t_{pr} = t_p + 0.25(\Delta t_r - \Delta t) \qquad (3.16)$$

where t_{pr} is the adjusted lag time in hours and Δt_r is the new unit hydrograph duration in hours.

Application of the method. The application of the Snyder unit hydrograph method should be preceded by an evaluation of t_p and C_p for the study area. This can be accomplished by analyzing unit hydrographs at other locations in the region. Snyder parameters can be derived for a number of gaged basins, and with these regional criteria can be developed for estimating parameters for similar ungaged basins. An example of regional criteria for the Upper Oconee River basin, in Georgia, is as follows:[6]

$$t_p = 1.8(LL_{ca})^{0.40}$$

Snyder unit hydrograph computations in HEC-1. The Snyder method does not define a complete unit hydrograph, so the HEC-1 program utilizes the Clark method in a trial-and-error procedure to complete the hydrograph. Initial Clark parameters are derived from the given Snyder parameters, t_p and C_p, and these are used to compute a Clark unit hydrograph. A new set of Snyder parameters computed from this hydrograph is compared with the given Snyder parameters, and this process is continued for 20 iterations or until the difference between the computed and given values is less than 1 percent.[7]

SCS method. The synthetic unit hydrograph method of the Soil Conservation Service is based on a dimensionless unit hydrograph developed from an analysis of a large number of unit hydrographs for small rural watersheds, representing a large number of geographic locations. A "dimensionless hydrograph" is a unit hydrograph for which the discharge is expressed as a ratio of discharge to peak discharge

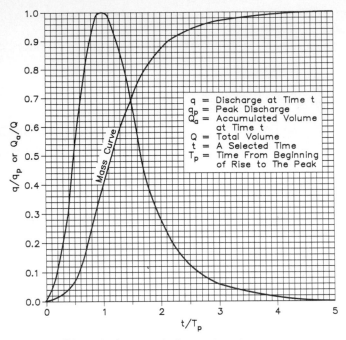

Figure 3.6 Dimensionless unit hydrograph and mass curve.

and the time by a ratio of time to lag time, thus eliminating the effect of basin size and much of the effect of basin shape.[2]

Concept. The SCS dimensionless unit hydrograph and mass curve are shown in Fig. 3.6. The ordinates of the curves are tabulated in Table 3.2. An equation for the peak discharge of a unit hydrograph is derived on the basis of the assumption that the dimensionless unit hydrograph can be represented by an equivalent triangular unit hydrograph (Fig. 3.7). The rising limb of the hydrograph accounts for 35.7 percent (0.357 in) of runoff. Assuming that this volume is represented by the area under the straight-line approximation, solving for the peak discharge yields the following:

$$q_p = \frac{484A}{T_p} \tag{3.17}$$

The Soil Conservation Service has found the factor "484" in Eq. (3.17) to vary from 600 in areas of steep terrain to 300 in very flat swampy areas.[5] For flat, high-water-table watersheds, this factor may be as low as 10 to 50.[4] However, to change this constant would require a revised dimensionless hydrograph. The fact that there is no

TABLE 3.2   Ratios for Dimensionless Unit Hydrograph and Mass Curve

Time ratios t/T_p	Discharge ratios q/q_p	Mass curve ratios Q_a/Q
0.0	0.000	0.000
0.1	0.030	0.001
0.2	0.100	0.006
0.3	0.190	0.012
0.4	0.310	0.035
0.5	0.470	0.065
0.6	0.660	0.107
0.7	0.820	0.163
0.8	0.930	0.228
0.9	0.990	0.300
1.0	1.000	0.375
1.1	0.990	0.450
1.2	0.930	0.522
1.3	0.860	0.589
1.4	0.780	0.650
1.5	0.680	0.700
1.6	0.560	0.751
1.7	0.460	0.790
1.8	0.390	0.822
1.9	0.330	0.849
2.0	0.280	0.871
2.2	0.207	0.908
2.4	0.147	0.934
2.6	0.107	0.953
2.8	0.077	0.967
3.0	0.055	0.977
3.2	0.040	0.984
3.4	0.029	0.989
3.6	0.021	0.993
3.8	0.015	0.995
4.0	0.011	0.997
4.5	0.005	0.999
5.0	0.000	1.000

provision in HEC-1 for changing the constant 484 poses a significant limitation on the use of this method.

The time to peak is found from Eq. (3.18). Lag L is estimated from the time of concentration with the equation $L = 0.6T_c$ or from other empirical equations developed by the SCS.[5]

$$T_p = \frac{\Delta t}{2} + L \tag{3.18}$$

where L is the lag time from the center of rainfall excess to the time of peak in hours.

Application.   A unit hydrograph for an ungaged basin can be obtained by means of the following steps:

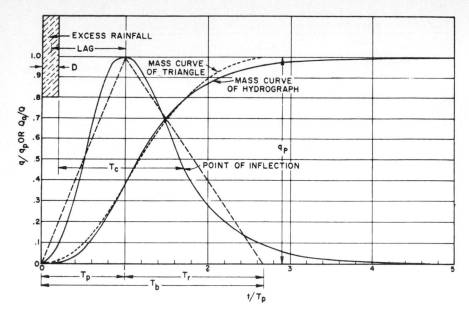

Figure 3.7 Dimensionless curvilinear unit hydrograph and equivalent triangular unit hydrograph.

1. Estimate T_c for the watershed from an analysis of travel times or some other means.
2. Determine lag from $L = 0.6 \ T_c$.
3. Determine time to peak T_p from Eq. (3.18).
4. Determine peak discharge q_p from Eq. (3.17).
5. Use values obtained for q_p and T_p in conjunction with ratios from Table 3.2 to define the unit hydrograph.

SCS unit hydrograph computations in HEC-1. The program uses data input for lag parameter L to compute peak flow and time to peak from Eqs. (3.17) and (3.18). The ordinates of the unit hydrograph are determined through interpolation of the dimensionless unit hydrograph curve at points defined according to the specified computational interval, which is also the duration of the rainfall excess. The duration of the unit hydrograph Δt should not be greater than $0.29L$.

Base Flow

HEC-1 simulates the base flow component of a flood hydrograph with three parameters: (1) STRTQ, the flow in the stream just prior to the start of the rising limb; (2) RTIOR, the exponential decay rate, which

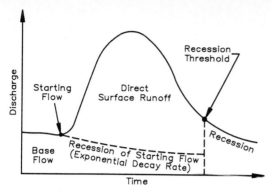

Figure 3.8 Base flow as defined by HEC-1.

is characteristic of the basin and defines base flow recession; and (3) QRCSN, the point on the falling limb called the recession threshold, beyond which the flow is assumed to recede at the exponential decay rate (Fig. 3.8). The starting flow is the result of the release of groundwater from previous storms, so its magnitude depends on the characteristics and timing of preceding storm events. The recession of the starting base flow is computed with Eq. (3.19):

$$Q = Q_0(\text{RTIOR})^{-n\Delta t} \tag{3.19}$$

where Q = base flow, ft^3/s
$\qquad Q_0$ = starting base flow, ft^3/s
$\quad$ RTIOR = variable name for recession decay rate
$\qquad n\Delta t$ = time since recession was initiated, h

The recession threshold and the exponential decay rate can be obtained from a semilog plot of observed flows vs. time (Fig. 3.9). The recession threshold is the point marking the beginning of the straight-line part of the curve. The decay rate RTIOR is equal to the slope of this straight line. It is also defined as the ratio of the recession limb flow at one point in time to the recession limb flow occurring 1 h later.[7]

In flood flows, the base flow contribution to the peak discharge or to total volume is generally minor, so errors associated with base flow simulation generally are not significant. In practice, the recession threshold is usually estimated; it is specified in HEC-1 input as a ratio of peak flow, typically in the range of 0.05 to 0.15.

The part of the streamflow hydrograph occurring prior to the time the recession threshold is reached is computed as the sum of direct surface runoff and base flow. After the recession threshold is reached, the falling limb of the hydrograph normally is computed with Eq.

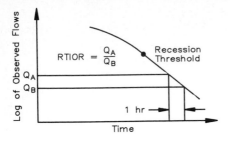

Figure 3.9 Determination of recession threshold and recession decay rate from plot of observed flows.

(3.19), and base flow and direct-flow components beyond that point are undefined. An exception occurs with a multiple-peak hydrograph, for example, a double-peaked hydrograph, when the computed direct surface runoff for the second peak rises above the recession limb of the first. When this happens, the second hydrograph is computed as the sum of the direct runoff for the second peak and the corresponding base flow defined by an extension of the original starting flow recession. Another recession threshold is established in the falling limb of the second hydrograph, and so forth (Fig. 3.10). Additional peaks are computed in a similar way.

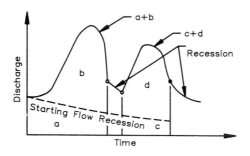

Figure 3.10 Base flow as defined by HEC-1 for a double-peaked hydrograph.

Workshop Problem 3.1: SCS Loss Rate and Clark Unit Hydrograph

Purpose

The purpose of this workshop problem is to provide an understanding of how the SCS loss-rate and Clark unit hydrograph methods are computed in HEC-1. Hand calculations are made first, and the results are compared with computed output from HEC-1.

Problem statement

A discharge hydrograph is needed for the area of the Rahway River basin above the Springfield gage for a storm of July 15, 1969. Hydrograph calculations are to be based on the following information:

DRAINAGE AREA: 25 mi^2

BASIN AVERAGE RAINFALL FOR JULY 15:

For hour ending at	Incremental rainfall, in
0300	0.08
0400	0.84
0500	1.36
0600	0.30

LOSS-RATE PARAMETERS:
  Initial abstraction: 0.5 in
  SCS curve number: 70

CLARK UNIT HYDROGRAPH PARAMETERS:
  Time of concentration T_c: 6.32 h
  Storage coefficient R: 8.42 h
  HEC-1 synthetic time-area curve

BASE FLOW: 20 ft^3/s (constant)

Tasks

1. Calculate rainfall excess amounts using the SCS curve number method. Prepare a table similar to Table P3.1 for your calculations.

TABLE P3.1   Rainfall Excess Calculations

Time, h	Incre- mental precipi- tation, in	Cumula- tive precipi- tation, in	Cumula- tive excess, in	Cumula- tive loss, in	Incre- mental loss, in	Incre- mental excess, in
0200						
0300						
0400						
0500						
0600						

2. Develop a translation hydrograph using Eqs. (3.11) and (3.12) for the HEC-1 synthetic time-area curve and a computational time interval of 1 h. Prepare a table patterned after Table P3.2 for tabulating the calculations.

TABLE P3.2   Translation Hydrograph Calculations

t, h	t/T_c	AI	Q, $(\text{mi}^2)(\text{in})/\text{h}$	Q, ft^3/s
0				
1				
2				
3				
4				
5				
6				
7				

3. Route the translation hydrograph through a linear reservoir using Eq. (3.7). Convert the resulting instantaneous unit hydrograph into a 1-h-duration unit hydrograph using Eq. (3.8). Calculate only the first 20 ordinates and tabulate the results in a table similar to Table P3.3.

TABLE P3.3   Unit Hydrograph Calculations

Time, h	Translation hydrograph, ft^3/s	Instantaneous unit hydrograph, ft^3/s	1-h unit hydrograph, ft^3/s
0			
1			
2			
.			
.			
.			
20			

4. Use the 1-h unit hydrograph and rainfall excess amounts to calculate the peak discharge (only). Determine the magnitude and time of occurrence of the peak discharge.

5. Compare the results of your hand computations with HEC-1 output (Fig. P3.1) based on the same data. Note and explain any differences.

6. Evaluate and discuss the suitability of the 1-h time interval used in the computations. Describe the effect on the calculated hydrograph that could be anticipated if a 3-h time interval were used.

9 LS   SCS LOSS RATE
       STRTL   .50   INITIAL ABSTRACTION
       CRVNBR 70.00   CURVE NUMBER
       RTIMP   .00   PERCENT IMPERVIOUS AREA

10 UC   CLARK UNITGRAPH
       TC   6.32   TIME OF CONCENTRATION
       R   8.42   STORAGE COEFFICIENT

SYNTHETIC ACCUMULATED-AREA VS. TIME CURVE WILL BE USED

***

UNIT HYDROGRAPH PARAMETERS
CLARK   TC 6.32 HR   R 8.42 HR
SNYDER TP 5.92 HR   CP .48

UNIT HYDROGRAPH
48 END-OF-PERIOD ORDINATES

82.	305.	615.	947.	1212.	1351.	1324.	1191.	1057.	939.
833.	740.	657.	583.	518.	460.	408.	363.	322.	286.
254.	225.	200.	178.	158.	140.	124.	110.	98.	87.
77.	69.	61.	54.	48.	43.	38.	34.	30.	27.
24.	21.	19.	16.	15.	13.	12.	10.		

Figure P3.1   HEC-1 output, hydrograph at the Springfield gage for storm of July 15, 1969.

```
************************************************************************************

                          HYDROGRAPH AT STATION    SPRGF

************************************************************************************
                                                      *
DA MON HRMN ORD  RAIN  LOSS EXCESS COMP Q             *  DA MON HRMN ORD  RAIN  LOSS EXCESS COMP Q
                                                      *
15 JUL 0200  1   .00   .00   .00    20.               *  15 JUL 1200  11   .00   .00   .00   845.
15 JUL 0300  2   .08   .08   .00    20.               *  15 JUL 1300  12   .00   .00   .00   755.
15 JUL 0400  3   .84   .80   .04    23.               *  15 JUL 1400  13   .00   .00   .00   673.
15 JUL 0500  4  1.36   .88   .48    71.               *  15 JUL 1500  14   .00   .00   .00   599.
15 JUL 0600  5   .30   .14   .16   204.               *  15 JUL 1600  15   .00   .00   .00   534.
15 JUL 0700  6   .00   .00   .00   402.               *  15 JUL 1700  16   .00   .00   .00   477.
15 JUL 0800  7   .00   .00   .00   621.               *  15 JUL 1800  17   .00   .00   .00   426.
15 JUL 0900  8   .00   .00   .00   807.               *  15 JUL 1900  18   .00   .00   .00   380.
15 JUL 1000  9   .00   .00   .00   915.               *  15 JUL 2000  19   .00   .00   .00   340.
15 JUL 1100 10   .00   .00   .00   919.               *  15 JUL 2100  20   .00   .00   .00   304.
                                                      *
************************************************************************************

TOTAL RAINFALL =   2.58, TOTAL LOSS =   1.90, TOTAL EXCESS =    .68

                                       MAXIMUM AVERAGE FLOW
PEAK FLOW    TIME
 (CFS)      (HR)             6-HR     24-HR    72-HR    19.00-HR
 919.       9.00
                   (CFS)     815.      483.     483.     483.
                 (INCHES)    .297      .557     .557     .557
                  (AC-FT)    404.      758.     758.     758.

                 CUMULATIVE AREA =   25.50 SQ MI
```

Figure P3.1 (*Continued*)

References

1. L. K. Sherman, "Stream Flow from Rainfall by the Unit Hydrograph Method," *Engineering News-Record,* vol. 108, Apr. 7, 1932, pp. 501–508.
2. Ven T. Chow (ed.), *Handbook of Applied Hydrology,* McGraw-Hill, New York, 1964.
3. Ven T. Chow, David R. Maidment, and Larry W. Mays, *Applied Hydrology,* McGraw-Hill, New York, 1988.
4. Phillip B. Bedient and Wayne C. Huber, *Hydrology and Flood Plain Analysis,* Addison-Wesley, New York, 1988.
5. Soil Conservation Service, *National Engineering Handbook,* Sec.4: *Hydrology,* U.S. Department of Agriculture, March 1985.
6. Hydrologic Engineering Center, *Upper Oconee River Basin, Hydrologic Study, Savannah District,* Special Projects Memo 456, U.S. Army Corps of Engineers, Davis, Calif., January 1976.
7. Hydrologic Engineering Center, *HEC-1 Flood Hydrograph Package, Users Manual,* U.S. Army Corps of Engineers, Davis, Calif., 1981 (rev. 1987).

Flood Routing Analysis

Introduction

Concept

A discharge hydrograph at the outlet of a basin reflects indirectly the rise and fall of a flood wave with time at that location. Defining the characteristics of a flood wave as it moves downstream through river reaches and reservoirs is essential in the hydrologic analysis of many basins. In a basin which has been subdivided, hydrographs at subbasin outlets are translated to the basin outlet or other locations of interest to determine water surface elevations, areas of inundation, and other impacts.

Flood routing may be defined as the technique of analyzing the movement of a flood wave through a river system. The course and character of the wave are traced by calculation as it progresses downstream. The effects of storage and flow resistance are reflected in the shape and timing of hydrographs of the wave at different locations. Diversions and tributary inflows must be accounted for, but these are usually handled separately from routing and added or subtracted at the ends of routing reaches.

Typically, it is necessary to obtain a discharge hydrograph at a point downstream from the location where another hydrograph has been observed or computed. For example, in Fig. 4.1 a hydrograph has been measured by a gage at A, located at the mouth of the basin, but a hydrograph also is needed at the ungaged location, B. To obtain the hydrograph at B, it is customary to determine with flood routing techniques how the flood wave is modified as it moves from A to B and then add the local inflow.

Purposes

One of the purposes of flood routing, as already noted, is to translate data on stage or discharge from one location downstream to another.

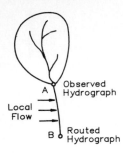

Local
Flow

Figure 4.1 Routing example, gage A to gage B.

This could be considered a descriptive purpose or application. Routing may also be used to predict streamflow at downstream locations resulting from the application of design storms and to determine the effects of constructed works, such as reservoirs and channel modifications, on streamflow.

Methods

General.   There are two general types of routing techniques: hydraulic and hydrologic. Hydraulic routing solves the differential equations of unsteady flow in open channels. Hydrologic routing, a simpler approach, generally employs the equation of continuity and the relationship between storage and discharge. However, there are a few hydrologic methods based on lagging of averaged hydrograph ordinates rather than on storage-discharge relationships.

A few of the most popular flood routing techniques are listed in Table 4.1. Except for the last one, these methods follow an order that ranges from the least empirical method at the top to the most empirical at the bottom. The last on the list, combining steady-state water surface profile computations and storage routing, is a hybrid technique that integrates the use of the hydrologic model, HEC-1, and the

TABLE 4.1   Routing Methods

Title	Type
Unsteady flow using Saint-Venant equations	Hydraulic method
Diffusion approximation of Saint-Venant equations	Hydraulic method
Kinematic-wave approximation of Saint-Venant equations	Hydraulic method
Modified Puls	Hydrologic method
Muskingum	Hydrologic method
Combination of steady-state backwater computations and modified Puls	Hydrologic method

hydraulic model, HEC-2. The HEC-2 water surface profile program is used to compute a storage-outflow function, and HEC-1 utilizes the storage-outflow data to do the routing and compute the peak discharge.

Hydraulic routing.   The basic equations that describe one-dimensional unsteady flow in open channels, the Saint-Venant equations, may be written in the form of an energy equation [Eq. (4.1)] and a continuity equation [Eq. (4.2)].[1] Solution of these equations, called the "dynamic-wave" equations, completely defines the flood wave with respect to distance along the channel and time.

For many open channel flow conditions, some of the terms in Eq. (4.1) can be neglected,[2] and simplified routing methods have been formulated by eliminating certain terms. If the three terms on the right side are neglected, the equation represents steady uniform flow. Coupling this equation with Eq. (4.2) constitutes the kinematic-wave approximation. Other approximations can be formulated by retaining more of the terms in Eq. (4.1).[3]

$$S_f = S_o - \frac{\partial y}{\partial x} - \frac{v}{g}\frac{\partial v}{\partial x} - \frac{1}{g}\frac{\partial v}{\partial t} \qquad (4.1)$$

$$A\frac{\partial v}{\partial x} + vB\frac{\partial y}{\partial x} + B\frac{\partial y}{\partial t} = q \qquad (4.2)$$

where S_f = friction slope
S_o = bed slope
y = depth of water
v = average velocity of water
x = distance in direction of flow (one-dimensional)
A = cross-sectional area
B = top width of flow
q = local inflow

Although the dynamic-wave equations provide the most accurate and comprehensive approach to flood routing, the data and computational time requirements are large. A typical set of input requirements includes (1) river cross sections, (2) Manning's n values, (3) an initial water surface profile, (4) an inflow hydrograph, and (5) a stage-discharge relationship for the downstream end of the routing reach. HEC-1 does not have the capability to solve the full unsteady-flow equations, but a few other programs are available with this capability.[4]

Hydrologic routing.   Hydrologic methods of flood routing, because of their computational simplicity, may be more suitable for use in simulation models. However, they do require the estimation of empirical coefficients that are difficult to evaluate in the absence of observed-flow data. Hydrologic methods available in HEC-1 include the Muskingum, modified Puls, working R and D, level-pool reservoir, and average lag. Two of the most popular of these, the modified Puls and the Muskingum, are discussed here.

Hydrologic Routing Methods in HEC-1

Modified Puls method

Concept of reservoir storage routing.   One of the simplest routing applications is the analysis of a flood wave that passes through an unregulated level-pool reservoir (Fig. 4.2a). The inflow hydrograph at A is known, and it is desired to find the outflow hydrograph for the reservoir at B. Assuming that the gate or spillway opening is fixed, a relationship is determined between storage and outflow. This relationship is derived from a hydraulic analysis of outflow vs. water level or stage, which is also linearly related to storage. Outflows for different water levels are computed on the basis of the hydraulics of the release features of the dam. Storage amounts for different water levels are computed from elevation-area measurements on a topographic map of the reservoir. Since both outflow and storage are directly related to water level elevation, they are, of course, related to each other (Fig. 4.2b).

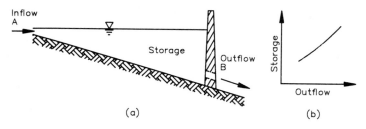

Figure 4.2   Reservoir storage routing.

An inflow hydrograph at A and an outflow hydrograph at B are shown in Fig. 4.3a. The peak is attenuated as a result of storage effects, and it is lagged by the travel time. The area under a hydrograph represents volume of discharge, so the area under A is the volume of inflow to the reservoir, and the area under B is the volume of outflow. The shaded areas between the curves for inflow and outflow account for changes in storage. When inflow is greater than outflow, storage is increasing; when outflow is greater than inflow, storage is decreasing. Thus, the storage volume peaks (Fig. 4.3b) at the point where the two hydrographs intersect, and since storage and outflow are related via a

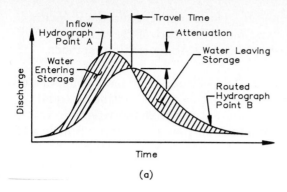

(a)

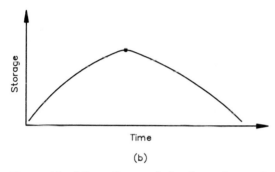

(b)

Figure 4.3 Attenuation and lagging of routed hydrograph. (a) Inflow and outflow hydrographs; (b) storage curve.

common water surface elevation, the outflow hydrograph also peaks at this point.

The equation defining storage routing based on the principle of conservation of mass can be written as inflow minus outflow equals change in storage:

$$I \, \Delta t - O \, \Delta t = \Delta S$$

or

$$I - O = \frac{\Delta S}{\Delta t}$$

This equation can be approximated for a routing period Δt with Eq. (4.3), where the subscripts "1" and "2" denote the beginning and end, respectively, of Δt.

$$\frac{O_1 + O_2}{2} = \frac{I_1 + I_2}{2} - \frac{S_2 - S_1}{\Delta t} \tag{4.3}$$

Given the complete inflow hydrograph from which I_1 and I_2 can be obtained and the initial conditions, O_1 and S_1, two unknowns remain to be found in Eq. (4.3): O_2 and S_2. Thus, another equation is required to solve for these two unknowns, and the nature of the storage-outflow relationship used distinguishes various storage routing methods.

Reservoir routing with the modified Puls method.   This method applied to a reservoir consists of a repetitive solution of the continuity equation [Eq. (4.3)] on the basis of the assumptions that the reservoir water surface remains horizontal and that outflow is a unique function of storage. The continuity equation can be manipulated so that the only unknowns for the first time step, O_2 and S_2, are located on one side.

$$\frac{S_2}{\Delta t} + \frac{O_2}{2} = \left(\frac{S_1}{\Delta t} + \frac{O_1}{2}\right) - O_1 + \frac{I_1 + I_2}{2} \qquad (4.4)$$

Since I is known for all time steps, and O_1 and S_1 are known for the first time step, the right-hand side of Eq. (4.4) can be calculated. Values of O_2 and S_2 are then used as input on the left-hand side, and the computation is repeated for the next time interval, and so on. A storage-indication curve is a plot of $[(S/\Delta t) + (O/2)]$ vs. O.

   Application.   The procedure for applying the modified Puls method is as follows:

1. Determine a discharge rating curve for the reservoir outlet.
2. Determine reservoir storage that goes with each elevation on the rating curve for reservoir outflow.
3. Construct a storage indication vs. outflow curve: $[(S/\Delta t) + (O/2)]$ vs. O.
4. Route the inflow hydrograph through the reservoir on the basis of Eq. (4.4).
5. Verify the results with data from historical events.

   Example of reservoir routing.   Tabulation of storage and outflow values under the headings shown in Table 4.2 is helpful in developing a storage indication vs. outflow curve. Corresponding values of storage and outflow are entered in columns (1) and (3), respectively. Column (2) is obtained by converting storage values in column (1) into units of cubic feet per second and dividing by the time interval that has been

TABLE 4.2   Storage Indication vs. Outflow Calculation

Storage, acre • ft (1)	$\frac{S}{\Delta t}$, ft^3/s (2)	Outflow, ft^3/s (3)	$\frac{\text{Outflow}}{2}$, ft^3/s (4)	$\frac{S}{\Delta t} + \frac{O}{2}$, ft^3/s (5)

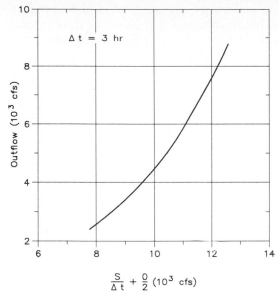

$$\frac{S}{\Delta t} + \frac{O}{2} \ (10^3 \ \text{cfs})$$

Figure 4.4   Storage-indication curve.

adopted. Column (4) is simply the outflow divided by 2, and the storage-indication value in column (5) is merely the sum of columns (2) and (4). The storage indication vs. outflow curve (Fig. 4.4) is obtained by plotting column (5) vs. column (3). The plot may be made on logarithmic graph paper if desired.

The routing can then be performed using a tabulation such as Table 4.3. All inflow amounts [column (2)] are known, as are the initial values of storage and outflow, columns (7) and (5), at time "0." Thus, the

TABLE 4.3   Storage Routing Calculation

Time, h (1)	Inflow, ft³/s (2)	Average inflow, ft³/s (3)	$\frac{S}{\Delta t} + \frac{O}{2}$, ft³/s (4)	Outflow, ft³/s (5)	$\frac{S}{\Delta t}$, ft³/s (6)	Storage, acre·ft (7)
0	3000		8600	3000	7100	1760
		3130				
3	3260		8730	3150	7155	1774
		3445				
6	3630		9025	3400	7325	1816
		3825				
9	4020		9450	3850	7525	1866
		4250				
12	4480		9850	4300	7700	1909
·						
·						

storage-indication value in column (4) at time "0" can be obtained from known values. The value in column (4) at the next time interval (3 h) is determined by applying Eq. (4.4). That is, the outflow at time "0" is subtracted from, and the mean inflow [column (3)] for the first period is added to, the value of $(S/\Delta t) + (O/2)$ at time "0" to yield $(S/\Delta t) + (O/2)$ at time "3 h." The outflow [column (5)] at time "3 h" is then determined from the curve in Fig. 4.4. This procedure is repeated for subsequent time periods until the desired outflow hydrograph is obtained. Storage values [column (7)] also may be obtained.

When the modified Puls method is used in HEC-1 to do reservoir routing, the storage and outflow values from columns (7) and (5) are required input data for the program. Storage-outflow input data may be specified in two ways: (1) as precomputed storage vs. elevation or discharge data and (2) as water surface area vs. elevation data. A conic method[5] is used to compute reservoir volumes if the latter option is used. An optional method of computing reservoir outflow from a description of the outlet works (low-level outlet and spillway) also is available in HEC-1.

Concept of storage routing in a channel. A "cascade of reservoirs" (Fig. 4.5) can be used to depict storage routing in a river. The river is divided into a series of routing reaches represented by reservoirs, with the outflow from the first reservoir being the inflow for the second, the outflow from the second being the inflow for the third, and so forth.

To illustrate the process of routing in terms of volume conservation and the effects of storage on the routed hydrograph, a fixed volume of water will be released though the reservoir systems A to D. The total volume of water to be routed is contained in reservoir A at the beginning. This water is released and flows into reservoir B. As the water

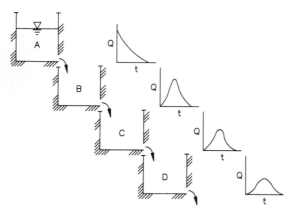

Figure 4.5 Cascade of reservoirs, depicting storage routing in a channel.

flows from A, the rate decreases as the level drops, as depicted by the hydrograph to the right of the reservoir. When all the water has drained from A, the volume that passed through the orifice is equal to the total volume originally contained in A. This is represented by the area beneath the hydrograph.

At reservoir B, the water flows in very rapidly at first, and the outflow from B is not as great as the inflow, so there is temporary storage of some of the water. The outflow increases with time as the level in B rises and then decreases as it surpasses the inflow from A, which has been decreasing. This rise and fall of outflow is reflected in the hydrograph for B.

A similar temporary storage phenomenon will occur at reservoirs C and D, but the inflows and outflows will be different, as reflected by the hydrograph shapes. Since a fixed amount of water was released through all reservoirs, the volumes under the hydrographs, A through D, are the same. The peaks of the hydrographs have been attenuated by the storage effects of the system in a manner similar to that which occurs in a stream channel.

To carry the analogy a step further, one of the reaches, C, is represented in Fig. 4.6. Reservoir C represents a reach of river and flood-

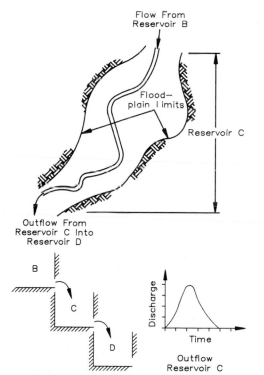

Figure 4.6 Reservoir representing channel routing reach.

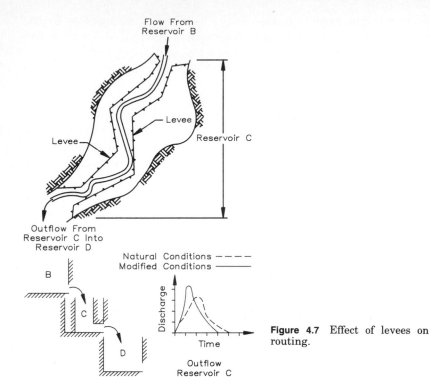

Figure 4.7 Effect of levees on routing.

plain, which typically would be several miles long. The magnitude of the time steps may be in the order of several hours. The number of time steps or reaches in an actual study is determined with calibration procedures explained later.

Determining the storage-outflow relationship for a reach is a critical part of the routing procedure. This can be clarified by considering the cascade-of-reservoirs analogy with a levee system added, as shown in Fig. 4.7. Adding levees has the effect of reducing the size of the reservoir and the amount of storage in the system. Comparing hydrographs "before" and "after" the levee is constructed in reservoir C, represented with dashed lines and solid lines, respectively, shows that both the attenuation and the travel time are less with the levees.

Additional information on storage routing in channels is provided in some general hydrology texts.[6]

A special case in which no attenuation of the peak occurs is the routing of a flood wave down a short, smooth reach of prismatic channel with a uniform cross section and a steep slope. This flood which is merely translated in time is called a "uniformly progressive wave" (Fig. 4.8).

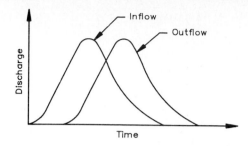

Figure 4.8 Uniformly progressive wave.

Channel routing with the modified Puls method.   When this method is applied to a river, a hydrograph at an upstream location, *A,* is routed to a downstream location, *B* (Fig. 4.9*a*). The storage is the volume in the channel under the water surface profile, and the outflow is the discharge in the channel at the downstream end of the reach at *B*. A storage-outflow relationship (Fig. 4.9*b*) must be defined which is similar to that for reservoir routing, but the approach is different. In river reaches, storage-outflow relationships are determined from one of the following: (1) steady-flow profile computations, (2) observed profiles, (3) normal-depth calculations, (4) storage calculations based on inflow and outflow hydrographs, and (5) optimization techniques applied to inflow and outflow hydrographs.

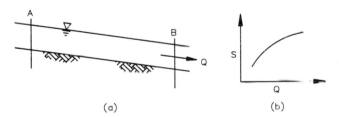

(a)   (b)

Figure 4.9 Storage-outflow curve for routing reach.

Determination of the storage-outflow relationship

Steady-flow profiles. Steady-flow water surface profiles, computed over a range of discharges, can be used to determine storage-outflow relationships in a river reach (Fig. 4.10*a*). In this illustration, a known hydrograph at *A* is routed to location *B*. The storage-outflow relationship required for routing is determined by computing a series of water surface profiles, WS_1 to WS_5, corresponding to discharges Q_1 to Q_5. This is done with a steady-flow water surface profile simulation program such as HEC-2.

The storage volumes are computed from the cross-sectional areas of the channel under the flow profiles and reach lengths. For a given water surface elevation, the storage is depicted by the area under the wa-

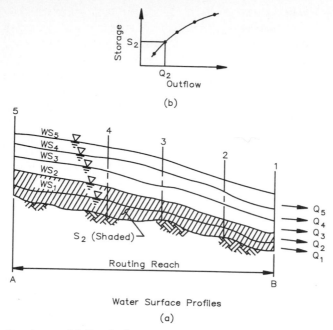

Figure 4.10 (*a*) Steady-flow water surface profiles; (*b*) storage-outflow curve.

ter surface profile, for example, the shaded area under profile WS_2. The volume under WS_2 corresponds to discharge Q_2 and is designated as S_2. These values for discharge and storage define one point on the storage-outflow curve (Fig. 4.10*b*) required for routing. Additional points on the curve can be obtained from the other water surface profiles in a similar way. Storage-outflow coordinates from this curve are input into HEC-1 for routing with the modified Puls method.

If the effects of channel or levee modification on flood routing are a concern, the cross sections in the reach represented in Fig. 4.10*a* can be redefined, the water surface profiles recomputed, revised storage-outflow values obtained, and a new routing reflecting the modifications computed.

Observed profiles. Observed profiles evidenced by high-water marks are used to compute storage-outflow relationships if sufficient data is available; however, it is unlikely that data would be available over the range of discharges needed for a stream reach. If some data points are available, they can be used to calibrate a steady-flow model for the reach with which storage-outflow data is computed as previously described.

Normal-depth calculations.   Normal depth associated with uniform flow does not exist in natural streams, but the concept is used to estimate water depth and storage in a stream reach if uniform-flow conditions can reasonably be assumed. For a typical cross section, Manning's equation is solved for a range of discharges, given appropriate n values and an estimated slope of the energy grade line. Under the assumption of uniform-flow conditions, the energy slope is considered equal to the average channel bed slope.

The normal-depth approach is programmed in HEC-1, as indicated in Fig. 4.11. Reach lengths, energy grade line slopes, roughness factors, and cross-sectional geometry are required as data inputs. From this data, the program computes the water surface elevation for a given discharge at each cross section. Then it computes the storage volumes from the cross-sectional areas under the water surface elevation and the reach lengths.

Storage calculations from inflow and outflow hydrographs.   Inflow and outflow hydrographs are used to compute channel storage by an inverse process of flood routing. If inflow and outflow are known, the change in storage can be computed. Tributary inflow, if any, also must be ac-

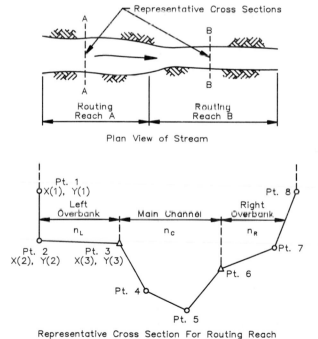

Figure 4.11   Normal-depth storage-volume computations.

counted for in this computation. The total storage is computed from some base-level storage at the beginning or end of the routing sequence.

Storage-outflow data from trial-and-error procedure. Inflow and outflow hydrographs are also used to compute routing criteria through a process of iteration in which an initial set of routing criteria is assumed, the inflow hydrograph is routed, and the results are evaluated. The process is repeated if necessary until a suitable fit of the routed with the observed hydrograph is obtained.

Determination of the number of routing steps in modified Puls. In reservoir routing, generally only one step is used under the assumption that the travel time through the reservoir is smaller than the computational time interval Δt. In channel routing, it may be necessary to define a number of routing steps to simulate flood-wave movement and changes. The number of steps or reach lengths affects the attenuation of the hydrograph and should be obtained by calibration. That is, it should be optimized or adjusted in a series of runs to replicate desired or observed hydrographs at the downstream end of the routing reach. An initial estimate of this parameter, represented by the variable NSTPS in HEC-1, is obtained by dividing the total travel time for the reach by the time interval Δt [Eq. (4.5)]. The time interval Δt is usually determined operationally to obtain a sufficient number of points to define the rising limb of the flood hydrograph.

$$K = \frac{L}{V_W}$$

$$\text{NSTPS} = \frac{K}{\Delta t} \qquad (4.5)$$

where K = travel time
L = reach length
V_W = velocity of flood wave
NSTPS = number of routing steps

Muskingum method

Looped storage-outflow relationship. In the modified Puls method of routing, a unique relationship between storage and outflow was needed. In routing a flood wave through a reach of river, such as from A to B in Fig. 4.12, storage vs. outflow is not a single-valued function. As the wave passes through the reach, the relationship changes. To illustrate this, assume that a control exists at B so that a unique relationship exists between depth and outflow, and for a given stage and

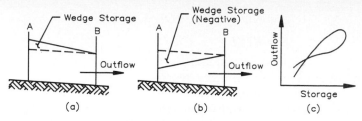

Figure 4.12 Looped storage-outflow relationship in the Muskingum method.

discharge, the steady-uniform-flow profile is represented by a line parallel to the streambed.

As a flood wave passes through the reach, at some time during the rising limb, the water surface will slope downward from A to B, as shown in Fig. 4.12a. The storage at that time will be the total volume under the sloped line. Later on, after the crest of the flood wave has passed through the reach, the water surface between A and B for the same discharge will change slope, as shown in Fig. 4.12b. The storage volume represented by the area under this water surface obviously is less than the storage volume associated with the rising limb. Thus, for each value of outflow there are two values of storage, one for the rising limb and one for the falling limb of the hydrograph, and this is shown in the looped storage-outflow curve in Fig. 4.12c. Storage is indicated on the lower part of the curve for the rising portion of the flood wave, and it is indicated on the upper part of the loop for the falling portion.

The Muskingum method was developed to accommodate this more complicated looped relationship of storage vs. outflow. Storage in a reach may be divided into two parts: prism storage and wedge storage. The prism storage is essentially the storage under the steady-flow water surface profile. The wedge storage is the additional storage under the actual water surface profile. So, in the rising stage of a flood wave the wedge storage is positive and added to the prism storage; in the falling stage the wedge storage is in effect negative and subtracted from prism storage.

The Muskingum routing equation. The total storage in a reach is equal to prism storage plus wedge storage [Eq. (4.6)]. The prism storage is computed as the routing coefficient K times the outflow, and the wedge storage is computed as K times a weighting coefficient X and the difference between inflow and outflow $(I - O)$. The coefficient K has units of time, corresponding to the travel time of the flood wave through the reach. The constant X is a dimensionless value expressing a weighting of the relative effects of inflow vs. outflow.

$$S = KO + KX(I - O)$$

$$S = K[XI + (1 - X)O] \tag{4.6}$$

where S = total storage in routing reach
O = rate of outflow from routing reach
I = rate of inflow to routing reach
K = travel time through routing reach
X = dimensionless constant, ranging from 0 to 0.5

The quantity in the brackets in the second form of the equation shown is considered an expression of weighted discharge. If $X = 0$, $S = KO$ from this equation, indicating prism storage only and a condition equivalent to level-surface reservoir routing. Routing in this case is unaffected by inflow. If $X = 0.5$, equal weight is given to inflow and outflow in the equation, and the condition is equivalent to a uniformly progressive wave discussed previously. Thus, "0" and "0.5" are limits on the value of X, and within this range the value of X determines the degree of attenuation of the flood wave through the routing reach. A value of "0" produces maximum attenuation, and "0.5" has no effect.

The Muskingum routing equation [Eq. (4.8)] is obtained by expressing the storage [Eq. (4.6)] at times 1 and 2 with Eq. (4.7), combining Eq. (4.7) with the continuity equation [Eq. (4.3)], and solving for O_2.

$$S_2 - S_1 = K[X(I_2 - I_1) + (1 - X)(O_2 - O_1)] \tag{4.7}$$

$$O_2 = C_0 I_2 + C_1 I_1 + C_2 O_1 \tag{4.8}$$

The coefficients C_0, C_1, and C_2 are routing coefficients defined in terms of Δt and calibrated parameters K and X.

$$C_0 = \frac{-KX + 0.5\,\Delta t}{K - KX + 0.5\,\Delta t} \tag{4.9}$$

$$C_1 = \frac{KX + 0.5\,\Delta t}{K - KX + 0.5\,\Delta t} \tag{4.10}$$

$$C_2 = \frac{K - KX - 0.05\,\Delta t}{K - KX + 0.05\,\Delta t} \tag{4.11}$$

Given an inflow hydrograph, a selected time interval Δt, and parameters K and X, the outflow hydrograph can be calculated. This is demonstrated in the example problem which follows.

Determination of Muskingum K and X for gaged reaches. These parameters are determined from observed inflow and outflow hydrographs in at least two ways. As demonstrated previously, a curve of storage vs.

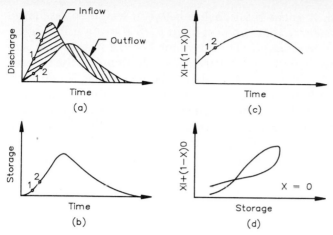

Figure 4.13 Determining Muskingum K and X from gaged data. (a) Measure storage vs. time; (b) plot accumulated storage vs. time; (c) plot weighted discharge vs. time; (d) plot weighted discharge vs. storage.

time can be obtained by measuring the area between the inflow and outflow hydrographs plotted on the same graph (Fig. 4.13a) and plotting the accumulated area (volume) vs. time on another graph Fig. 4.13b. For a given value of X, weighted discharge vs. time is plotted in Fig. 4.13c. The values of I and O for different times are obtained from Fig. 4.13a to compute and plot the points for the weighted discharge. Two points, 1 and 2, are shown in the figure for illustration. A fourth curve of weighted discharge vs. storage (Fig. 4.13d) is obtained by taking points from Fig. 4.13b and c for common times and plotting them.

Two examples of weighted discharge vs. storage curves for different values of X are shown in Fig. 4.14. The object of constructing these

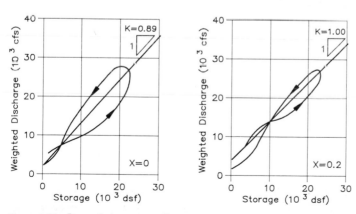

Figure 4.14 Looped storage-outflow curves.

curves is to obtain a linear relationship between weighted discharge and storage, so the curve that can best be fit with a straight line is the one selected for determining K. The curve for $X = 0.2$ in this case produces the closest fit. With $X = 0.2$, the value of K, found as the inverse slope of the straight line, is 1.00. It should be noted that K defines a linear relationship between storage and outflow in the Muskingum method. This is in contrast to the unique nonlinear relationship in the modified Puls method. This is a disadvantage of the Muskingum method, particularly as applied to the routing of flows through reaches with wide overbank areas exhibiting a nonlinear relationship of storage and outflow.

Another technique for estimating K and X from observed inflow and outflow hydrographs is to measure the time K as the interval between certain points on the two hydrographs and then use a trial-and-error computational procedure for obtaining X. The travel time K can be measured as the elapsed time between centroids of areas of the two hydrographs, between the hydrograph peaks, or between midpoints of the rising limbs. These alternatives are shown in Fig. 4.15. After K has been determined, a value of X is assumed, and the inflow hydrograph is routed with these parameters. The routed hydrograph is compared with the observed-outflow hydrograph for fit. If the fit is not suitable, another value of X is assumed, and the inflow hydrograph is again routed. This procedure is repeated until a suitable fit is obtained.

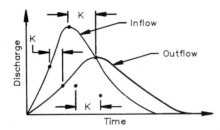

Figure 4.15 Alternative measurements of K between observed inflow and outflow hydrographs: peaks, centroids of areas, and midpoints of rising limbs.

To determine the Muskingum routing parameters K and X with the automatic parameter estimation routine in HEC-1, it is necessary to provide the inflow hydrograph, the outflow hydrograph, and a pattern hydrograph for distributing local inflow with time. The program estimates routing parameters with the same routine it uses for estimating unit hydrograph and loss-rate parameters.

Determining K and X for ungaged reaches.   A number of techniques have been developed for estimating Muskingum routing parameters at ungaged locations.[7] One is similar to the approach used in the modified Puls method in which a storage-outflow relationship is derived from water surface profile analysis. Channel geometry is measured and water surface profiles computed according to the procedures previously discussed. A plot of weighted discharge vs. storage is derived from the results, and the parameter K is measured as the slope of this curve. This approach entails considerable work, and unless the water surface profiles analysis is needed anyway, an alternative method of determining the routing parameters is probably warranted.

An estimate of K may be obtained by applying Seddon's principle [Eq. (4.12)], in which the velocity of a flood wave is approximated from a discharge rating curve. Manning's equation is applied to a cross section representative of the reach to obtain the rating curve. The slope of this curve at a discharge representative of the magnitude of flows which will be routed through the reach is equal to dQ/dy. After the wave velocity is determined from Eq. (4.12), K is found as the ratio of reach length to V_W with Eq. (4.13).

$$V_W = \frac{1}{B} \frac{dQ}{dy} \tag{4.12}$$

where V_W = wave velocity, ft/s
$\quad\quad B$ = channel width, ft

$\dfrac{dQ}{dy}$ = slope of discharge rating curve

$$K = \frac{L}{V_W} \tag{4.13}$$

where L is the reach length in feet.

An examination of X values for similar streams may yield values that can be transferred to ungaged reaches, but it is not a straightforward procedure because allowances must be made for differences in the effects of reach length, travel time, and time interval of the ungaged reach.

Selecting the number of subreaches. The Muskingum equation has a constraint related to the relationship between the parameter K and the time interval Δt [Eq. (4.14)]. Ideally, the two should be equal, but Δt should not be less than $2KX$ to avoid negative coefficients and instabilities in the routing.

$$2KX < \Delta t \leq K \tag{4.14}$$

A long routing reach should be subdivided into subreaches so that the travel time through each subreach is approximately equal to the routing interval Δt. That is,

$$\text{Number of subreaches} = \frac{K}{\Delta t}$$

This assumes that factors such as channel geometry and roughness have been taken into consideration in determining the length of the routing reach and the travel time K.

Muskingum method example. The inflow hydrograph shown in Table 4.4 is to be routed down a 2.66-mi reach of river with the Muskingum routing method. The average flood-wave velocity in this reach is estimated to be 3 ft/s, on the basis of channel geometry and bed roughness. The inflow hydrograph with a 1-h time interval is shown in column (1) of Table 4.4.

The Muskingum parameter K is determined to be 1.3 h with Eq. (4.11), and parameter X is estimated to be 0.1, on the basis of the large number of small ponds and reservoirs along the routing reach. The computational time interval of 1 h is found to be within the bounds set by criterion (4.14). The routing coefficients C_0, C_1, and C_2 are calculated to be 0.2216, 0.3772, and 0.4012, respectively.

The initial inflow and outflow values are assumed to be equal. The routing equation [Eq. (4.8)] is solved for successive time increments, with O_2 for one routing period becoming O_1 for the next period. For example, for the time interval that ends at 0900, $Q_1 = 11$ ft^3/s, $I_1 = 11$ ft^3/s, $I_2 = 26$ ft^3/s, and O_2 is computed to be 14 ft^3/s. For the next time interval that ends at 1000, $Q_1 = 14$ ft^3/s, $I_1 = 26$ ft^3/s, $I_2 = 117$ ft^3/s,

TABLE 4.4   Muskingum Routing Calculation

Time, h	Inflow, ft^3/s (1)	C_0I_2 (2)	C_1I_1 (3)	C_2O_1 (4)	Outflow, ft^3/s (5)
0800	11				11
0900	26	6	4	4	14
1000	117	26	10	6	42
1100	297	66	44	17	127
1200	473	105	112	51	268
1300	649	144	178	108	430
1400	865	192	245	173	610
1500	964	214	326	245	785
1600	814	180	364	315	859
1700	577	128	307	345	780
1800	427	95	217	313	625
1900	316	70	161	251	482
2000	211	47	119	193	359
2100	140	31	80	144	255
2200	96	21	53	102	176
2300	93	21	36	71	128
2400	91	20	35	51	106

and O_2 is calculated to be 42 ft^3/s. A complete tabulation of the routing calculations is shown in Table 4.4.

Applicability of routing methods

With a multitude of routing methods available, questions have arisen as to the applicability of various methods to different routing problems. One concerns the conditions under which it is reasonable to use one of the hydrologic methods, such as those in HEC-1, and when an unsteady-flow model should be used. Another, more specific question regarding one of the methods in HEC-1 concerns the applicability of the modified Puls method to channel routing. There is a difference of opinion on this question due to the lack of a physical basis for defining the number of steps or routing reaches to be used. The number of steps chosen affects the attenuation of the hydrograph, and proponents of the method would probably agree that this parameter should be verified with observed data whenever possible, through the use of model calibration procedures.

Three investigations that have been conducted to evaluate the accuracy and reliability of the modified Puls method are briefly outlined below. The conclusions of these investigations are too involved to be reviewed in detail. The reader should refer to the reports cited for more information.

One comparative study investigated conditions under which the modified Puls technique yields results close to the complete one-dimensional unsteady-flow solution. The influence of channel slope, roughness, cross-sectional configuration, base flow, and inflow hydrograph peak and rise time were analyzed. The modified Puls method was found to give good results under a number of limiting conditions.[8]

Testing of the modified Puls method in the field in another study revealed certain conditions for which the method is not suitable. In the reach studied, the modified Puls method using a storage-outflow relationship based on steady-flow profiles did not produce satisfactory results. Storage was very loosely related to outflow, and the shape of the flood wave and storage volumes were found to be quite different from those indicated by the steady-flow profile. These results were attributed to a small channel slope and the broad, flat floodplain in which the channel meanders. Unsteadiness caused by the small channel slope greatly affected the depth gradient, which was very large relative to other forces affecting the flow. The effect of depth on storage was highly exaggerated by the broad, flat floodplain, and the storage-outflow function was highly event-dependent.[9]

A third study[10] compared the performance of selected unsteady-flow models with the modified Puls method in HEC-1 applied to dam-break analyses. Comparisons were made between the results of computations of maximum water surface elevation, travel time, and maximum discharge by HEC-1 and three unsteady-flow models. The findings indicate that HEC-1 produces results which are quite comparable to the unsteady-flow models in computing maximum water surface elevations for the conditions studied. In this case, there were steep valley slopes, averaging approximately 30 ft/mi. Under these conditions, the bed slope is the dominant term in the Saint-Venant equations, and a procedure such as the modified Puls, utilizing normal-depth storage-outflow relationships, could be expected to perform well.

Since the evaluation of unsteady-flow methods is beyond the scope of this book, only a general suggestion is offered relative to their use vs. the use of the methods in HEC-1. Application of the empirical methods in HEC-1 to large events outside the range of events with which the model has been calibrated is subject to question. Empirical methods have been found to work reasonably well in "descriptive applications" within the range of observed data. The routing of dam-break floods and other large events that go well beyond the range of the observed data should probably be analyzed with the more physically based hydraulic methods.

Workshop Problem 4.1: Flood Routing in the Rahway River Basin

Purpose

The purpose of this workshop problem is to provide an understanding of the Muskingum and modified Puls methods of flood routing applied to channels.

Problem statement

The map of the Rahway River basin shown in Fig. P4.1 shows the reaches of river for which routing is to be performed.

Part 1: Muskingum channel routing

Description: The reach from Orange Reservoir to Millburn on the West Branch has numerous small reservoirs and ponds along its length. The reach has the following hydraulic characteristics:

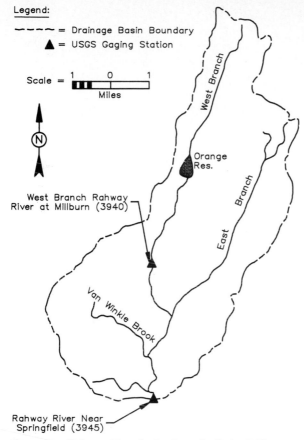

Figure P4.1   Rahway River basin above the Springfield gage.

Channel reach length = 2.6 mi

Channel slope = 40 ft/mi

Hydraulic radius = 2 ft

Composite Manning's $n = 0.07$

Task

Estimate the Muskingum routing parameters K and X for this reach.

Part 2: modified Puls routing, Millburn to confluence with East Branch

Description: The reach from Millburn, on the West Branch, to the confluence with the East Branch has the following channel characteristics:

Channel reach length = 1.4 mi

Channel slope = 30 ft/mi

Hydraulic radius = 1.5 ft

Composite Manning's n = 0.07

Storage-discharge data developed from steady water surface profile computations is as follows:

Cross section location	Discharge, ft³/s	Accumulated volume, acre • ft
Confluence at East Branch	0	0
	2100	1285
	3500	2545
	6905	3460
Millburn	0	0
	2100	1405
	3500	2725
	6905	3900

The average flood velocity is assumed to be the same as the average channel velocity computed with Manning's equation.

Tasks

Route the flood hydrograph from Millburn to the East Branch confluence using the modified Puls method, and determine the peak discharge at the confluence. Neglect tributary inflow. Develop and plot a storage-indication curve on 2 × 2 cycle logarithmic graph paper. Use a tabulation patterned after the one illustrated in Table 4.2 to compute points for the curve. Route the hydrograph using a tabulation similar to that shown in Table 4.3 for the routing computations. Plot the inflow and outflow hydrographs for the reach on the same graph, and determine if the routed hydrograph peak has been attenuated. Discuss any unexpected results noted in the plot. Determine the number of time steps used, and evaluate the validity of the 1-h time step used in the computations.

References

1. F. M. Henderson, *Open Channel Flow,* MacMillan, New York, 1966.
2. P. S. Eagleston, *Dynamic Hydrology,* McGraw-Hill, New York, 1970.
3. Phillip B. Bedient and Wayne C. Huber, *Hydrology and Flood Plain Analysis,* Addison-Wesley, New York, 1988.

4. D. L. Fread, *National Weather Service Operational Dynamic Wave Model,* National Weather Service, Silver Springs, Md., April 1978.
5. Hydrologic Engineering Center, *HEC-1 Flood Hydrograph Package, Users Manual,* U.S. Army Corps of Engineers, Davis, Calif., 1981 (rev. 1987).
6. Ray K. Linsely, Jr., Max A. Kohler, and Joseph L. H. Paulhus, *Hydrology for Engineers,* 3d ed., McGraw-Hill, New York, 1986.
7. *Routing of Floods through River Channels,* U.S. Army Corps of Engineers. Engineering Manual EM 1110-2-1408, March 1960.
8. Hydrologic Engineering Center, *Comparative Analysis of Flood Routing Methods,* U.S. Army Corps of Engineers, Davis, Calif., 1980.
9. Hydrologic Engineering Center, *Modified-Puls Routing in Chiquatonchee Creek,* Research Document 23, U.S. Army Corps of Engineers, Davis, Calif., September 1980.
10. Hydrologic Engineering Center, *Guidelines for Calculating and Routing a Dam Break Flood,* HEC Research Note 5, U.S. Army Corps of Engineers, Davis, Calif., January 1977.

The HEC-1 Flood Hydrograph Package and Parameter Estimation

Overview of the HEC-1 Program Capabilities

Introduction

The HEC-1 computer program has been developed for use in analyzing the hydrologic processes of flood events in river basins varying in size and complexity from a small urban catchment to a large multibasin river system. It is used as a basic tool to determine runoff from synthetic as well as historical events, and it has several analytical options.

Choice of metric or English units

Either metric or English units of measure can be used in the program. English units are used unless otherwise specified in input. The selection of metric units is indicated on an IM line of input immediately in front of the first KK line in the data input file. Lines of input, such as the IM and KK lines, and the composition of an input file are explained later in this chapter. The metric units used in HEC-1 are as follows:

Drainage area: square kilometers

Precipitation depth: millimeters

Length, elevation: meters

Flow: cubic meters per second

Storage volume: cubic meters

Surface area: square meters

Temperature: degrees Celsius

Since the metric unit of length (meter) used by the program has a significantly greater magnitude than the English unit (foot) used by the program, it is important to be aware of this in evaluating lengths, areas, and volumes in program output. It is also important to make sure that units used for data input are consistent with the system of units selected for use in the program computations.

Data storage system

The data storage system HECDSS, including a set of supporting utility programs, has been developed by the Hydrologic Engineering Center to manage data sets commonly used in water resources applications. This system has the capability to organize, store, and transfer blocks of continuous data. It allows "(1) storing and maintaining data in a centralized location, (2) preparing input to and storing output from application programs [such as HEC-1], (3) transferring data between applications programs, and (4) displaying the data in graphs or tables for interpretation."[1]

A PC version of HECDSS, including the utility programs DSSUTL and DSPLAY, is under development and is expected to be released in the fall of 1988. Subsequent to this release, the PC version of HEC-1 will have the capability to read data from and write data to HECDSS, and the utility programs DSSUTL and DSPLAY will provide capability for editing and displaying the information in this data base.

Basic modeling capability

Fundamentally, the HEC-1 program has the capability to simulate the precipitation-runoff process and compute flood hydrographs at desired locations in a basin. The physical characteristics of the river basin are represented by an interconnected system of geographic and hydrologic components. The basin boundaries are delineated, and the land area is divided into subbasins on the basis of study objectives and hydrologic characteristics. The model is structured so that lumped parameters can be used to simulate the precipitation-runoff process. Channel routing reaches provide the connecting link between subbasins. Thus, the basic hydrologic components of the model include land-surface runoff from each subbasin, channel and reservoir routing, and the combining of hydrographs at confluences.

Major program options

In addition to the basic rainfall-runoff simulation capability, the program has several significant optional features. These include computations for snowfall and snowmelt simulation, dam safety analysis, pumping and diversion schemes, parameter estimation, multiple-flood and multiple-plan analyses, and simulation of precipitation depth–area relationships.

Snowfall and snowmelt. Snowfall and snowmelt can be simulated in HEC-1 for up to 10 different elevation zones in each subbasin. A lapse rate is used to translate temperature data input for the lowest zone to higher zones, and a base temperature indicating freezing conditions determines if precipitation is falling as rain or snow. The program has an optional capability to use either the degree-day method or an energy budget method to compute snowmelt in each zone.

Dam safety analysis. In dam safety investigations, overtopping conditions can be analyzed. If the dam has the potential of eroding under these conditions, the formation of a breach also can be simulated through the use of a trapezoidal or triangular weir configuration that changes with time.

Pumping. A pumping option is particularly useful in solving interior drainage problems, such as the ponding of water behind a levee. In this option, pumping at a constant rate is simulated with the capability to automatically turn pumping on and off at preselected pool levels. Pumped flows may be retrieved at any downstream analysis point, i.e., at a subbasin outlet or at the end of a routing reach.

Diversions. Diversions can also be simulated at any analysis point in a stream network. Diversion rates are computed as a function of flow in the main channel on the basis of rating curves or the hydraulic characteristics of the diversion structure. Flows can be diverted out of the basin permanently or retrieved at a downstream point.

Estimation of rainfall-runoff parameters. Estimation of parameters is an essential part of the modeling process, and HEC-1 has an optimization capability for automatically estimating unit hydrograph and

loss-rate parameters for individual basins. The optimization technique determines a set of parameters that best reproduces an observed
runoff hydrograph from an observed precipitation event. Unit
hydrograph and loss-rate parameters can be determined individually
or in combination. Initial estimates of the parameters are specified in
data input or are selected automatically by the program. Details on
the theory and operation of this technique are covered later in this
chapter.

Multiple-flood–multiple-plan analyses.   Several different sizes of floods
are analyzed in one run with the multiple-flood capability of HEC-1.
Each flood is computed as a ratio of a base flood event. Thus, floods
ranging from those which are a fraction of a single base event to ones
several times larger are computed in one run. This capability is particularly useful in comparing the effects of events of different sizes in
floodplain studies.

The multiple-plan option modifies a basin model to reflect changes
such as channel improvements, the addition of reservoirs, and new
land uses. The hydrologic impact of several alternative flood control
plans or basin development scenarios can be analyzed in a single run.
These options are discussed in detail in Chap. 9.

Simulation of the precipitation depth–area relationship.   The computation
of flood hydrographs at a number of different locations in a river basin
on the basis of a design storm would normally require runoff simulations for storms centered at several different points upstream from
each location of interest. This need arises because the average depth
of precipitation over the area covered by a storm generally decreases
as the size of the area increases. Design storms are derived on the basis of precipitation amounts observed on a given area. A storm centered so that it produces a flood with a certain frequency at one downstream location does not produce a flood with the same frequency at
other downstream locations. Conversely, a design storm centered at different points in a complex watershed does not produce consistent
hydrographs at the same downstream location.

To avoid the many executions of the model that normally would be
required to compute a consistent flood of a specified frequency and duration at numerous locations throughout a large river basin, a precipitation depth–area computational capability has been incorporated in
HEC-1. HEC-1 simultaneously simulates runoff from several rainfall
quantities, each of which corresponds to a specific subbasin size and to
a specific precipitation depth–drainage area relationship. The depth-
area computation uses index hydrographs derived from a range of pre-

cipitation depth–drainage area values, a time distribution of rainfall pattern, and appropriate parameters for precipitation loss and unit hydrographs.[2] A description of the computational procedure and an example application are presented in Chap. 8.

Structuring the model

Definition of river basin components.   One of the first steps in developing a precipitation-runoff model is to define the boundaries of the basin and, typically, divide the basin into a number of smaller subbasins on the basis of study objectives and other factors. To illustrate, subbasins are delineated with dashed lines and identified with numbers in small boxes in Fig. 5.1a.

In subdividing the basin, it is important to identify points where runoff information is needed. The model can be structured to produce hydrographs at any location desired, and a virtually unlimited number of components can be used. Study objectives largely determine the level of detail that is appropriate. Since different areas of a large basin can be expected to have different hydrologic response characteristics, it is also important to select an appropriate computational time interval and subdivide the area so that lumped parameters provide a reasonable characterization of conditions.

An HEC-1 model has three basic building blocks: a subbasin runoff component, a routing component, and a hydrograph combining component. Only the first of these is employed in a single-basin model. A more complex model with subbasins requires all three. The components for the basin depicted in Fig. 5.1a are represented with symbols in Fig. 5.1b.

The simulation process.   To provide a better understanding of how the components function in the model, the simulation process is described with reference to Fig 5.1. The first step in the simulation is the computation of runoff for one of the headwater subbasins, such as number 30. This produces a hydrograph at the outlet of 30, which is then routed to combining point 20. Next, the runoff hydrograph for subbasin 40 is computed at its outlet, which coincides with combining point 20. In a similar manner the hydrograph for subbasin 10 is computed and routed to combining point 20, and the hydrograph for subbasin 20 is computed. Thus, there are four hydrographs at combining point 20: two that have been routed from headwater subbasins and two that have been computed as local inflow at point 20.

To continue the simulation downstream, these four hydrographs are combined, and the combined hydrograph is routed to combining point 50, where it is combined with local inflow from subbasin 50. This com-

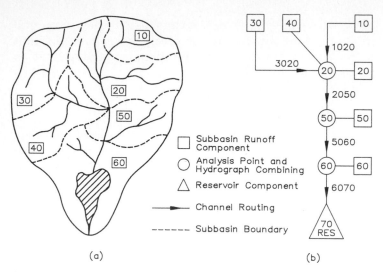

Figure 5.1 River basin simulation. (*a*) Map of subdivided basin; (*b*) network diagram of model components.

bined hydrograph is routed to combining point 60, and so on, until a hydrograph for the entire basin is computed at the outlet of reservoir 70. Any number and arrangement of components for a complex river basin can be simulated in this manner.

Data Input Requirements

Model parameters

Each of the subbasins in the model must be defined with appropriate parameters representing the precipitation-runoff processes. These include parameters for precipitation loss, the unit hydrograph or kinematic-wave transformation method, and base flow. The model must also have precipitation data input and routing parameters, and in some cases it may be appropriate to enter a known discharge hydrograph as input.

Data input file

Since HEC-1 utilizes batch processing of data, an input file in a prescribed format must be prepared for entering the data. Input files are structured to reflect the topography and hydrology of the basin; the sequence of the data prescribes how the basin is to be simulated.

Input files are composed of lines of data laid out in 10 fields of eight columns each. Each line of input, sometimes referred to as a "record"

TABLE 5.1   HEC-1 Input Data Identification Scheme

Data category	Record identification	Description of data
Loss-rate data	LE	Exponential rainfall loss rate function
	LM	Exponential Snowmelt function
	LU	Initial and uniform rates
	LS	SCS curve number
	LH	Holtan's function

or a "card," contains a specific set of data identified by the characters entered in the first two columns of the line. Input variables are entered in the remaining fields of a line, one per field. If decimal points are not indicated, all numbers must be right-justified within the field. Blank fields are read as zeros.

The two identification characters are the basis of a two-tier classification system of data input. The first character identifies a data category, and the second indicates a type of data or option within that category. For example, an "L" in the first column (Table 5.1) indicates precipitation loss data, which has five subcategories, as indicated by the second character in the line identifiers—LE, LM, LU, LS, and LH. A list of data categories and detailed information on different lines of input are presented in the Appendix A of the user's manual.[2] Line input descriptions and variable definitions also are provided in help files of the text editor COED, described in Chap. 1.

Consistent with the standard format for input files, an 80-column, 10-field heading (Fig. 5.2) is used in illustrations of input data throughout the text.

ID.....1.......2.......3.......4.......5.......6.......7.......8.......9......10

Figure 5.2   Standard format for lines of input: 80 columns, 10 fields, line identifiers in columns 1 and 2.

Input file organization

There are two general types of data in HEC-1 input files: input control and simulation data. Input control lines, identified with an asterisk in the first column, identify file format and control optional diagnostic output. Simulation data lines identify and activate simulation processes and provide required data inputs.

Input control.   Input control commands include *FREE, *FIX, *LIST, *NOLIST, *, and *DIAGRAM. The commands *FREE and *FIX provide options to use either free-format or fixed-format input structures, or both, in an input file. If neither command is entered, the program defaults to fixed-format mode, and all lines of data must be structured in ten 8-column fields. Lines of data following a free-format command, *FREE, are entered in the file in free format, i.e., with data fields separated by commas or blanks and successive commas used to indicate blank spaces in a line. A change from one mode to the other—i.e., from free format to fixed, and vice versa—can be implemented at any line in the input file by including either the *FREE or the *FIX command as appropriate.

A preprocessor in the program converts free-format data into the fixed-format structure and lists the fixed-format input file at the beginning of the output. This "echo print" of input can be turned off or turned back on again at any line in the input file by entering a *NOLIST line or a *LIST line, respectively. An "*" in the first column followed by a blank in the second column of a line indicates an alphanumeric comment, which can occupy the remaining columns, 3 through 80. Comments can be entered at any location in the input file, but they are printed only with an input echo listing, i.e., when the *LIST option is implemented, either by default or by direction. A stream network diagram is printed at the beginning of output when the *DIAGRAM command is used. This is particularly useful in verifying that the sequence of computations conforms to the intended analysis of subbasins and routing reaches. Components of the model inadvertently omitted may be identified by reviewing this network diagram.

Simulation data. Three broad subdivisions of simulation data are shown in Table 5.2. These follow a specific sequence in an input file;

TABLE 5.2   Subdivisions of Simulation Data

Job control	Hydrology and hydraulics	Economics and end of job
I __, job initialization	K __, job step control	E __, etc., economics data
V __, variable-output summary	H __, hydrograph transformation	ZZ, end of job
O __, optimization	Q __, hydrograph data	
J __, job type	B __, basin data	
	P __, precipitation data	
	L __, loss (infiltration) data	
	U __, unit graph data	
	M __, melt data	
	R __, routing data	
	S __, storage data	
	D __, diversion data	
	W __, pump withdrawal data	

the job control data comes first, followed by the hydrology and hydraulics data, and then the economics data, if included. The job control data, beginning with an ID line, contains an alphanumeric description of the job and sets the job type, output control, time interval and time span, and type of units to be used. The hydrology and hydraulics data includes all the data necessary to simulate the various physical processes in the river basin. Each hydrograph calculation begins with a KK line, followed by other lines providing data specifying how the hydrograph is to be calculated. The economics data, which is optional, begins with an EC line; it is used when expected annual damages are calculated. A ZZ line is the last line in an input file, signaling to the program the end of the job.

The sequence of lines of data for a typical job is shown in Table 5.3a, with KK groups for subbasin runoff and routing shown in Table 5.3b and c.

TABLE 5.3   Input File Structure

Line identification	Description of data
(a) For a Typical Job File	
ID	Job identification
IT	Time specification
I__*	Additional initialization data
J__*	Job type
O__*	Optimization
VV*, VS*	Variable-output summary tables
KK	Hydrograph computation identification
.	KK-record groups describing *runoff, routing, combining,* etc., components are repeated as necessary to simulate the processes and connectivity of a river basin.
.	
.	
.	
.	
EC*	Economic data identification
ZZ	End-of-job record
(b) For Hydrograph Computation	
KK	Hydrograph computation identification
BA	Basin area
BF*	Base flow data
P__	Precipitation data
L__	Loss data
U__	Unit graph or kinematic-wave data
(c) For Routing Computation	
KK	Hydrograph computation identification
R__	Routing option
S__*	Reservoir data or dam-break analysis

*Optional records.

```
                         HEC-1 INPUT DESCRIPTION
                     JOB INITIALIZATION (I Records)


    2.2 ** IT RECORD - TIME SPECIFICATION


        The IT record is used to define time interval, starting date and time, and
    length of hydrographs calculated by the program.


    FIELD    VARIABLE    VALUE    DESCRIPTION


    Col 1+2  ID          IT       Record identification.

      1      NMIN        +        Integer number of minutes in tabulation
                                  interval.  Minimum value is one minute.

      2      IDATE*      +        Day, month, and year of the beginning of the
                                  first time interval (e.g., 17MAR78 is input for
                                  March 17,1978).  Required to specify pathname
                                  part D when using DSS.

      3      ITIME*      +        Integer number for hour and minute of the
                                  beginning of the first time interval (e.g., 1645
                                  is input for 4:45 pm).

      4      NQ          +        Integer number of hydrograph ordinates to be
                                  computed (300 max).  If end date and time are
                                  specified in Fields 5 and 6, NQ will be
                                  computed from the beginning and end dates and
                                  times.

      5      NDDATE      +        Day, month, and year of last ordinate (used to
                                  compute NQ).

      6      NDTIME      +        Integer number for time of last ordinate (used to
                                  compute NQ).


    *CAUTION: IDATE and ITIME are the time of initial flow conditions.  No runoff
    calculations are made from precipitation preceding this time.

    Use 3-character code for month: JAN, FEB, MAR, APR, MAY, JUN, JUL, AUG, SEP,
    OCT, NOV, DEC.  Use of any other code for month means this is not a date, and
    days will be numbered consecutively from the given day.  Default is day = 1.


    **REQUIRED
```

Figure 5.3 Input description for the IT line.

Time data for HEC-1 is entered on IT and IN lines. One time interval and one time span are used for calculating runoff hydrographs, and these are defined with variables specified on the IT line. The input description for the IT line, typical of input line descriptions in the appendix of the user's manual,[2] is shown in Fig. 5.3. The computational time interval has the variable name NMIN; the starting date and time of the simulation are designated as IDATE and ITIME; the number of hydrograph ordinates to be computed can be specified with NQ or determined by the program from the ending date and time of the simulation, NDDATE and NDTIME, if they are specified. It should be noted that 300 is the maximum number of ordinates that can be used, including the one at the starting date and time.

Time-series data may be input into HEC-1 with different time parameters than those used in the simulation. For example, the precipitation data can have different starting and ending dates and times and a different time interval. These are specified on an IN line located in the file just before the precipitation data. The precipitation data is read by the program and interpolated to fit the simulation time conditions that have been specified on the IT line.

Certain physical characteristics of several subbasins in a model may be the same, and in a multiple-plan analysis, characteristics for some subbasins may not change between plan 1 and subsequent plans. To avoid the repetition of data input for these conditions, HEC-1 has the capability to automatically repeat data for certain variables when no change is indicated (Table 5.4). This does not apply to routing data.

TABLE 5.4   Data Repetition Options

Data types which are automatically repeated	Line identification
Rainfall	P
Infiltration	L
Base flow	BF
Snowmelt	M
Unit hydrograph*	US, UC, UD
Kinematic wave*	UK, RK†

*Not recommended.
†Only in multiplan analysis when all lines of input remain unchanged.

Data input options

To facilitate the preparation of input files, tables have been prepared identifying options for data input. Several options for precipitation data are shown in Table 5.5. The data can be input as basin average rainfall, gage data, synthetic storm data, etc. Data lines to be used

TABLE 5.5   Precipitation Data Input Options

Type of storm data	Line identification
Basin average storm depth and time series	PB and/or (PI or PC)
Recording and nonrecording gages	PG for all nonrecording gages PG and (PI or PC) for all recording gages PR, PW, PT, PW for each subbasin
Synthetic storm from depth-duration data	PH
Probable maximum storm	PM
Standard project storm	PS
Depth-area with synthetic storm	JD, PH, or (PI or PC)

with the various options are shown. Reference should be made to the user's manual[2] for similar tables on hydrograph computation, hydrograph optimization, channel and reservoir routing, and other model processes.

Example input file

An example data input file in fixed format for a single basin is shown in Fig. 5.4a. A listing in free format is shown in Fig. 5.4b. An example of an input file for a three-subbasin model, including routing and combining components, is presented in Chap. 6. The first three lines in Fig. 5.4 contain job control data, and the following lines down to the ZZ line contain hydrology and hydraulics data.

The lines from the file containing job control data are shown in Fig. 5.5. The first two lines are "ID" lines on which titles can be written without regard to fields. The IT line is for date and time specifications. It might be noted that the simulation time interval of 30 min in the first field of the IT line contrasts with a precipitation data interval of 60 min on the IN line farther down in the file. Whenever there is a difference like this, the data is interpolated automatically to conform to the time interval specified on the IT line. Thus, the 60-min rainfall amounts on the PI lines are automatically interpolated to conform to the 30-min interval. Zeros appear in fields 2 and 3 of the IT line in this example. Normally, the starting date and time of the simulation would be entered in these fields. The number of hydrograph ordinates to be computed, 51, appears in field 4.

```
ID.....1.......2.......3.......4.......5.......6.......7.......8.......9......10

ID  FROG CREEK BASIN EXAMPLE
ID  CALCULATE RUNOFF FOR A SINGLE BASIN
IT    30      0      0      51
KK  FROG
KM  CALCULATE RUNOFF AT FROG CREEK
BA   5.8
BF  -2.3    -.10    1.03
PB
IN    60
PI     0     .10     .22     .15      0     .08     .13     .09     .20     .33
PI   .29     .36     .51     .42     .13     .05     .18
LU   0.8    0.10
US   2.3     0.6
ZZ


                               (a)


ID  FROG CREEK BASIN EXAMPLE
ID  CALCULATE RUNOFF FOR A SINGLE BASIN
*FREE
IT  30,0,0,51
KK  FROG
KM  CALCULATE RUNOFF AT FROG CREEK
BA  5.8
BF  -2.3,-.10,1.03
PB
IN  60
PI  0,.10,.22,.15,0,.08,.13,.09,.20,.33,.29,.36,,51,.42,.13,.05,.18
LU  0.8,0.10
US  2.3,0.6
ZZ

                               (b)
```

Figure 5.4   Example data input file. (a) Fixed format; (b) free format.

```
ID.....1.......2.......3.......4.......5.......6.......7.......8.......9......10

ID  FROG CREEK BASIN EXAMPLE
ID  CALCULATE RUNOFF FOR A SINGLE BASIN
IT    30      0      0      51
```

Figure 5.5   Job control data.

```
ID.....1.......2.......3.......4.......5.......6.......7.......8.......9......10

KK  FROG
KM  CALCULATE RUNOFF AT FROG CREEK
BA    5.8
BF  -2.3    -.10    1.03
```

Figure 5.6 Hydrologic data: station name and description, basin area, and base flow parameters.

The first four lines of the hydrologic group of data are shown in Fig. 5.6. The first line in this group is a KK line containing the station name FROG in the first field. This identifier is limited to six characters in length. The KM line that follows is a message line containing alphanumeric data describing the computation. The BA line indicates basin area in the first field. Three base flow parameters—STRTQ, QRCSN, and RTIOR—are entered on the BF line. The minus signs on the first two indicate that an optional method of specifying these parameters is being used. The value for STRTQ, when negative, is multiplied by the basin area to obtain the starting base flow. The value for QRCSN, when negative, is a ratio that is applied to the peak discharge to obtain the base flow recession threshold. RTIOR is a ratio used in defining the rate of recession.

The next four lines in the file (Fig. 5.7) describe the precipitation input. The PB line without any data indicates that the basin average precipitation for the total storm will be computed from the incremen-

```
ID.....1.......2.......3.......4.......5.......6.......7.......8.......9......10

PB
IN    60
PI     0    .10    .22    .15     0    .08    .13    .09    .20    .33
PI   .29    .36    .51    .42    .13    .05    .18
```

Figure 5.7 Hydrologic data: precipitation data.

tal values specified on the PI lines. The IN line indicates the time interval of the incremental precipitation data entered on the PI lines.

The next two lines of the input file (Fig. 5.8) contain precipitation loss function and unit hydrograph parameters. The LU line indicates

```
ID.....1.......2.......3.......4.......5.......6.......7.......8.......9......10

LU   0.8    0.10
US   2.3    0.6
ZZ
```

Figure 5.8 Hydrologic data: rainfall loss and unit hydrograph parameters.

that the initial and uniform loss method will be used with an initial loss of 0.8 in and a uniform loss of 0.10 in/h. The US line specifies the Snyder unit hydrograph parameters t_p and C_p in fields 1 and 2, respectively. The final line of input in the file has the identifier ZZ with no other entries. This is the end-of-job line that causes the summary computations and printout to occur.

Program Output

General

Output consists of an input file listing, intermediate simulation results, summary results, and error messages. The extent and degree of detail obtained in the output are controlled by data entries on the IO line; however, these output specifications can be overridden by KO lines for specific KK groups. The input file listing in fixed format and several of the summary tables are written to scratch files[2] that may be saved by renaming or copying to other files.

Input file listing

The input file is listed in fixed format in output. Input data, whether in fixed or free format, is written to a working file in fixed format and assigned line numbers before processing by the program.

If a *DIAGRAM command is included in input, a schematic diagram of the stream network is printed following the data listing. An example network for the multiple-subbasin model described in Chap. 6 is shown in Fig. 5.9. The program scans the line identification codes in the input file to generate this diagram, so a review of this schematic provides a good check on the completeness of the file. The diagram also provides a check on the sequence of the lines of data by showing how the hydrographs are stored in computer memory. As a new branch is added to the diagram, a new hydrograph is added to storage. Except for the diversion hydrographs, which are stored in a separate file, each new hydrograph replaces in computer memory the one printed above it.

Intermediate simulation results

As controlled by the IO line in general, or the KO line for a specific KK group, the data used in each hydrograph computation, as well as each computed hydrograph, can be included in output. The input data upon which the calculations are based is printed together with the line identification codes and input file line numbers immediately before the computed hydrograph. An examination of this listing provides

```
                SCHEMATIC DIAGRAM OF STREAM NETWORK
INPUT
LINE      (V) ROUTING            (--->) DIVERSION OR PUMP FLOW

NO.       (.) CONNECTOR          (<---) RETURN OF DIVERTED OR PUMPED FLOW

 13     RED RI
          .
          .
 23       .      EAST10
          .        .
          .        .
 37       .        .-------> DIVERT
 34       .      EAST10
          .        .
          .        .
 40     RED10............
           V
           V
 44     10TO20
          .
          .
 52       .        .<------- DIVERT
 50       .      LOSTBR
          .        .
          .        .
 53       .        .      LOSTBR
          .        .        .
          .        .        .
 63       .      LOSTBR............
          .        V
          .        V
 66       .      E.DAM
          .        .
          .        .
 73       .        .      WEST20
          .        .        .
          .        .        .
 83     RED20.......................
           V
           V
 86     20TO30
          .
          .
 91       .      RED30
          .        .
          .        .
101     RED30............
```

Figure 5.9 Schematic diagram of stream network.

a valuable check to ensure that intended data is being used in the computations.

The hydrographs may be output in both tabular form and printer plots with the date, time, and sequence number of each ordinate included. Rainfall, rainfall loss, and rainfall excess also are included in the tables and plots. When observed hydrographs are input for particular locations, they are included in the tables and plots, thus facilitating comparisons with computed hydrographs.

If the parameter optimization option is used, the output contains values for each of the variables optimized and the objective function at each iteration of the process. This output can be reviewed to determine when and why values are changed in the optimization process and verify that results are reasonable. An example of optimization output is discussed in Chap. 6.

Summary output

Standard output includes a summary table showing peak flow, time of peak, average peak flows of various durations, accumulated drainage area, maximum stage, and time to maximum stage. Summaries for multiplan-multiflood analyses include peak flows and stages and corresponding times for each plan and ratio in the computation. User-defined tables of precipitation, precipitation loss, precipitation excess, observed and calculated flows, storage, and stage data can be specified on VS and VV lines.

Example output

Output for the one-basin example in the data input section above, excluding the input listing, is shown in Fig. 5.10. A heading at the top of the printout indicates the version of the program. Title data from the ID lines is printed after the banner and followed by hydrograph time data from the IT line and an indication of the units used in the output. Data used in the hydrograph computation is shown next; the line numbers and identifiers associated with the data are shown in the left margin. After the basin name listed on the KK line, the computational data is listed by variable name and value. For example, under "BASE FLOW CHARACTERISTICS," the variable STRTQ, representing starting flow, has a value of 13.34 ft^3/s. Under the heading "PRECIPITATION DATA," basin total precipitation is 3.24 in. Since total precipitation was not indicated on the PB line of input, this total was computed by summation of the incremental precipitation amounts on the PI lines. The incremental precipitation pattern shown

```
************************************
*                                  *
*  FLOOD HYDROGRAPH PACKAGE (HEC-1) *
*         FEBRUARY 1981             *
*       REVISED 6 FEB 87            *
*                                  *
*  RUN DATE 04/26/1988 TIME 15:35:58 *
*                                  *
************************************
```

```
************************************
*                                  *
*  U.S. ARMY CORPS OF ENGINEERS     *
*  THE HYDROLOGIC ENGINEERING CENTER *
*        609 SECOND STREET          *
*      DAVIS, CALIFORNIA 95616      *
*  (916) 440-3285 OR (FTS) 448-3285 *
*                                  *
************************************
```

            FROG CREEK BASIN EXAMPLE
            CALCULATE RUNOFF FOR A SINGLE BASIN

IT   HYDROGRAPH TIME DATA
          NMIN   30   MINUTES IN COMPUTATION INTERVAL
          IDATE   1   0   STARTING DATE
          ITIME   0000   STARTING TIME
          NQ   51   NUMBER OF HYDROGRAPH ORDINATES
          NDDATE   2   0   ENDING DATE
          NDTIME   0100   ENDING TIME
          ICENT   19   CENTURY MARK

          COMPUTATION INTERVAL   .50 HOURS
          TOTAL TIME BASE   25.00 HOURS

ENGLISH UNITS
    DRAINAGE AREA   SQUARE MILES
    PRECIPITATION DEPTH   INCHES
    LENGTH, ELEVATION   FEET
    FLOW   CUBIC FEET PER SECOND
    STORAGE VOLUME   ACRE-FEET
    SURFACE AREA   ACRES
    TEMPERATURE   DEGREES FAHRENHEIT

*** ***
```
```

```
***************
*             *
*    FROG     *
*             *
***************

4 KK        FROG

            CALCULATE RUNOFF AT FROG CREEK

9 IN    TIME DATA FOR INPUT TIME SERIES
            JXMIN      60   TIME INTERVAL IN MINUTES
            JXDATE      1    0   STARTING DATE
            JXTIME      0        STARTING TIME

        SUBBASIN RUNOFF DATA

6 BA    SUBBASIN CHARACTERISTICS
            TAREA     5.80   SUBBASIN AREA

7 BF    BASE FLOW CHARACTERISTICS
            STRTQ     13.34    INITIAL FLOW
            QRCSN      -.10    BEGIN BASE FLOW RECESSION
            RTIOR   1.03000    RECESSION CONSTANT

        PRECIPITATION DATA

8 PB    STORM     3.24   BASIN TOTAL PRECIPITATION

10 PI   INCREMENTAL PRECIPITATION PATTERN
            .00    .00    .05    .05    .11    .08    .07    .00    .00
            .04    .04    .0E    .06    .04    .10    .10    .17    .16
            .14    .14    .18    .18    .26    .21    .21    .07    .07
            .03    .02    .09    .09    .25

12 LU   UNIFORM LOSS RATE
            STRTL      .80   INITIAL LOSS
            CNSTL      .10   UNIFORM LOSS RATE
            RTIMP      .00   PERCENT IMPERVIOUS AREA

13 US   SNYDER UNITGRAPH
            TP       2.30   LAG
            CP        .60   PEAKING COEFFICIENT

        SYNTHETIC ACCUMULATED-AREA VS. TIME CURVE WILL BE USED
```

Figure 5.10   Example HEC-1 output.

APPROXIMATE CLARK COEFFICIENTS FROM GIVEN SNYDER CP AND TP ARE TC= 5.21 AND R= 4.63 INTERVALS

UNIT HYDROGRAPH PARAMETERS

CLARK TC= 2.60 HR,   R= 2.32 HR
SNYDER TP= 2.31 HR,   CP= .60

UNIT HYDROGRAPH
28 END-OF-PERIOD ORDINATES

87.	315.	612.	861.	970.	896.	729.	587.	473.	381.
307.	247.	199.	160.	129.	104.	84.	67.	54.	44.
35.	28.	23.	18.	15.	12.	10.	8.		

****

HYDROGRAPH AT STATION   FROG

****

DA	MON	HRMN	ORD	RAIN	LOSS	EXCESS	COMP Q
1		0000	1	.00	.00	.00	13.
1		0030	2	.00	.00	.00	13.
1		0100	3	.00	.00	.00	13.
1		0130	4	.05	.05	.00	13.
1		0200	5	.05	.05	.00	13.
1		0230	6	.11	.11	.00	12.
1		0300	7	.11	.11	.00	12.
1		0330	8	.08	.08	.00	12.
1		0400	9	.07	.07	.00	12.
1		0430	10	.00	.00	.00	12.
1		0500	11	.00	.00	.00	12.
1		0530	12	.04	.04	.00	11.
1		0600	13	.04	.04	.00	11.
1		0630	14	.06	.06	.00	11.
1		0700	15	.06	.06	.00	11.
1		0730	16	.05	.05	.00	11.
1		0800	17	.04	.04	.00	11.
1		1300	27	.26	.05	.21	649.
1		1330	28	.21	.05	.16	757.
1		1400	29	.21	.05	.16	865.
1		1430	30	.07	.05	.02	944.
1		1500	31	.07	.05	.02	964.
1		1530	32	.03	.03	.00	915.
1		1600	33	.02	.02	.00	814.
1		1630	34	.09	.05	.04	690.
1		1700	35	.09	.05	.04	577.
1		1730	36	.00	.00	.00	492.
1		1800	37	.00	.00	.00	427.
1		1830	38	.00	.00	.00	371.
1		1900	39	.00	.00	.00	316.
1		1930	40	.00	.00	.00	261.
1		2000	41	.00	.00	.00	211.
1		2030	42	.00	.00	.00	172.
1		2100	43	.00	.00	.00	140.

	TIME				
1	0830	.10	.07	.04	13.
1	0900	.10	.05	.05	26.
1	0930	.17	.05	.12	57.
1	1000	.16	.05	.11	117.
1	1030	.14	.05	.09	202.
1	1100	.14	.05	.09	297.
1	1130	.18	.05	.13	390.
1	1200	.18	.05	.13	473.
1	1230	.25	.05	.20	554.

	TIME							
1	2130	44	*	.00	.00	.00	114.
1	2200	45	*	.00	.00	.00	96.
1	2230	46	*	.00	.00	.00	95.
1	2300	47	*	.00	.00	.00	93.
1	2330	48	*	.00	.00	.00	92.
2	0000	49	*	.00	.00	.00	91.
2	0030	50	*	.00	.00	.00	89.
2	0100	51	*	.00	.00	.00	88.

****

TOTAL RAINFALL = 3.24, TOTAL LOSS = 1.63, TOTAL EXCESS = 1.61

PEAK FLOW	TIME		MAXIMUM AVERAGE FLOW			
(CFS)	(HR)		6-HR	24-HR	72-HR	25.00-HR
+		(CFS)				
+ 964.	15.00		723.	262.	252.	252.
		(INCHES)	1.158	1.680	1.684	1.684
		(AC-FT)	358.	520.	521.	521.

CUMULATIVE AREA = 5.80 SQ MI

1

RUNOFF SUMMARY
FLOW IN CUBIC FEET PER SECOND
TIME IN HOURS, AREA IN SQUARE MILES

OPERATION	STATION	PEAK FLOW	TIME OF PEAK	AVERAGE FLOW FOR MAXIMUM PERIOD			BASIN AREA	MAXIMUM STAGE	TIME OF MAX STAGE
				6-HOUR	24-HOUR	72-HOUR			
+ HYDROGRAPH AT									
+	FROG	964.	15.00	723.	262.	252.	5.80		

*** NORMAL END OF HEC-1 ***

Figure 5.10 (Continued)

after the basin total in the output has been interpolated to a 30-min time interval consistent with the simulation time interval.

The Snyder unit hydrograph parameters entered on the US line are listed in output. HEC-1 uses the Clark method to fit a hydrograph to the peak defined by the Snyder parameters. Recomputed Snyder parameters based on the Clark approximation are shown below the Clark parameters in the printout. The unit hydrograph ordinates are listed next.

The computed hydrograph for the basin is tabulated with values for rainfall, rainfall loss, and rainfall excess, corresponding to each computed discharge ordinate. The ordinates are numbered and spaced according to the simulation time interval, with the first ordinate located at the starting time of simulation.

A summary of the data for the basin, shown at the end of the tabulation, gives the totals for rainfall, rainfall loss, and rainfall excess; the time and amount of peak flow; maximum average flows for durations of 6, 24, and 72 h; and the average for the storm, which was 25 h in duration in this example. The value of 252 ft^3/s for the 25-h duration also appears for the duration of 72 h because of the short duration of this event. The average flows are shown in three different units: cubic feet per second, inches, and acre-feet.

The runoff summary at the end of the printout summarizes data for the entire simulation. It would include data for other subbasins if others were included in the model.

The last line of output is "NORMAL END OF HEC-1," indicating a normal termination of the program. If this line is not printed, an error may exist, and an investigation is warranted. Some errors do not terminate a run; the program may print an error message and continue processing the data. However, such errors may result in erroneous output, and their source and impact need to be understood. A list of error messages together with brief explanations is contained in the user's manual.[2]

Parameter Estimation

Introduction

General. Calibration of a model with historical data is an essential part of model development. It is a procedure of adjusting parameters until the model satisfactorily simulates the observed physical processes in the hydrologic system and produces reliable results. The parameter optimization option in HEC-1 estimates unit hydrograph, rainfall loss function, and routing parameters using gaged precipitation and runoff data from the same event. The program automatically determines a set of parameters that "best" replicates an observed run-off hydrograph for a subbasin.

The calibration concept. As a consequence of precipitation from a storm, a hydrograph is observed at a streamflow gaging station at the outlet of a subbasin. Utilizing observed precipitation data for the event in a rainfall-runoff model with specified unit hydrograph and rainfall loss parameters produces a computed hydrograph at the same outlet.

The goal of calibration is to obtain a set of parameters so that the model will respond like the physical system it represents. The model must compute a hydrograph that is essentially the same as an observed one, not only for the calibration event but for others as well. This is normally accomplished with a trial-and-error procedure in which a hydrograph is computed with an initial set of parameters and compared with the observed hydrograph. The parameters are adjusted on the basis of the comparison, and the procedure is repeated until a suitable fit is obtained. A schematic diagram of the trial-and-error procedure used in parameter estimation is shown in Fig. 5.11.

Parameter estimation in HEC-1

Optimization methodology. Since HEC-1 computes all the ordinates of the discharge hydrograph, a numerical index of closeness of fit of ob-

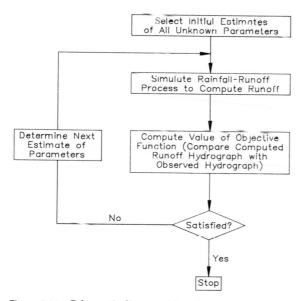

Figure 5.11   Schematic diagram of parameter estimation.

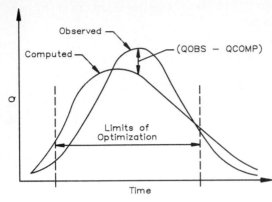

Figure 5.12   Index of closeness of fit of observed and computed hydrograph ordinates.

served and computed ordinates (QOBS minus QCOMP) is used over the full range of the optimization (Fig. 5.12). The objective function STDER in Eq. (5.1) is a root mean square error computation. It provides an index of replication of the observed hydrograph based on a variety of factors, including peak discharge, volume, time to peak, and shape.

The objective function is weighted [Eq. (5.2)] to emphasize the closeness of fit of flows higher than average. Thus, the reduction of differences of the highest flows yields the greatest reduction of the objective function. This emphasis on flows closest to the peak is consistent with the importance attached to peak flows in flood hydrograph analysis.

$$\text{STDER} = \sqrt{(1/n) \sum_{i=1}^{n} (\text{QOBS}_i - \text{QCOMP}_i)^2 * \text{WT}_i} \quad (5.1)$$

where STDER = root mean square error
$\quad$ n = number of observed hydrograph ordinates
$\quad$ QOBS_i = observed hydrograph ordinate for time period i
$\quad$ QCOMP_i = computed ordinate for period i
$\quad$ WT_i = weight for the squared difference between ordinates for i

$$\text{WT}_i = \frac{\text{QOBS}_i + \text{QAVE}}{2 * \text{QAVE}} \quad (5.2)$$

where QAVE is the average observed discharge.

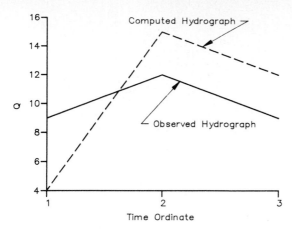

Figure 5.13 Simplified hydrographs for example computation of the objective function.

Computation of the objective function.   An example of the computation of the objective function [Eq. (5.1)] is shown with simplified hydrographs in Fig. 5.13 and tabulated in Table 5.6. The observed and computed discharge values are listed in columns (2) and (3), respectively. The weights in column (4) are obtained with Eq. (5.2). The objective function, measuring how well the computed hydrograph replicates the observed hydrograph, is computed with Eq. (5.1). The result is 3.75. How good is this match? If it were a perfect match, the summation of the differences between ordinates would be zero, and the objective function would equal zero.

The search for optimal parameters.   A plot of solutions to the objective function for a model with two parameters, the Clark unit hydrograph

TABLE 5.6   Objective Function Computation

Ordinate (1)	Observed discharge, ft^3/s (2)	Computed discharge, ft^3/s (3)	Weight (4)	Squared difference between observed and computed discharge (5)	Weighted squared difference (6)
1	9	4	0.95	25	23.75
2	12	15	1.25	9	9.90
3	9	12	0.95	9	8.55
Sum					42.20

Objective function = [column (6)/number of ordinates]$^{1/2}$ = $(42.20/3)^{1/2}$ = 3.75.

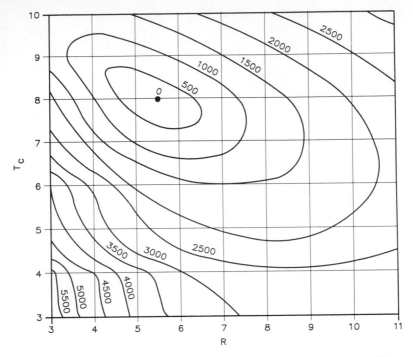

Figure 5.14 Illustration of search procedure for finding the optimal combination of two parameters, T_c and R.

parameters T_c and R, is shown in Fig. 5.14. Contours represent equal values of the objective function. The plot can be visualized as a topographic map with a low point of zero. In the calibration process, the program searches for the minimum value to determine an optimal combination of parameters. With a complete view of the contours, this task appears trivial; however, for the program to, in effect, have such a view is impractical because of the extensive computations required.

In a simple trial-and-error procedure, the parameters would have to be divided into parts or increments as represented by the grid in Fig. 5.14, and the objective function would have to be computed for each possible combination of values. This would require the computation of each point on the grid. The precision of this approach is limited by the size of the increments. Also, the cost may be prohibitive if the increments are extremely small or the model has a large number of parameters.

Univariate gradient search procedure in HEC-1. A univariate gradient search procedure is used in HEC-1 to determine the optimal parameter estimates. In this search procedure the optimum of the objective function occurs at the root of the first partial derivative of the function with respect to each of the parameters. These derivatives are difficult to solve analytically, so numerical approximations are obtained in a process of iteration. A single parameter is varied in each iteration, the derivatives are estimated numerically, and Newton's technique is used to improve parameter estimates.[3]

The range of feasible values of parameters is bounded by physical and numerical limitations represented by the constraints shown in Table 5.7. These constraints are imposed automatically by the program; other constraints which may be appropriate under certain circumstances would have to be imposed from outside the program.

TABLE 5.7   Constraints on Unit Hydrograph and Rainfall Loss Parameters In HEC-1

Clark Unit Graph Parameters	
$TC \geq 1.03\,\Delta t$	
$R \geq 0.52$	
Δt = computation interval	

Loss-Rate Parameters	
Exponential:	SCS:
ERAIN ≤ 1.0	$0 \leq CN \leq 100$
RTIOL ≥ 1.0	
Snowmelt:	
RTIOK ≥ 1.0	
$-1.11°C \leq FRZTP \leq 3.33°C$	
Uniform:	Holtan:
STRTL ≥ 0	FC ≥ 0
CNSTL ≥ 0	GIA ≥ 1.0
	BEXP ≥ 0

This optimization procedure does not guarantee a "global optimum" solution of the objective function; a local minimum might be found instead. Graphical and statistical output of the program comparing computed and observed hydrographs can be used to judge the accuracy of the optimization results. Sometimes a global optimum can be found and the results improved by varying the starting values of the parameters in a number of optimization runs.[2]

HEC-1 input for parameter estimation

Execution line.   An OU line is required in an input file to execute the parameter estimation option of HEC-1 and to define a time window of optimization if a limited number of ordinates are to be compared. Including an OU line in the file with blank fields will cause all the ordinates in the simulation to be compared. If the comparison of observed and computed ordinates is to be limited, the values of the first and last ordinates to be used in computing the objective function are specified as IFORD and ILORD in the first and second fields of the OU line. Such limits are shown with vertical dashed lines in Fig. 5.12.

Specifying the parameters to be optimized. The unit hydrograph and loss function parameters that are to be estimated are indicated in the appropriate fields of the lines of input on which the parameters are normally specified. For example, if the t_p and C_p of Snyder's unit hydrograph are to be estimated, the values are entered in fields 1 and 2 of the US line. The values specified can be either initial estimates of the parameters with minus signs affixed or -1s, indicating that default values in the program are to be used.

An extract from an input file is shown in Fig. 5.15. The limits of optimization are set at ordinates 10 and 50 on the OU line. The initial estimates of 1.2 for starting loss and 0.2 for uniform loss rate on the

```
ID.....1.......2.......3.......4.......5.......6.......7.......8.......9......10

OU    10      50
LU   -1.2    -0.2
UC    -1      -1
```

Figure 5.15 Lines of input data required for implementing parameter estimation.

LU line are to be optimized as indicated by the minus signs. The -1 values on the UC line indicate that default initial estimates of the Clark unit hydrograph parameters are to be used. The default values for unit hydrograph and loss-rate parameters in the program are shown in Table 5.8.

Output

The first and last ordinates in the optimization region, as specified on the OU line of input, are listed after the time and units data in the first section of output. The output normally includes a log of iterations used in the optimization computations, a summary of optimization results, and comparative statistics of the computed and observed hydrographs. See the example of output for estimating Clark unit hydrograph parameters TC and R and exponential loss-rate parameters STRKR and DLTKR in Fig. 5.16. In this example, loss-rate parameters ERAIN and RTIOL are constant.

TABLE 5.8   Default Initial Estimates for Unit Hydrograph and Rainfall Loss Parameters in HEC-1

	Parameter	Initial value
	Unit Graph	
Clark	TC + R	$(TAREA)^{1/2}$
	R/(TC + R)	0.50
	Loss Rates	
Exponential	COEF	0.07
	STRKR	0.20
	STRKS	0.20
	RTIOK	2.00
	ERAIN	0.50
	FRZTP	0.00
	DLTKR	0.50
	RTIOL	2.00
Initial and uniform	STRTL	1.00
	CNSTL	0.10
Holtan	FC	0.01
	GIA	0.50
	SA	1.00
	BEXP	1.40
Curve number	STRTL	1.08
	CRVNBR	65.00

Log of iterations in optimization computations.   Initial estimates for the optimization variables are shown at the beginning of the optimization output. Because of the interrelationship of TC and R, the program uses values for TC + R and R/(TC + R) in the computations but produces the "best estimate" of TC and R in the results.

In the tabulation of iterations, intermediate values of the optimization variables are shown. The first line reflects an adjustment of the volume of the computed hydrograph to within 1 percent of the volume of the observed hydrograph. This adjustment is made if the option to adjust rainfall loss function parameters has been selected. After the volume adjustment, the program optimizes the objective function for one variable at a time while holding the others constant. The optimized value for each variable is identified with an asterisk (*). A variable which does not improve the value for the objective function under this procedure remains unchanged and is identified with a plus sign (+). However, none occurs in this example. The first set of values for the optimized variable is 6.890 for TC + R, .500 for R/(TC + R), .448 for STRKR, and 1.119 for DLTKR. The program repeats this proce-

			INITIAL ESTIMATES FOR OPTIMIZATION VARIABLES			
	TC+R	R/(TC+R)	STRKR	DLTKR	RTIOL	ERAIN
	6.16	.50	.20	.50	1.00	.50
			INTERMEDIATE VALUES OF OPTIMIZATION VARIABLES			
			(*INDICATES CHANGE FROM PREVIOUS VALUE)			
			(+INDICATES VARIABLE WAS NOT CHANGED)			
OBJECTIVE FUNCTION	TC+R	R/(TC+R)	STRKR	DLTKR	RTIOL	ERAIN
VOL. ADJ.	6.156	.500	.448*	1.119*	1.000	.500
349.3	6.890*	.500	.448	1.119	1.000	.500
346.8	6.890	.521*	.448	1.119	1.000	.500
344.4	6.890	.521	.438*	1.119	1.000	.500
339.3	6.890	.521	.438	.984*	1.000	.500
339.1	6.919*	.521	.438	.984	1.000	.500
335.8	6.919	.546*	.438	.984	1.000	.500
335.1	6.919	.546	.443*	.984	1.000	.500
328.3	6.919	.546	.443	.812*	1.000	.500
327.0	7.014*	.546	.443	.812	1.000	.500
326.8	7.014	.550*	.443	.812	1.000	.500
324.5	7.014	.550	.453*	.812	1.000	.500
311.1	7.014	.550	.453	.541*	1.000	.500
309.9	7.100*	.550	.453	.541	1.000	.500
309.9	7.100	.551*	.453	.541	1.000	.500
305.6	7.100	.551	.465*	.541	1.000	.500
293.4	7.100	.551	.465	.361*	1.000	.500
288.2	7.100	.551	.465	.241*	1.000	.500
286.2	7.100	.551	.465	.160*	1.000	.500
281.7	7.100	.551	.478*	.160	1.000	.500
281.7	7.100	.551	.477*	.160	1.000	.500
281.2	7.044*	.551	.477	.160	1.000	.500
VOL. ADJ.	7.044	.551	.487*	.164*	1.000	.500

Figure 5.16 Log of optimization iterations in program output.

dure of optimizing one variable at a time until four sets of computations have been completed. The resulting values for the variables are 7.100, .551, .465, and .361.

Next, the program optimizes the objective function for the variable that most improved the value of the objective function in its last change. It then optimizes the objective function for the variable that had the next greatest impact in its last change, and so on, until no single change in a variable yields an improvement in the objective function of more than 1 percent. In the example, this leads to new values for STRKR and DLTKR of .477 and .160, respectively. Following this, one more complete search of all variables is made for improving the value of the objective function. This results in a minimum value of the objective function of 281.2 and a new value for TC + R of 7.044. Finally, STRKR and DLTKR are adjusted slightly to again bring the volume of the computed hydrograph within 1 percent of the volume of the observed hydrograph.

Summary of optimization results.   The results of the optimization for the Clark unit hydrograph parameters and the exponential loss-rate parameters are shown in Fig. 5.17. Equivalent parameters for the other unit hydrograph methods in HEC-1 and for uniform rainfall loss also are presented.

Comparative statistics.   Peak flow, time of peak, and other key variables are tabulated for the computed and observed hydrographs. Comparative statistics include the "standard error," defined as the root mean squared sum of the differences between observed and computed hydrographs; the "average percent absolute error," defined as the average of the absolute values of percent difference between computed and observed hydrograph ordinates; and other error estimates.

Guidelines for estimating parameters

Effective parameter estimation involves more than just using the parameter optimization capabilities of HEC-1. The use of engineering judgment in selecting the parameters and fixing them in a sequence of steps has been found to produce the best results. Due to differences in data availability and other factors, the sequence of steps varies from study to study. A set of guidelines for calibrating exponential loss function parameters for a subbasin based on experience gained in modeling studies is as follows.[3]

1. For each storm used in the calibration (several storm events should be used), determine the three base flow parameters: STRTQ, the starting flow from antecedent runoff; QRCSN, the discharge at which recession flow begins on the recession limb of the hydrograph; and RTIOR, the recession coefficient. These parameters are event-dependent and are not calibrated with the parameter optimization capabilities of HEC-1.

2. For each storm, estimate all unknown unit hydrograph and rainfall loss function parameters for gaged subbasins.

3. If ERAIN, the exponent that reflects the influence of precipitation rate on basin average loss characteristics, is to be estimated, deter-

```
*****************************************************
*                                                   *
*               OPTIMIZATION RESULTS                *
*                                                   *
*****************************************************
*                                                   *
*      CLARK UNITGRAPH PARAMETERS                   *
*          TC    3.16                               *
*          R     3.88                               *
*                                                   *
*      SNYDER STANDARD UNITGRAPH PARAMETERS         *
*          TP    2.99                               *
*          CP     .52                               *
*                                                   *
*      LAG FROM CENTER OF MASS OF EXCESS            *
*      TO CENTER OF MASS OF UNITGRAPH    5.36       *
*                                                   *
*      UNITGRAPH PEAK    4333.                      *
*      TIME OF PEAK      3.00                       *
*                                                   *
*****************************************************
*                                                   *
*      EXPONENTIAL LOSS-RATE PARAMETERS             *
*          STRKR    .49                             *
*          DLTKR    .16                             *
*          RTIOL   1.00                             *
*          ERAIN    .50                             *
*                                                   *
*      EQUIVALENT UNIFORM LOSS RATE     .444        *
*                                                   *
*****************************************************
```

```
*********************************************************************************************************
*                                                                                                       '
*                            STATISTICS BASED ON OPTIMIZATION REGION                                    '
*                                    (ORDINATES 1 THROUGH 61)                                           '
*                                                                                                       '
*********************************************************************************************************
*                                                                                                       '
*                                                  TIME TO      LAG                                     '
*                    SUM OF     EQUIV     MEAN     CENTER     C.M. TO    PEAK     TIME OF                '
*                    FLOWS      DEPTH     FLOW     OF MASS      C.M.     FLOW      PEAK                  '
*                                                                                                       '
*   PRECIPITATION EXCESS          .937               4.13                                               '
*                                                                                                       '
*   COMPUTED HYDROGRAPH   84787.   .867    1390.     8.51      4.38     3622.     7.00                  '
*   OBSERVED HYDROGRAPH   84787.   .867    1390.     8.16      4.03     3540.     7.00                  '
*                                                                                                       '
*   DIFFERENCE               0.   .000       0.      .35       .35       82.       .00                  '
*   PERCENT DIFFERENCE       .00                              8.66      2.30                            '
*                                                                                                       '
*              STANDARD ERROR    270.             AVERAGE ABSOLUTE ERROR        207.                    '
*           OBJECTIVE FUNCTION   283.      AVERAGE PERCENT ABSOLUTE ERROR      27.24                    '
*                                                                                                       '
*********************************************************************************************************
```

Figure 5.17 Summary of optimization results.

mine a regional value on the basis of the results of all the storm events used in step 2.

4. Repeat the estimation of the unknown parameters for all gaged subbasins using the same storm events, but with ERAIN fixed at the value determined in step 3. Determine a regional value for RTIOL if RTIOL is unknown.

5. With ERAIN and RTIOL fixed, again estimate the unknown parameters for all gaged subbasins, determining a value of STRKR for each storm being used in the procedure. Check the values for regional consistency.

6. With ERAIN, RTIOL, and STRKR fixed, reestimate the remaining parameters for all gaged subbasins, and determine a generalized (average) value of DLTKR for each subbasin, if desired. It should be noted, however, that this parameter is considered to be relatively event-dependent.

7. With all the other parameters set, reestimate the variables TC + R and R/(TC + R) for all gaged subbasins, and determine TC and R for each subbasin. In determining TC and R, a regional value for R/(TC + R), based on an average of estimated values for the gaged subbasins, and a generalized value of TC + R for each subbasin are generally used.

8. After all the parameters have been estimated, verify the values by simulating the response of the gaged subbasins to other events.

Estimating routing parameters

The HEC-1 parameter optimization option can be used for estimating routing criteria for three of the hydrologic routing techniques: the Tatum, the straddle-stagger, and the Muskingum. Observed inflow and outflow hydrographs and a pattern local inflow hydrograph for the river reach are required inputs. The assumed pattern hydrograph for local inflow accounts for the volume difference between observed inflow and outflow hydrographs for the routing reach, and its shape can have a significant effect on the estimates of the routing criteria. In the optimization procedure, the squared sum of the deviations between the observed hydrograph and the computed hydrograph is minimized. This procedure is essentially the same as that used to optimize unit hydrograph and loss-rate parameters.

Workshop Problem 5.1: Developing a Single-Basin Model for the Rahway River above the Gage near Springfield

Purpose

The purpose of this workshop problem is to provide an understanding of how a single basin is modeled with HEC-1.

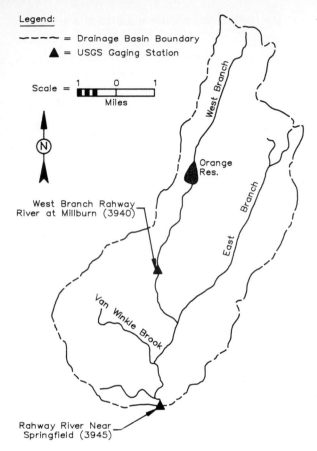

Legend:

- – – – – = Drainage Basin Boundary
- ▲ = USGS Gaging Station

Scale = [scale bar] Miles

Orange Res.

West Branch

East Branch

West Branch Rahway River at Millburn (3940)

Van Winkle Brook

Rahway River Near Springfield (3945)

Figure P5.1 Rahway River basin above the Springfield gage.

Problem statement

An HEC-1 rainfall-runoff model is to be developed for the Rahway River basin above the USGS gaging station near Springfield (Fig. P5.1). The entire 25.5-mi^2 area is to be modeled as a single basin. The storm of August 27 and 28, 1971, which produced a basin average total precipitation of 8.84 in, is to be used in the simulation. Hourly rainfall amounts for this storm recorded at the Springfield recording rain gage are given in Table P5.1. These are to be used as a pattern to distribute the total storm depth in time. The following parameters are to be used in the model:

TABLE P5.1   Hourly Rainfall at Springfield for the Storm of August 27 and 28, 1971

Date	Time, h	Rainfall, in	Date	Time, h	Rainfall, in
Aug. 27	0100		Aug. 27	1600	0.06
	0200			1700	0.04
	0300			1800	0.09
	0400	0.04		1900	
	0500	0.12		2000	
	0600	0.06		2100	
	0700	0.10		2200	
	0800	0.08		2300	
	0900	0.43	Aug. 27	2400	
	1000	0.76	Aug. 28	0100	
	1100	0.37		0200	0.10
	1200	0.70		0300	0.76
	1300	1.15		0400	1.53
	1400	1.36		0500	1.12
	1500	0.18		0600	0.37
				0700	0.03

SCS loss rate	CRVNBR = 65
	STRTL = 4.0
	RTIMP = 26
Clark unit hydrograph	TC = 6.32
	R = 8.42
Base flow	STRTQ = 2.0
	QRCSN = − 0.04
	RTIOR = 1.10

Tasks

1. Set up the model and run it with the given storm data.

2. Verify by hand calculation the results of the HEC-1 computation of rainfall losses and excess during the first 6 h of the simulation.

3. Use the unit hydrograph ordinates and rainfall excess amounts printed in the HEC-1 output to verify by hand calculations the HEC-1 computed discharge at the end of the nineteenth hour *only*. Note that the rainfall excess values have been truncated to two decimal places. The answers from hand calculations will be slightly smaller than the HEC-1 answers because of the number of significant digits used in the computations.

Workshop Problem 5.2: Parameter Estimation, Rahway River Basin

Purpose

The purpose of this workshop problem is to provide an understanding of the automatic model calibration capability of HEC-1 and the techniques of estimating model parameters.

Problem statement

Unit hydrograph and loss function parameters are needed for an HEC-1 model to predict runoff in the Rahway River basin above the Springfield gage. To estimate these parameters, the automatic parameter estimation capability of HEC-1 is used with observed data for the storms of August 27 and 28, 1971, and July 13–15, 1975.

Tasks

1. Make a list of line-identification characters for lines of input corresponding to steps *a* through *h* below:
   a. Indicate that parameters are to be estimated, and identify the ordinates to be used in the computations.
   b. Specify the rain gages and weights to be used for the computation of basin average rainfall depth.
   c. Specify the rain gages and weights to be used for the computation of the basin average hyetograph.
   d. Specify the discharge hydrograph ordinates to be used for parameter estimation.
   e. Specify the time step and starting date and time of the observed hydrograph to be entered as input.
   f. Indicate which parameters of the Clark unit hydrograph are to be estimated, and provide initial estimates of those parameters.
   g. Indicate which parameters of the initial and uniform-rate loss function are to be estimated, and provide initial estimates of those parameters.
   h. Specify the time step and starting date and time for computations.
2. Complete the input file shown in Fig. P5.2 to estimate the parameters for the Clark unit hydrograph and the initial and uniform rainfall loss functions for the storm of August 27 and 28, 1971. The observed rainfall and discharge data are already included in the file. Since this exercise is for the same subbasin and storm used in the preceding workshop problem, the file from that workshop problem may be revised for use in this problem. Use the following base

```
ID    RAHWAY RIVER BASIN, NEW JERSEY
ID    PARAMETER ESTIMATION - RAHWAY RIVER NEAR SPRINGFIELD
ID    STORM OF AUGUST 27-28 1971
IT    60 27AUG71      0      85
PB  8.84
PI    0       0       0     0.04   0.12   0.06    0.1    0.08   0.43   0.76
PI  0.37     0.7     1.15   1.35   0.18   0.06    0.04   0.09    0      0
PI    0       0       0      0      0     0.1     0.76   1.53   1.12   0.37
PI  0.03     0       0       0      0      0       0      0      0      0
QO    2       2       3      3      4      9      16     22     39     105
QO   200     310     425    543    660    800    925    1080   1300   1700
QO  1725    1620    1480   1365   1294   1200   1085   1054   1300   1675
QO  2160    2620    3080   3400   3350   3060   2820   2450   2160   1975
QO  1760    1525    1300   1160   1010   920    820    720    609    500
QO   410     320     225    177    140    115    90     80     75     74
QO    70      60      55     50     50     50     48     47     45     40
QO    38      34      31     29     28     27     25     24     23     23
QO    22      20      20     19     18      0      0      0      0      0
ZZ
```

Figure P5.2 Incomplete data input file for the storm of August 27 and 28, 1971.

flow parameters: STRTQ = 2 ft^3/s, RTIOR = 1.0718, and QRCSN = 4 percent of the peak discharge. Use the program default initial estimates for the unit hydrograph and rainfall loss parameters.

3. Prepare another input file to estimate parameters for the storm of July 13–15, 1975. Rainfall and discharge data for this storm are provided in Fig. P5.3. Use the following base flow parameters: STRTQ = 30 ft^3/s, RTIOR = 1.0718, and QRCSN = 5 percent of the peak discharge. Use the program default initial estimates for the unit hydrograph and rainfall loss parameters.

4. Execute the HEC-1 program with each of the input files, and before examining the output, answer the following questions:

   a. Which of the unit hydrograph and loss function parameters affect the volume of runoff?

   b. Could the parameters of the SCS rainfall loss method be estimated automatically with HEC-1? If so, how would your input file have to be modified to accomplish this? Would you expect the parameter estimation procedure to yield the same values of T_c and R?

   c. If different initial estimates of the parameters were specified, would you expect the program to identify the same "optimal" parameters? Explain your answer.

   d. To estimate model parameters automatically, you must specify a stimulus and a corresponding response. In rainfall-runoff modeling, the stimulus is the rainfall, and the response is the

```
ID    RAHWAY RIVER BASIN, NEW JERSEY
ID    PARAMETER ESTIMATION - RAHWAY RIVER NEAR SPRINGFIELD
ID    STORM OF 13-15 JULY 1975
IT    60 13JUL75        0      85
PB  7.22
PI    0        0     0.03      0       0       0       0    0.15     0.7    0.11
PI  0.07     0.12     0.0      0       0      0.6      1    0.14    0.06      0
PI    0        0       0       0       0       0       0       0       0       0
PI    0        0       0       0    0.31    1.45     .95     .96     .02     .02
PI    0        0       0       0       0       0       0       0       0       0
PI    0       .1      .87     .95     .08     .13    .123      0      .03     .01
PI  0.01       0       0       0       0       0    0.05       0    0.01       0
QO   23       23      31      32      29      19      21      19      48     246
QO  412      489     542     548     493     415     664    1120    1270    1250
QO 1250     1260    1230    1150    1060     943     827     716     576     542
QO  300       97     149      59     120     336    1180    1820    2510    2900
QO 3080     3090    3010    2810    2470    2160    1850    1600    1400    1250
QO 1100      951     915    1250    1570    1650    1650    1640    1580    1480
QO 1370     1270    1150    1020     900     791     696     586     479     396
QO  315      252     202     165     142     128     116     108     103      97
QO   92       89      86      83      81       0       0       0       0       0
ZZ
```

Figure P5.3 Incomplete data input file for the storm of July 13–15, 1975.

discharge. The automatic calibration option of HEC-1 can also estimate the parameters of the Muskingum routing method. What are the stimulus and response in that case, and what are the lines of input (indicate with the two-character line identifiers) used to specify data for the stimulus and response? What line of input is included to indicate that routing parameters are to be estimated? On what line of input are the initial estimates of Muskingum K and X specified?

5. Examine the program output from the two runs. If you are not satisfied with the results, adjust the input and reexecute the program for either or both of the storms, as is appropriate. Prepare a summary table patterned after Table P5.2, and complete columns (2) and (3).

6. As you might expect from calibration with two different storms, the resulting "optimal" estimates of the parameters are different. Engineering judgment is required to resolve the differences and select values to be used for predicting runoff from other storms. Select the "best" estimates of the parameters for the Rahway basin on the basis of the calibration results for the two events recorded in columns (2) and (3) of Table P5.2, and complete column (4). Reexecute HEC-1 for each of the two storms using your "best" estimates of parameters listed in column (4), and complete columns (5) and (6).

TABLE P5.2 Results of Parameter Estimation

Results (1)	Calibration Aug. 27 and 28, 1971 (2)	Calibration July 13–15, 1975 (3)	Calibration "Best" estimates (4)	Verification Aug. 27 and 28, 1971 (5)	Verification July 13–15, 1975 (6)
R, h				X	X
TC, h				X	X
Initial loss, in				X	X
Uniform rate, in/h				X	X
Final objective function			X		
Differences:					
Sum of flow, ft³/s			X		
Peak flow, ft³/s			X		
Time of peak, h			X		

References

1. Hydrologic Engineering Center, *HECDSS Users Guide and Utility Program Manuals,* U.S. Army Corps of Engineers, Davis, Calif., November 1987.
2. Hydrologic Engineering Center, *HEC-1 Flood Hydrograph Package, Users Manual,* U.S. Army Corps of Engineers, Davis, Calif., 1981 (rev. March 1987).
3. Hydrologic Engineering Center, *Corps of Engineers' Experience with Automatic Calibration of a Precipitation-Runoff Model,* Technical Paper 70, U.S. Army Corps of Engineers, Davis, Calif., 1980.

6

River Basin Modeling
with HEC-1

Modeling Concepts

Introduction

River basin models are developed for a variety of engineering and management purposes, including the development of discharge-frequency relationships for use in planning studies and project design, the analysis of the effects of urban development and other changes on runoff response, and the evaluation of the hydrologic impacts of alternative water development plans.

A river basin precipitation-runoff model, frequently called a watershed model, is a network of computational components programmed to simulate surface runoff and compute discharge hydrographs at locations of interest. An HEC-1 model of a complex river basin has basic components for subbasin runoff, channel and reservoir routing, and hydrograph combining. Optional components include diversions and pumping. The number and location of components depend on basin conditions and modeling objectives.

Model components

Subbasin runoff. The subbasin runoff component represents runoff response from a subdivided part of the basin. Subbasin boundaries are delineated so that lumped parameters defining precipitation and precipitation loss give reasonable results. Input includes precipitation data, precipitation loss function parameters, and precipitation-runoff transformation function parameters (unit hydrograph or kinematic wave). The computed output consists of a discharge hydrograph at the subbasin outlet.

Channel and reservoir routing. The routing component represents flood-wave movement through a channel reach or unregulated reservoir. Input consists of an inflow hydrograph and parameters defining the routing characteristics of the reach or reservoir. Output consists of an outflow hydrograph at the reservoir outlet or downstream end of the reach.

Hydrograph combining. The hydrograph combining component, essential to the operation of the system as a whole, performs the function of linking or combining the hydrographs from the other components.

Subdividing a basin

General. Subdivision of the area of a large basin is necessary because of the size and complexity of the physical system, including the heterogeneity of basin characteristics and storms. A basin with major tributaries and a diversity of topography and land use must be broken down into smaller components to fit the constraints and assumptions in the model. Subdivision also may be necessary to obtain the information needed in terms of level of detail and location as dictated by the objectives of the study.

Sometimes it may be advantageous to subdivide the area in two phases: the first for parameter estimation and the second for developing an operational model of the complete system. The primary purpose of subdivision for calibration is to establish the conditions necessary for estimating parameters, including taking advantage of gage locations with historical data and defining areas with homogeneous conditions above the gages. In the operational model, other purposes related to the study objectives may come into play, and these may require altering subbasin boundaries to provide the output desired.

Basin characteristics and model constraints

Stream gage locations. Existing data should be utilized as much as possible in developing a model. If a subbasin boundary is located so that the subbasin outlet coincides with a stream gage location, the program can compare observed hydrographs from the gage with computed hydrographs in estimating model parameters.

Hydrologic integrity. The subdivision should be consistent with the topography and the hydrologic system that exists in the basin. Subbasin boundaries must coincide with hydrologic boundaries, and the interrelationships of major hydrologic elements in the system need to be considered in deciding where boundaries are to be drawn. For example, a subbasin encompassing more than one major tributary is not likely to provide the degree of resolution needed. Under these conditions, subbasin characteristics may not be adequately represented by a unit hydrograph, which is based on the assumption of a single homogeneous unit. Furthermore, information that might be useful regarding the contributions of the individual tributaries to the runoff process would be lacking.

Size restrictions imposed by lumped parameters. The uniformity of conditions in a subbasin is key to the validity of using lumped parameters in the model, and this generally limits the size of subbasins that can be used. Variation in loss rates, base flow, and other components of the runoff process obviously tends to increase with the size of the area involved.

Channel slope. Slope is a significant factor in the runoff response of a basin. The slope of the main watercourse is frequently used in regression equations relating unit hydrograph parameters to basin characteristics. Thus, it is desirable to divide a basin into areas of relatively uniform slope, both for effective modeling and for facilitating the analysis of ungaged subbasins using regression relationships.

Soil types and cover. Since infiltration rates and other runoff response characteristics of a basin vary with soil type and cover, a model may be enhanced by the separation of areas with different soil and cover conditions. Generally, forested areas should be separated from unforested areas, and areas with permeable soils should be separated from areas with impermeable soils, other factors being equal.

Elevation differences. Since mountains affect storm patterns and intensity due to orographic effects, it may be appropriate to separate areas of high elevation from areas of lower elevation in some river basins.

Basin shape. The shape of a basin affects the configuration and timing of the runoff hydrograph. Because of time-area differences, a long, narrow basin generates a different hydrograph than a relatively square one of the same area. Since unit hydrograph methods, such as the Clark method, are based on typical basin shapes, irregular basins should be subdivided to conform to the assumed shapes if practical. For example, a basin area composed of a wide section and a long, narrow section normally would be divided to separate the narrow part from the wide part.

Study scope and objectives

Study scope. The level of detail in a modeling study set by the scope and budget of the project generally dictates the number of subbasins to be analyzed. A reconnaissance-level study with a relatively low budget requires much less detail than a project design study, for example. The greater the number of subbasins, the higher the costs for computer time and other resources required to develop and run the model.

Sometimes it may be feasible to develop a model in stages, starting with only a few subbasins as required for a preliminary study and refining the model by further subdivision later to meet the needs of the project as it advances.

Reservoir locations. Usually, subdivisions are made at reservoir locations, both existing and planned, so that inflows to reservoirs can be computed and reservoir routing effects determined.

Channel improvements. To evaluate the results of channel improvements, it may be necessary to isolate an improved section of a river as a routing reach, thus dictating where subbasin boundaries are located.

Urban development. If an area of a basin is in the process of urbanization or future development is anticipated, it may be appropriate to separate this area in the model to facilitate revising basin parameters later to reflect changed conditions.

Damage centers. River basin models are frequently developed to evaluate potential damage and the effects of damage reduction measures at key damage centers, such as highly urbanized locations. Subdividing a basin to obtain essential data at such locations is an important consideration. When set up properly, the model can be run under different sets of conditions to predict peak flows and stages at these locations. This hydrologic data is translated into expected annual damage figures and benefits for evaluating alternative development plans.

Jurisdictional boundaries. Occasionally political boundaries influence the subdivision of a basin for modeling.

Example of subdivision of the Rahway River basin above the Springfield gage. The Rahway River basin is divided into five subbasins for a flood control study above the Springfield gage (Fig. 6.1). Subbasin boundaries shown with dashed lines are delineated to produce hydrographs at major points of interest, including the outlet from the Orange Reservoir, the gage at Millburn, the confluence of the two ma-

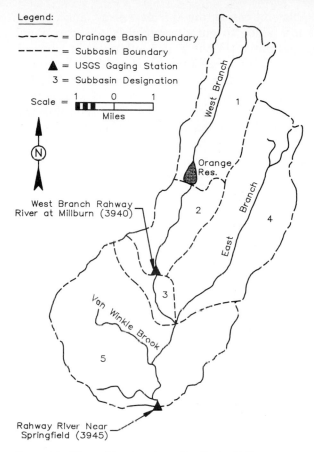

Legend:

~~~~ = Drainage Basin Boundary
----- = Subbasin Boundary
▲ = USGS Gaging Station
3 = Subbasin Designation

Scale = |1 0   1|
          Miles

Figure 6.1 Five subbasins above the Springfield gage.

jor tributaries, and the outlet to the basin at Springfield. The delineation of these subbasins is based on watershed boundaries taken from a
topographic map and other criteria mentioned above.

Estimation of model parameters

Calibration of a model normally has two phases. In the first phase,
unit hydrograph and loss function parameters for gaged subbasins are
optimized with HEC-1 using observed precipitation data and observed
discharge data from selected historical storm events. Only gaged
headwater subbasins are normally calibrated in this phase. Sometimes a simplified subdivision of the basin area may be used to take
maximum advantage of existing stream gages.

In the second phase, the complete model is calibrated. The number and configuration of subbasins are changed if necessary, and routing and other parameters are added to put the model in an operational mode. The parameters estimated for the gaged subbasins in phase 1 are transferred to the remaining or redefined ungaged subbasins in phase 2 with regression or regionalization techniques that relate the parameters to basin characteristics.

Calibration in the second phase is accomplished by running the model with data from observed storm events and comparing computed hydrographs with observed hydrographs at gaged locations. If the fit is not satisfactory, the parameters are adjusted and the computation is repeated until a suitable fit between observed and computed hydrographs is obtained.

Developing Input for a River Basin Model

Introduction

In this section, the organization of model components and the preparation of data input to represent runoff from a subdivided basin are discussed. Basic model components representing subbasin runoff, channel and reservoir routing, and hydrograph combining are defined and linked in a network representing the hydrologic system as a whole. The HEC-1 user's manual[1] or help files in the text editor COED, normally installed in the computer with HEC-1, should be used as a reference for details on data input.

Input for hydrologic components

Input requirements are discussed for the basic components of subbasin runoff, routing, and combining, as well as for an optional diversion component. Data input for each component consists of groups of lines of input beginning with a KK line that assigns a label, followed by lines containing parameters and data required for the computations. These groups of lines are referred to as K groups.

Subbasin runoff. A set of lines of input data for the subbasin runoff component is represented in Fig. 6.2. The label "21" on the KK line is assigned to the discharge hydrograph computed for this subbasin. The KM line provides alphanumeric data (comments) describing the component. The rest of the lines are used to specify parameters and data for the computations as indicated. Some of these, such as the BF line, would not have to be included here if the data had been specified in a

Subbasin Runoff

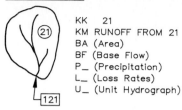

KK   21
KM RUNOFF FROM 21
BA (Area)
BF (Base Flow)
P__ (Precipitation)
L__ (Loss Rates)
U__ (Unit Hydrograph)

Figure 6.2   Subbasin runoff data.

previous group. The data would be automatically repeated from the previous specification.

The precipitation, loss-rate, and unit hydrograph lines have optional forms of input, as indicated by the space after the first letter of the line identifiers: P__, L__, and U__. For example, precipitation may be entered as basin average storm depth and time-series data on PB and PI or PC lines, or as gage data on PG, PI or PC, PR, PW, and PT lines, and so forth.

Combining. In Fig. 6.3, the runoff hydrographs from two subbasins are combined at a common outlet point. After the two K groups of data for computing runoff for subbasins 12 and 13, the data input file contains a KK line labeled "113" for the combining component. The combining of two hydrographs at this point is specified with the "2" on the HC line.

Combining

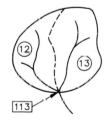

KK   12
KM RUNOFF FROM 12
    :
    :
KK   13
KM RUNOFF FROM 13
    :
    :
KK 113
KM TOTAL RUNOFF AT 113
HC   2

Figure 6.3   Hydrograph combining data.

Routing. One of several optional methods of routing can be specified in HEC-1. An inflow hydrograph at location 142 is routed to location 143 with the Muskingum method in Fig. 6.4. The reach between 142 and 143 is divided into two subreaches, the travel time between the two locations represented by the parameter K is 3.6 h, and the weighting coefficient X is equal to 0.2. These values are specified on the RM line in fields 1, 2, and 3, as shown. The KK line for this component assigns it the label "143."

Routing

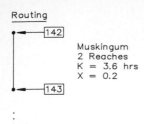

KK 143
KM ROUTE FROM 142 TO 143
RM   2   3.6   0.2

Figure 6.4 Muskingum routing data.

An optional routing method, the modified Puls, is illustrated in Fig. 6.5. With this method, a storage-outflow relationship for the reach, tabulated in the figure, must be input into the program. These values are specified in corresponding fields on SV and SQ lines. The routing parameters are specified on the RS line, with the first field indicating the number of routing reaches. Fields 2 and 3 indicate starting conditions for the storage routing. The conditions can be specified in terms of discharge, stage, or storage. In this example, discharge is selected, and the " − 1" in field 3 indicates that the outflow equals inflow as a starting condition.

Routing

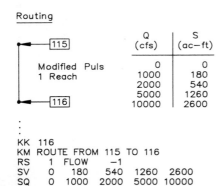

KK 116
KM ROUTE FROM 115 TO 116
RS   1   FLOW   −1
SV   0   180   540   1260   2600
SQ   0   1000   2000   5000 10000

Figure 6.5 Modified Puls routing data.

Diversion. A diversion out of the system at location 132 is illustrated in Fig. 6.6. The diverted hydrograph, D6, is computed as a function of inflow in the channel represented by the tabulation of Q_{in} vs. Q_{div}. The "132" on the KK line is the label assigned to the remainder of the flow in the channel after the diverted flow is removed. The label for the diversion hydrograph is indicated on the DT line. The DI and DQ lines indicate corresponding values of inflow and diverted flow, respectively, taken from the tabulation defining their relationship.

Diversion

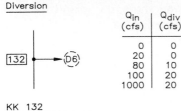

	Q_{in} (cfs)	Q_{div} (cfs)
	0	0
	20	0
	80	10
	100	20
	1000	20

```
KK  132
KM  DIVERSION D6
DT  D6
DI     0    20    80   100  1000
DQ     0     0    10    20    20
```

Figure 6.6   Diversion data.

A diverted hydrograph can be retrieved at a location as shown in Fig. 6.7. The KK line contains the label assigned to the flow in the channel after the retrieved hydrograph is added to channel inflow. Labeling this hydrograph "D6" is purely arbitrary; another label could have been used. The label "D6" on the DR line indicates the hydrograph being retrieved.

Retrieve Diversion

```
KK   D6
KM  RETRIEVE DIVERSION D6
DR   D6
```

Figure 6.7   Retrieve diversion data.

Hydrograph comparison.   An observed hydrograph can be read into the program for input or for comparison only. The option of reading in a hydrograph for comparison is illustrated for location 107 in Fig. 6.8. The KK line provides the label of the observed hydrograph, and the QO lines contain the observed hydrograph ordinates. HEC-1 will plot

Hydrograph Comparison

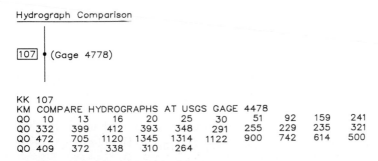

```
KK  107
KM  COMPARE HYDROGRAPHS AT USGS GAGE 4478
QO   10    13    16    20    25    30    51    92   159   241
QO  332   399   412   393   348   291   255   229   235   321
QO  472   705  1120  1345  1314  1122   900   742   614   500
QO  409   372   338   310   264
```

Figure 6.8   Hydrograph comparison data.

the observed and computed hydrographs on the same graph for this location if given the appropriate output instructions.

Computational sequence in HEC-1 (the stack principle)

In a multiple-subbasin model, calculated discharge hydrographs are stored in a "stack" in the program. Each time a new hydrograph is calculated, it is placed on top of the stack. When routing is performed, the hydrograph on top of the stack is treated as the inflow hydrograph, and the calculated hydrograph replaces the inflow hydrograph on top of the stack. When x hydrographs are combined, the x hydrographs on top of the stack are combined. These are replaced on top of the stack by the single combined hydrograph. When a diversion hydrograph is calculated, the hydrograph of flow remaining after the diversion is subtracted is placed on top of the stack. The diverted hydrograph is not placed on the stack; it is stored in a scratch file. However, when a diverted hydrograph is retrieved, it is placed on top of the stack.

It is very important to keep the "stack principle" in mind in organizing the data input file. To provide for the proper sequence of computations, the components must be arranged in an upstream to downstream order until a confluence is reached. All flows above a confluence must be computed and routed to the confluence and combined before computations are continued below the confluence.

Example of a river basin model

Basin layout.   This example problem of the Red River basin involves a basic stream network model structured to demonstrate a wide range of features and options in HEC-1. The problem is to prepare the data input for the model, defined as having five subbasins, a dam, a diversion, and a gaged outlet (Fig. 6.9a).

Model schematic.   One of the first steps in developing input is the preparation of a schematic diagram (Fig. 6.9b) that can be used as a guide in organizing the data. The schematic will help clarify the correct sequence of subbasin runoff, routing, and combining computations.

Data input file.   Discussion of the building of the input file in this example is broken down into steps corresponding to the components in the model.

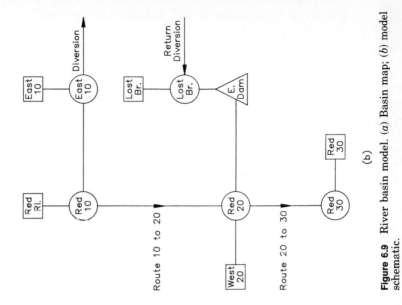

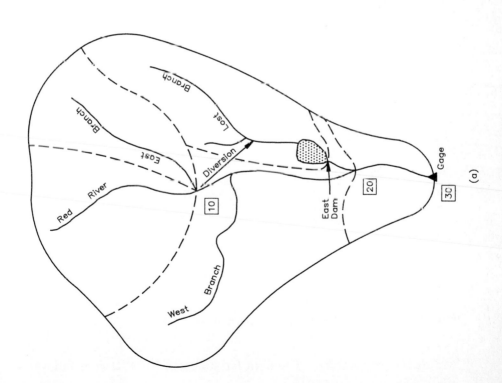

Figure 6.9 River basin model. (*a*) Basin map; (*b*) model schematic.

1. *Job control and basin precipitation data:*   Total storm precipitation for four gages and hourly amounts for one recording gage are available for input.

                    Total storm data
        Gage no.   *Storm depth, in*
            60   4.68
            61   4.65
            62   4.85
            63   4.90
            64   5.10

Hourly precipitation data
STARTING TIME: 7:15 a.m.
DATE: June 12, 1968
STATION No. 400
HOURLY INCREMENTAL RAINFALL:
.04, .35, .01, .03, .73, .21, .02, .01, .03, .01

The first two lines of the input file (Fig. 6.10) provide job title information; the next line calls for a network diagram to be printed in output; and the IT line specifies simulation time data and the number of ordinates. It should be noted that the IN line is included before the incremental rainfall data to indicate a time interval of 60 min for rainfall, which is different from the 15-min simulation time interval specified on the IT line. This signals the program to interpolate the rainfall data.

```
LINE            ID.......1.......2.......3.......4.......5.......6.......7.......8.......9......10

1       ID          EXAMPLE PROBLEM NO. 1
2       ID          STREAM NETWORK MODEL
        *DIAGRAM
3       IT      15 12JUN68      715      58
4       IO       5
5       PG      60    4.68
6       PG      61    4.65
7       PG      62    4.85
8       PG      63    4.90
9       PG      64    5.10
10      PG     400      0
11      IN      60 12JUN68      715
12      PI     .04      .35     .01      .03     .73      .21     .02      .01     .03      .01
```

Figure 6.10   Job control and precipitation input data.

2. *Subbasin RED RI:*   The data for subbasin RED RI is as follows:

AREA = 0.82 mi^2

PRECIPITATION GAGE WEIGHTING:

Gage	Weight
400	1.0 (pattern wt.)
60	.75 (total depth wt.)
61	.25 (total depth wt.)

LOSS RATE: SCS, curve no. = 80

UNIT HYDROGRAPH: SCS, lag = 1.47 h

BASE FLOW: STRTQ = 10.0, QRCSN = − .25, RTIOR = 1.2

The corresponding section of the input file is shown in Fig. 6.11. Precipitation is input as weighted gage data, with the station numbers and corresponding weights specified on the PT and PW lines, respectively. The time distribution of rainfall is identical with that for recording gage 400, as specified on the PR and PW lines. SCS loss-rate and unit hydrograph parameters are specified on the LS and UD lines.

```
LINE        ID.......1.......2.......3.......4.......5.......6.......7.......8.......9......10

 13     KK  RED RI
 14     KO      4
 15     KM          SCS RUNOFF COMPUTATION
 16     BA     .82
 17     BF    10.0      -.25     1.2
 18     PR     400
 19     PW       1
 20     PT      60        61
 21     PW     .75       .25
 22     LS                80
 23     UD    1.47
```

Figure 6.11   Red River subbasin-runoff input data.

3. *Subbasin EAST10:*   Precipitation data is handled the same way in this subbasin as in the preceding one, except that three stations are used in lieu of two (Fig. 6.12). Exponential loss rates and Snyder unit hydrograph methods are used in this subbasin; the parameters for these are entered on the LE and US lines. Since the base flow parameters for this subbasin are the same as for the preceding subbasin in the model, the BF line could have been omitted. The parameters would have been repeated from the previous subbasin.

4. *Diversion and hydrograph combining at EAST10:*   Under the assumption that the diversion takes place immediately upstream of the confluence, it must be accounted for before hydrographs are combined at EAST10. The diversion data is as follows:

Channel inflow, ft³/s	Diverted flow, ft³/s
0	0
100	25
300	100
600	180
900	270

```
LINE          ID......1.......2.......3.......4.......5.......6.......7.......8.......9......10

 24           KK  EAST10
 25           KO   1      2
 26           KM        SNYDER UNIT GRAPH COMPUTATION-EXPONENTIAL LOSS RATE
 27           BA   .66
 28           BF  10.0    -.25    1.2
 29           PR  400
 30           PW   1
 31           PT   61     62      63
 32           PW   .6     .3      .1
 33           LE   .6     1.0     1.0     0
 34           US  1.3     .8
```

Figure 6.12 East branch subbasin-runoff input data.

The corresponding file entries are shown in Fig. 6.13. The diverted hydrograph is labeled DIVERT, as indicated on the DT line, and is computed from the inflow–diverted flow data shown on the DI and DQ lines. The remaining hydrograph is labeled EAST10. It is combined with the hydrograph RED RI in the next part of the input file with a "2" entered on the HC line.

```
LINE          ID......1.......2.......3.......4.......5.......6.......7.......8.......9......10

 35           KK  EAST10
 36           KM        DIVERT FLOW TO LOSTBR
 37           DT  DIVERT
 38           DI   0     100     300     600     900
 39           DQ   0      25     100     180     270
 40           KK  RED10
 41           KM        COMBINE HYDROGRAPHS FROM SUBBASINS EAST10 AND RED RI
 42           HC   2
```

Figure 6.13 Diversion and hydrograph-combining input data.

5. *Channel routing 10TO20:* After the two hydrographs are combined at RED10, the combined hydrograph is routed downstream from location 10 to location 20. The storage volume and outflow data required for storage routing is as follows:

Volume, acre · ft	Outflow, ft^3/s	Volume, acre · ft	Outflow, ft^3/s
0	0	110	2600
18	500	138	3000
36	1000	174	3450
54	1500	228	4000
84	2150	444	6000

The storage volume and corresponding outflow data is entered on SV and SQ lines. The RS line indicates one routing reach and an initial flow condition of outflow equal to inflow. The routed hydrograph will have the label 10TO20, as entered on the KK line in Fig. 6.14.

6. *Retrieval of diverted flow:* At this point in the model, there are three alternatives for the next step: (1) compute runoff for the subbasin containing the West Branch, (2) compute runoff for the subbasin containing the Lost Branch, or (3) retrieve the diverted flow above East Dam. The third alternative is chosen in the example, and the diverted hydrograph labeled DIVERT is retrieved with the next three lines in the input file (Fig. 6.14). The DR line is used to retrieve the previously diverted flow. Routing of the diverted flow is neglected in this example for simplicity, but it could have been included.

```
LINE          ID......1......2......3......4......5......6......7......8......9......10

  43      KK  10TO20
  44      KO      1      2
  45      KM       ROUTE FLOWS FROM STATION RED10 TO RED20
  46      RS      1    FLOW     -1
  47      SV      0      18      36      54      84     110     138     174     228     444
  48      SQ      0     500    1000    1500    2150    2600    3000    3450    4000    6000
  49      KK  LOSTBR
  50      KM       RETRIEVE DIVERSION FROM EAST10
  51      DR  DIVERT
```

Figure 6.14 Routing and diversion-retrieval input data.

7. *Computing runoff from Lost Branch and combining with the diverted flow:* The next steps are to compute runoff for the Lost Branch subbasin, using the initial and uniform loss functions and the Clark unit hydrograph, and to combine this hydrograph with the diverted hydrograph to determine inflow to East Dam. The data for the Lost Branch subbasin is as follows:

AREA = 0.36 mi^2

PRECIPITATION GAGE WEIGHTING:

Gage	Weight
400	1.0 (pattern wt.)
62	.5 (total depth wt.)
63	.5 (total depth wt.)

LOSS RATE: initial and uniform, STRTL = .3, CNSTL = .04

UNIT HYDROGRAPH: Clark, TC = .80 h, R = 1.2

The input file for computing the runoff from Lost Branch and combining this flow with the diverted flow is shown in Fig. 6.15.

```
LINE          ID.......1.......2.......3.......4.......5.......6.......7.......8.......9......10

 52       KK  LOSTBR
 53       KM          CLARK UNIT GRAPH COMPUTATION - INITIAL AND UNIFORM LOSS RATES
 54       BA   .36
 55       BF  10.0    -.25    1.2
 56       PR   400
 57       PW    1
 58       PT    62      63
 59       PW    .5      .5
 60       LU    .3      .04
 61       UC   .80     1.2
 62       KK  LOSTBR
 63       KM          COMBINE RUNOFF FROM LOSTBR WITH DIVERTED FLOW
 64       HC    2
```

Figure 6.15 Input data for Lost Branch subbasin-runoff and two-hydrograph combining.

8. *Reservoir routing at East Dam:* The modified Puls routing method is used to route flows through the reservoir at East Dam. The data for reservoir routing is as follows:

Reservoir: E.Dam

Low-level outlet

Invert elevation = 851.2 ft., mean sea level
(downstream end)
Cross-sectional area = 12 ft^2
Discharge coefficient = .6
Head exponent = .5

Spillway

Crest elevation = 856 ft., mean sea level
Width = 60 ft
Weir coefficient = 2.7
Head exponent = 1.5

Volume-elevation data

Volume, acre • ft	Elevation, ft
21	850.0
100	851.5
205	853.3
325	856.5
955	858.0

The storage routing parameters are specified on the RS line, in this case indicating a starting reservoir elevation of 851.2 ft (Fig. 6.16). The dam has both a low-level outlet and an overflow spillway, and the hydraulic characteristics of these outlet structures are specified on the SL and SS lines. Reservoir volume and corresponding elevation data is specified on the SV and SE lines. From this data, the program computes the storage-outflow relationship required for the modified Puls routing.

```
LINE        ID......1.......2.......3.......4.......5.......6.......7.......8.......9......10

 65         KK   E.DAM
 66         KM        ROUTE FLOWS THROUGH DAM
 67         RS    1    ELEV    851.2
 68         SV    21    100    205    325    955
 69         SE  850.0  851.5  853.3  856.5  858.0
 70         SL  851.2   12     .6     .5
 71         SS   856    60    2.7    1.5
```

Figure 6.16   Reservoir routing input data.

9. *Runoff from WEST20, combining, and routing:*   The next calculations in the model consist of computing runoff from subbasin WEST20, combining this hydrograph with the outflow hydrograph from E.DAM and the routed hydrograph 10TO20, and routing this combined hydrograph to location 30 (Fig. 6.17). The Holtan loss rate and the SCS unit hydrograph methods are demonstrated in computing runoff from WEST20. Routing is by the modified Puls method.

10. *Runoff at RED30, combining, and comparing:*   The final calculations consist of computing runoff at RED30, combining this hydrograph with the routed hydrograph 20TO30, and comparing computed and observed hydrographs at gage location 30 (Fig. 6.18). Runoff at RED30 is computed with the SCS curve number loss rate (LS line) and the SCS dimensionless unit hydrograph (UD line). Observed-flow data for comparison is specified on QO lines, and the output control line KO, containing a "1" in the

```
LINE          ID......1.......2.......3.......4.......5.......6.......7.......8.......9......10

 72      KK   WEST20
 73      KM        SCS RUNOFF COMPUTATION-HOLTAN LOSS RATE
 74      KO     1      2
 75      BA    .80
 76      BF   10.0    -.25    1.2
 77      PR    400
 78      PW     1
 79      PT    63      64
 80      PW    .6      .4
 81      LH    .04     .4      .3     1.4
 82      UD    .94
 83      KK   RED20
 84      KM          COMBINE RUNOFF FROM WEST20, OUTFLOW FROM E.DAM AND REACH 10T020
 85      HC     3
 86      KK   20T030
 87      KM        ROUTE FLOWS FROM RED20 TO RED30
 88      RS     1    FLOW     -1
 89      SV     0     17      42      67     100     184     274     386     620
 90      SQ     0     500    1000    1500    2000    3000    4000    5000    7000
```

Figure 6.17 Subbasin runoff, combining, and routing data.

```
LINE          ID......1.......2.......3.......4.......5.......6.......7.......8.......9......10

 91      KK   RED30
 92      KM        RUNOFF BY THE SCS METHOD
 93      BA    .19
 94      BF   10.0    -.25    1.2
 95      PR    400
 96      PW     1
 97      PT    64      63
 98      PW    .65     .35
 99      LS            79
100      UD   1.03
101      KK   RED30
102      KM         COMBINE RUNOFF FROM RED30 AND OUTFLOW FROM REACH 20T030
103      HC     2
104      KK   GAGE
105      KO     1
106      KM        COMPARE COMPUTED AND OBSERVED HYDROGRAPHS AT RED30
107      IN    15 12JUN68    715
108      QO    10      13      16      20      25      30      51      92     159     241
109      QO   332     399     412     393     348     291     255     229     235     321
110      QO   472     705     921    1120    1255    1345    1373    1314    1228    1122
111      QO   996     900     817     742     668     614     549     500     444     409
112      QO   388     372     359     348     338     328     321     310     300     291
113      QO   282     274     267     277     252     240     231     224
114      ZZ
```

Figure 6.18 Subbasin runoff, combining, and hydrograph-comparing data.

first field, requests complete output, including the comparative tabulation and plot shown in Fig. 6.19a and b.

Regionalization of Unit Hydrograph and Loss Function Parameters

Introduction

The lack of meteorologic and hydrologic data is a common problem in hydrologic analyses. There are few precipitation stations and even fewer streamflow measuring stations in many basins. To cope with this problem, techniques have been developed for predicting runoff at locations without sufficient data.

The purpose of regional analysis is to identify mathematical relationships between runoff parameters and physical characteristics of gaged basins so that these relationships can be used to determine parameters for ungaged basins. The gaged basins do not have to be within the same watershed as the ungaged basins, but all of the basins must be in a region that is hydrologically and meteorologically homogeneous. The region must be large enough to obtain a good sampling of gaged basins.

Basic steps in regional analysis

A regional study to develop unit hydrograph and loss function relationships for a basin generally consists of the following steps:[2]

1. Collect precipitation data and discharge hydrograph records for the gaged subbasins in the region. Data should be obtained for several flood events.

2. Develop an HEC-1 flood hydrograph model for estimating unit hydrograph and loss function parameters for these gaged subbasins. In this model each subbasin is optimized separately; routing and combining components are omitted.

3. Run the model developed in step 2 with the data from step 1 to estimate unit hydrograph and loss function parameters for each gaged subbasin.

4. Correlate the estimated parameters from step 3 with subbasin characteristics to develop regional relationships.

5. Use the regional relationships from step 4 to determine parameters for ungaged subbasins. Sometimes further subdivision of the basin is appropriate at this stage of model development to reflect other

```
*********************************************************************
*                                                                   *
*                                                                   *
*            COMPARISON OF COMPUTED AND OBSERVED HYDROGRAPHS         *
*                                                                   *
*********************************************************************
*                                                                   *
*                                                         TIME TO   LAG                              *
*                     SUM OF   EQUIV   MEAN    CENTER     C.M. TO   PEAK    TIME OF                    *
*                     FLOWS    DEPTH   FLOW    OF MASS    C.M.      FLOW    PEAK                       *
*                                                                   *
* COMPUTED HYDROGRAPH 27066.   3.705   467.    7.63       7.63      1331.   6.50                       *
* OBSERVED HYDROGRAPH 26768.   3.664   462.    7.75       7.75      1373.   6.50                       *
*                                                                   *
* DIFFERENCE          298.     .041    5.      -.12       -.12      -42.    .00                        *
* PERCENT DIFFERENCE  1.11                                -1.59     -3.09                              *
*                                                                   *
*                  STANDARD ERROR      21.     AVERAGE ABSOLUTE ERROR        18.                       *
*                  OBJECTIVE FUNCTION  22.     AVERAGE PERCENT ABSOLUTE ERROR 18.56                    *
*                                                                   *
*********************************************************************
```

HYDROGRAPH AT STATION   GAGE

DA MON HRMN	ORD	COMP Q	OBS Q	RESIDUL
12 JUN 0715	1	38.	10.	28.
12 JUN 0730	2	37.	13.	24.
12 JUN 0745	3	37.	16.	21.
12 JUN 0800	4	38.	20.	18.
12 JUN 0815	5	40.	25.	15.
12 JUN 0830	6	46.	30.	16.
12 JUN 0845	7	64.	51.	13.
12 JUN 0900	8	106.	92.	14.
12 JUN 0915	9	178.	159.	19.
12 JUN 0930	10	270.	241.	29.
12 JUN 0945	11	359.	332.	27.
12 JUN 1000	12	418.	399.	19.
12 JUN 1015	13	435.	412.	23.
12 JUN 1030	14	417.	393.	24.
12 JUN 1045	15	378.	348.	30.
12 JUN 1100	16	330.	291.	39.
12 JUN 1115	17	289.	255.	34.
12 JUN 1130	18	261.	229.	32.
12 JUN 1145	19	270.	235.	35.
12 JUN 1200	20	358.	321.	37.

DA MON HRMN	ORD	COMP Q	OBS Q	RESIDUL
12 JUN 1215	21	536.	472.	64.
12 JUN 1230	22	731.	705.	26.
12 JUN 1245	23	937.	921.	16.
12 JUN 1300	24	1119.	1120.	-1.
12 JUN 1315	25	1249.	1255.	-6.
12 JUN 1330	26	1318.	1345.	-27.
12 JUN 1345	27	1331.	1373.	-42.
12 JUN 1400	28	1290.	1314.	-24.
12 JUN 1415	29	1210.	1228.	-18.
12 JUN 1430	30	1100.	1122.	-22.
12 JUN 1445	31	987.	996.	-9.
12 JUN 1500	32	890.	900.	-10.
12 JUN 1515	33	806.	817.	-11.
12 JUN 1530	34	731.	742.	-11.
12 JUN 1545	35	662.	668.	-6.
12 JUN 1600	36	602.	614.	-12.
12 JUN 1615	37	551.	549.	2.
12 JUN 1630	38	502.	500.	2.
12 JUN 1645	39	458.	444.	14.
12 JUN 1700	40	425.	409.	17.

DA MON HRMN	ORD	COMP Q	OBS Q	RESIDUL
12 JUN 1715	41	401.	388.	13.
12 JUN 1730	42	382.	372.	10.
12 JUN 1745	43	365.	359.	6.
12 JUN 1800	44	350.	348.	2.
12 JUN 1815	45	336.	338.	-2.
12 JUN 1830	46	324.	328.	-4.
12 JUN 1845	47	312.	321.	-9.
12 JUN 1900	48	301.	310.	-9.
12 JUN 1915	49	290.	300.	-10.
12 JUN 1930	50	280.	291.	-11.
12 JUN 1945	51	270.	282.	-12.
12 JUN 2000	52	261.	274.	-13.
12 JUN 2015	53	252.	267.	-15.
12 JUN 2030	54	243.	277.	-34.
12 JUN 2045	55	235.	252.	-17.
12 JUN 2100	56	227.	240.	-13.
12 JUN 2115	57	220.	231.	-11.
12 JUN 2130	58	213.	224.	-11.

(a)

Figure 6.19 Comparative output. (a) Tabulations.

```
                    (I) INFLOW,  (O) OUTFLOW,  (*) OBSERVED FLOW

         0.      200.     400.     600.     800.    1000.    1200.    1400.      0.       0.       0.       0.       0.
DAIHRMN PER
120715  11*0
120730  21*0
120745  31*0
120800  41*0
120815  51*0
120830  6I *
120845  7I  *
120900  8I    *
120915  9I     *0.
120930  10I      *0
120945  11I       *0
121000  12I         *0
121015  13I          .*0
121030  14I          *0
121045  15I        * 0.
121100  16I        * 0
121115  17I       *0
121130  18I      * 0
121145  19I       * 0
121200  20I        * 0
121215  21I            * .0.
121230  22I               * 0
121245  23I                   *0
121300  24I                       *
121315  25I                          0*
121330  26I                           0*
```

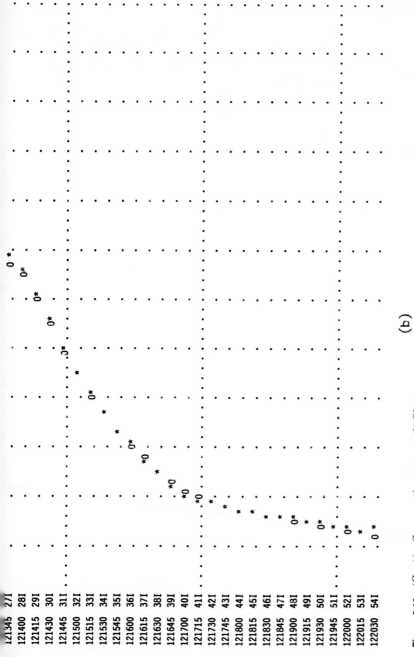

Figure 6.19 *(Cont.)* Comparative output. *(b)* Plot.

(b)

157

factors that come into play in the final model. See the discussion on delineating subbasins at the beginning of this chapter, in the section on developing an HEC-1 model.

6. After the HEC-1 model is fully developed, complete with routing and combining components, perform an analysis of runoff from the entire basin, including both gaged and ungaged subbasins. Compare computed hydrographs at gaged locations with observed hydrographs, and adjust parameters for unit hydrographs, rainfall loss functions, and routing if necessary to obtain suitable replications of the observed hydrographs.

Basin characteristics used in regionalization

In correlating precipitation-runoff parameters with basin characteristics, it is important to select basin characteristics that can be readily measured. This facilitates both the development of regional relationships and their application to ungaged basins. Some suggested parameters that might be used are as follows:[2]

DA	Drainage area
S	Representative (average) slope of the longest watercourse
L	Length of the longest watercourse, measured from the basin outlet to the upper limit of the watershed boundary
L_{ca}	Length of the main watercourse, measured from the basin outlet to a point on the watercourse nearest the centroid of the basin
I	Index of impervious cover as a percentage of total land area

Other characteristics that may influence runoff include soil type, vegetation, land use, percentage of land area occupied by lakes, and percentage of an urban area with sewers.

Correlation techniques

Correlation analyses are performed graphically or numerically to determine which basin characteristics significantly affect runoff response. Linear relationships are analyzed numerically with simple linear regression or multiple linear regression techniques. Nonlinear relationships can sometimes be analyzed with linear techniques if the variables are transformed with logarithms or other factors.

Basin characteristics are analyzed alone and in various combinations with respect to selected parameters. In developing the functional relationships to be analyzed, it is important to remember that they must be reasonable. It is possible to obtain spurious relationships with statistical analysis.

Linear regression. In graphical correlation, coordinate points representing corresponding values of two variables are plotted on graph paper to obtain a "scatter diagram." One of the variables is considered "independent" and the other "dependent." If the scatter diagram is plotted on rectangular coordinate paper and the points can be approximated by a straight line, a linear relationship exists. Linear relationships and nonlinear relationships, approximated with a straight line plotted on semilog or log-log graph paper, can be represented by the following equation:

$$Y = a + bX \tag{6.1}$$

where Y = dependent variable or its logarithm
    a = regression constant
    b = regression coefficient
    X = independent variable or its logarithm

If the scatter diagram of log Y vs. X shows a linear relationship, the equation of the curve is

$$\log Y = a + bX \tag{6.2}$$

or, in exponential form,

$$Y = ab^X \quad \text{(exponential curve)}$$

If the scatter diagram of log Y vs. log X shows a linear relationship, the equation is

$$\log Y = a + b \log X \tag{6.3}$$

or

$$Y = aX^b \quad \text{(geometric curve)}$$

Eq. (6.4) is an example of a geometric regression relationship developed from studies in the Lehigh River basin in Pennsylvania.

$$t_p = 2.10(LL_{ca})^{0.30} \tag{6.4}$$

where t_p is the Snyder unit hydrograph parameter and LL_{ca} is the product of two basin-length characteristics. Numerous other equations used in curve fitting, including quadratic, cubic, and other polynomial equations, can also be used.

Coefficients a and b that will allow the linear regression equation [Eq. (6.1)] to give a "best-fitting line" to the data can be computed with "the method of least squares," described in most statistics texts. Programs for making these "least-squares" computations with computers or calculators are readily available.

Multiple regression. Watershed response usually depends on several watershed characteristics, so a multiple regression equation with several independent variables is required for correlation analysis:

$$Y = a + b_1X_1 + b_2X_2 + \cdots + b_nX_n \qquad (6.5)$$

where Y = dependent variable
a = regression constant
$b_1, b_2,\ldots, b_n$ = regression coefficients
$X_1, X_2,\ldots, X_n$ = independent variables

Just as there is a least-square regression line that provides a "best-fitting line" approximating data points in a two-dimensional scatter diagram, so also is there, in essence, a least-square regression surface for approximating data points in an n-dimensional relationship. Because of the complexity and extent of the computations involved, solution of a multiple regression equation generally requires a computer program such as the stepwise multiple regression program MLRP.[3] Multiple regression analytical capability also is available in most statistics software packages.

Multiple linear regression program—MLRP. When several basin characteristics are being analyzed, the ones that have little effect on the dependent variable should be dropped. In the "stepwise regression" employed in MLRP, developed at the Hydrologic Engineering Center, the analysis is made in a series of iterations, starting with all the specified independent variables included. In each iteration of the computation, the least significant independent variable is deleted on the basis of a comparison of the adjusted partial determination coefficients (r^2). This coefficient indicates the incremental increase of unexplained variance that will result if the variable with which it is associated is deleted from the regression equation.

MLRP also permits combining and transforming variables. A variable such as DA $S^{0.5}$ (basin area DA times the square root of the channel slope S) can be analyzed if the area and slope have been specified in program input. And individual variables may be transformed (by taking the square root, reciprocal, or logarithm) to make the relationships more linear. For example, the equation $Z = cX^dY^e$ is equivalent to log Z = log c + d log X + e log Y through a logarithmic transformation. After such a transformation, multiple linear regression can be used to determine the coefficients that best fit the data.

Criteria for accepting the results of regression analysis. The results of regression analysis are evaluated with statistics indicating the "goodness of fit" between the regression equation and the data. The statistical measures of correlation typically used are the coefficient of determination, the partial determination coefficient, and the standard error of estimate.[4,5]

The coefficient of multiple determination R^2 and the unbiased coefficient of determination $\bar{R}^2$ provide a measure of the percentage of variance in the dependent variable explained by the independent variables in the regression equation. The magnitude of these coefficients varies from 0 to 1. The closer the value is to unity, the greater the reliability of the estimate.

The partial determination coefficient r^2 provides a measure of the loss in correlation or increase in variance that would result from omitting a particular variable from the regression equation. The standard error of estimate S_e provides a measure of scatter of data relative to the computed regression. It is the standard deviation of the differences between the observed dependent values and the values computed with the regression equation.

Example regression analysis. Unit hydrograph parameters and basin characteristics for six gaged basins in the region of the Rahway River basin are shown in Table 6.1. The multiple linear regression program MLRP was used to correlate T_c and R with various physical characteristics of the basin. Imperviousness was included as a variable be-

TABLE 6.1   Selected Data Used in Multiple Regression Analysis

Gage location	DA, mi^2	S, ft/mi	I, %	$T_c + R$, h	T_c, h
1: Rahway River near Springfield, N.J.	25.50	14.6	26.0	12.4	4.9
2: Rahway River at Rahway, N.J.	40.90	8.8	24.0	29.2	16.5
3: Chester Creek near Chester, Pa.	61.10	22.6	9.5	11.9	8.1
4: Green Brook at Plainfield, N.J.	9.75	49.2	25.0	4.7	2.6
5: Robinsons Branch Rahway River at Rahway, N.J.	21.60	13.3	19.0	9.9	5.0
6: Elizabeth River at Elizabeth, N.J.	18.00	19.5	45.0	5.4	2.7

TABLE 6.2   Results of Multiple Regression Analysis

	Standard error of estimate S_e	Correlation coefficient R	Coefficient of determination R^2
$T_c = 26.19I^{-0.53}S^{-0.29}(\text{DA})^{0.23}$	.0495	.9710	.9428
$T_c = 19.84I^{-0.50}(\text{DA}/S)^{0.26}$	.0358	.9849	.9701
$T_c = 8.29K^{-1.28}(\text{DA}/S)^{0.28}$	.0269	.9915	.9831
$T_c = 4.14(\text{DA}/S)^{0.39}$	.1296	.7800	.6084
$(T_c + R) = 122.64I^{0.42}S^{-0.55}(\text{DA})^{0.09}$	.1442	.6844	.4684
$(T_c + R) = 15.69I^{-0.21}(\text{DA}/S)^{0.34}$	.1161	.8094	.6552
$(T_c + R) = 11.52K^{-0.67}(\text{DA}/S)^{0.33}$	.1054	.8461	.7159
$(T_c + R) = 7.98(\text{DA}/S)^{0.39}$	.1093	.8333	.6944

T_c = time of concentration
I = percentage of imperviousness
K = 1.0 + 0.03I
DA = drainage area
S = slope
R = Clark unit hydrograph parameter

cause of the high degree of development in parts of the basin. Results of the regression analysis are shown in Table 6.2. Equation (6.6) was adopted on the basis of the partial determination coefficients for the individual parameters and a comparison of the standard errors of estimate for the various expressions.[6]

$$T_c = 8.29(1.0 + 0.03I)^{-1.28}\left(\frac{\text{DA}}{S}\right)^{0.28} \tag{6.6}$$

where T_c = time of concentration, h
I = imperviousness, %
DA = drainage area, mi^2
S = slope, ft/mi

Evaluation of residuals. The residuals, which are the differences between the computed and observed values in regression analysis, can be used for adjusting the variables computed for ungaged subbasins. They can also be used as a check for consistency of the results of the analysis. The residuals are plotted on a map of the study region at the centroids of the subbasins, and isometric lines of residuals are drawn. The variables computed for an ungaged subbasin are adjusted according to the location and values of the isometric lines relative to the centroid of the ungaged subbasin.

Substantial variation in residuals between different areas of the region may suggest the need for further subdivision of the area in the regression analysis. For example, if the residuals in the upper half of the region are all positive and those in the lower half are all negative, division of the region into two parts according to this pattern would yield more consistent results.

A part of a residuals map developed in a regional frequency study[7] is shown in Fig. 6.20. "Drainage area" was determined to be the basin characteristic that could explain most of the variation in the mean of annual flood peaks. The adopted regression equation was as follows:

$$\log Q_m = a + 0.87 \log A \qquad (6.7)$$

where Q_m = geometric mean of annual flood peaks, ft^3/s
$\quad\quad a$ = a constant
$\quad\quad A$ = drainage area, mi^2

The residual effects of the basin characteristics that were either not considered or not quantifiable in the regional analysis were evaluated by computing the difference between observed values and values computed with Eq. (6.7). The map in Fig. 6.20 was developed with isometric lines representing a modified regression coefficient C_m that in-

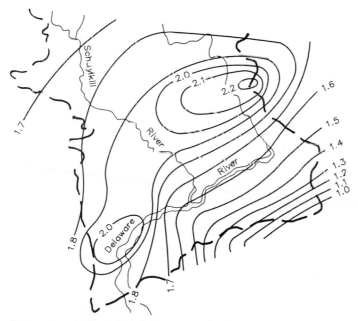

Figure 6.20 Residuals map: isometric lines representing a modified regression coefficient.

cludes the influence of the residuals as well as the constant a used in Eq. (6.7). Thus, Eq. (6.8) can be used in conjunction with Fig. 6.20 for a more accurate computation of annual flood peaks.

$$\log Q_m = C_m + 0.87 \log A \qquad (6.8)$$

where C_m is a map coefficient for the mean log of annual peak.

Workshop Problem 6.1: Regionalization of Unit Hydrograph Parameters in the Region of the Rahway River Basin

Purpose

The purpose of this workshop problem is to provide an understanding of the application of multiple linear regression analysis to develop regional relationships for the Clark unit hydrograph parameters, T_c and R. Use of the Multiple Linear Regression Program, MLRP, is described.

Problem description

Regional regression relationships of T_c and R vs. selected basin characteristics are to be derived for estimating parameters for ungaged subbasins in the Rahway River basin. In the Workshop Problem 5.2, "best" estimates of T_c and R were determined for the subbasin above the Springfield gage. The other gage in the subbasin is not suitable for parameter estimation because of the reservoir in the watershed.

Tasks

1. Unit hydrograph parameters that have been developed at other nearby gages are listed, together with associated basin characteristics, in Table P6.1. An input file prepared with this data for the multiple regression program MLRP is shown in Fig. P6.1. The values of T_c and R for the Springfield gage are omitted and should be added. Use the values estimated in Workshop Problem 5.2. The text editor COED provides a convenient mechanism for making this file change.

2. Formulate several possible functional relationships, including appropriate transformations and combinations of variables. Select two or three of these relationships, and perform a multiple linear regression analysis to determine a suitable equation for estimating parameters at ungaged locations. Running the Multiple Linear Regression Program, MLRP, with the file generated in task 1 is a convenient and fast way to perform this analysis.

TABLE P6.1 Basin Characteristics and Unit Hydrograph Parameters for Basins in the Region of the Rahway River

USGS station number	Station name and location	Drainage area, mi² (1)	Slope, ft/mi (2)	Length, mi (3)	Lake storage, % (4)	Imperviousness, % (5)	T_c, h (6)	R, h (7)
01379000	Passaic River near Millington, N.J.	55.40	30.30	11.00	19.1	7.8	18.2	1.03
01380500	Rockaway River at Boonton, N.J.	116.00	11.90	28.50	7.7	7.6	15.0	32.00
01381500	Whippany River at Morristown, N.J.	29.40	49.00	11.80	2.0	12.6	8.2	19.00
01391500	Saddle River at Lodi, N.J.	54.60	16.60	18.80	5.0	16.0	18.3	10.00
01392000	Weasel Brook at Clifton, N.J.	4.45	95.00	2.70	1.4	30.0	1.2	2.20
01392500	Second River at Belleville, N.J.	11.60	48.50	6.30	1.9	35.0	1.6	2.80
01393500	Elizabeth River at Elizabeth, N.J.	18.00	19.50	8.20	1.1	45.0	2.7	2.70
01394500	Rahway River near Springfield, N.J.	25.50	14.60	9.10	1.4	26.0		
01395000	Rahway River at Rahway, N.J.	40.90	8.84	7.60	1.5	24.0	15.0	15.00
01396000	Robinsons Branch Rahway River at Rahway, N.J.	21.60	13.30	7.10	6.6	19.0	5.0	4.90
01403500	Green Brook at Plainfield, N.J.	9.75	49.20	7.69	2.0	25.0	2.6	2.10

```
T1 REGIONALIZATION OF UNIT HYDROGRAPH PARAMETERS
T2 BY USE OF THE MULTIPLE LINEAR REGRESSION PROGRAM (MLRP)
T3 CLARK'S TC AND R VS. SELECTED BASIN CHARACTERISTICS
J1     7
NM          AREA   SLOPE  LENGTH STORAGE  IMPERV     TC      R
TR
DT  3790    55.4   30.3    11.0    19.1     7.8     18.2    1.03
DT  3805    116.   11.9    28.5     7.7     7.6     15.0    32.0
DT  3815    29.4   49.0    11.8     2.0    12.6      8.2    19.0
DT  3915    54.6   16.6    18.8     5.0    16.0     18.3    10.0
DT  3920    4.45   95.0     2.7     1.4    30.0      1.2     2.2
DT  3925    11.6   48.5     6.3     1.9    35.0      1.6     2.8
DT  3935    18.0   19.5     8.2     1.1    45.0      2.7     2.7
DT  3945    25.5   14.6     9.1     1.4    26.0
DT  3950    40.9   8.84    17.6     1.5    24.0     15.0    15.0
DT  3960    21.6   13.3     7.1     6.6    19.0      5.0     4.9
DT  4035    9.75   49.2    7.69     2.0    25.0      2.6     2.1
ED
NJ
T
```

Figure P6.1   Incomplete input file for the multiple regression program MLRP. Parameters from five regional subbasins are used.

3. Select a regression equation, or equations, from your analysis to estimate the parameters for the ungaged basins delineated in Fig. 6.1. Explain the reasons for your choice.

4. Apply the selected regression equation(s) to the five Rahway subbasins shown in Fig. 6.1, with the subbasin characteristics provided in Table P6.2. A partially completed input file containing this data is shown in Fig. P6.2. Modify the TR and RP lines to reflect your selected equation, and then execute the program MLRP using this file as input. Tabulate the results of this analysis, and examine the values of T_c and R for reasonableness. Making some additional runs with other equations formulated in task 2 may provide additional information for making this evaluation.

Workshop Problem 6.2: Development of a Multiple-Subbasin Model for the Rahway River Basin

Purpose

The purpose of this workshop problem is to provide an understanding of the procedures used in formulating and executing a river basin model with several subbasins and routing reaches.

TABLE P6.2 Basin Characteristics for Subbasins in the Rahway River Basin

Subbasin number	Description	Drainage area, mi^2 (1)	Slope, ft/mi (2)	Length, mi (3)	Lake storage, % (4)	Impervi- ousness, % (5)	T_c, h (6)	R, (7)
1	West Branch above Orange Reservoir	4.62	115	4.6	2.5	22		
2	Downstream Orange Reservoir to Millburn gage	2.51	143	3.1	1.2	10		
3	Millburn gage to junction of East Branch	1.15	160	2.4	1.1	25		
4	East Branch Rahway River	7.53	37	6.0	1.1	35		
5	Junction East and West branches to Springfield	9.68	48	4.6	1.2	27		

```
T1 REGIONALIZATION OF UNIT HYDROGRAPH PARAMETERS
T2 BY USE OF THE MULTIPLE LINEAR REGRESSION PROGRAM (MLRP)
T3 ESTIMATION OF PARAMETERS AT SUBBASINS 1 - 5
J1    7
NM         AREA   SLOPE  LENGTH STORAGE  IMPERV    TC     R
TR                                                        -1
RP
DT    1    4.62   115.    4.6    2.5     22.      1.    1.
DT    2    2.51   143.    3.1    1.2     10.      2.    2.
DT    3    1.15   160.    2.4    1.1     25.      3.    3.
DT    4    7.53   37.     6.0    1.1     35.      4.    4.
DT    5    9.68   48.     4.6    1.2     27.      5.    5.
ED
NJ
T1
```

Figure P6.2 Incomplete input file for MLRP. Parameters from five Rahway subbasins are used.

Problem description

A floodplain study is to be conducted for the Rahway River basin upstream from the gage near Springfield. The basin, divided into five subbasins, is shown in Fig. 6.1. A partially completed input file for a five-subbasin model is shown in Fig. P6.3. This model is to be completed and calibrated; some of the data developed in previous workshops and additional information provided here are to be used for this assignment.

The percentage of imperviousness used in the model was determined from a previous study.[8] Base flow parameters for all the subbasins are estimated to be as follows: QRCSN = 4 percent of peak flow; RTIOR = 1.10; and, for the event being simulated, STRTQ = 0.08 ft^3/(s)(mi^2). There are three routing reaches and one reservoir in the study area. The routing parameters, including two sets derived from a previous workshop, are shown in Table P6.3. The discharge-storage data for the Orange Reservoir is provided in Table P6.4. The basin average rainfall pattern for this storm from the preceding parameter estimation workshop should be used. The average total rainfall for each subbasin is as follows:

Subbasin	Total storm rainfall, in
1	8.00
2	9.00
3	9.35
4	8.69
5	9.25

```
ID        RAHWAY RIVER BASIN
ID        MODEL CALIBRATION
ID        STORM OF 26-28 AUGUST 1971
IT    15       0       0     240
IO     2       1
KK   101
KM        INFLOW TO THE ORANGE RESERVOIR
BA  4.62
PB  8.00
IN    60
PI     0       0       0     .04     .12     .06     .10     .08     .43     .76
PI   .37     .70    1.15    1.36     .18     .06     .04     .09      0       0
PI     0       0       0       0       0     .10     .76    1.53    1.12     .37
PI   .03
LU  4.82    0.50      22
UC   1.9     3.5
KK   102
BA  2.51
PB  9.00
LU  4.82    0.50      10
UC   2.8     5.2
KK   103
BA  1.15
PB  9.35
LU  4.82    0.50      25
UC   1.0     1.9
KK   104
KM        EAST BRANCH OF THE RAHWAY RIVER IMMEDIATELY UPSTREAM OF CONFLUENCE
BA  7.53
PB  8.69
LU  4.82    0.50      35
UC   2.1     3.9
KK   105
KM        LOCAL AREA - CONFLUENCE TO THE SPRINGFIELD GAGE
BA  9.68
PB  9.25
LU  4.82    0.50      27
UC   2.1     3.9
KKSPRFLD
KM        TOTAL FLOW OF THE RAHWAY RIVER AT THE SPRINGFIELD GAGE
HC     2
KKSPRFLD
KO     1       2
KM        COMPARE COMPUTED AND OBSERVED HYDROGRAPHS AT THE SPRINGFIELD GAGE
QO     2       2       3       3       4       9      16      22      39     105
QO   200     310     425     543     660     800     925    1080    1300    1700
QO  1725    1620    1480    1365    1294    1200    1085    1054    1300    1675
QO  2160    2620    3080    3400    3350    3060    2820    2450    2160    1975
QO  1760    1525    1300    1160    1010     920     820     720     609     500
QO   410     320     225     177     140     115      90      80      75      74
QO    70      60      55      50      50      50      48      47      45      40
QO    38      34      31      29      28      27      25      24      23      23
QO    22      20      20      19      18       0       0       0       0       0
ZZ
```

Figure P6.3 Incomplete input file for five-subbasin model of the Rahway River above the Springfield gage.

TABLE P6.3   Routing Reach Data

Reach 101-102: Orange Reservoir to Millburn Gage			
Muskingum	$X = 0.10$	$K = 1.30$ h	

Reach 102-103: Millburn Gage to East-West Confluence			
Modified Puls using one step:			
Storage, acre • ft	0	120	180	440
Outflow, ft^3/s	0	2100	3500	6905

Reach 104-105: East-West Confluence to Springfield Gage					
Modified Puls using two steps:					
Storage, acre • ft	0	660	1320	1775	2220	3830
Outflow, ft^3/s	0	2400	4850	6500	8125	14000

TABLE P6.4   Discharge-Storage Data for the Orange Reservoir

Head on crest of spillway, ft	Coefficient C	Discharge $Q = CLH^{3/2}$, ft^3/s	Storage, acre • ft
0.0		0	0
1.0	3.36	240	65
2.0	3.36	680	135
3.0	3.36	1250	200
3.4	3.36	1510	230
4.0		2600	270
5.0		5900	340

The flood flows of August 27 recorded at the Springfield gage also were given in a previous workshop.

Tasks

1. Construct an HEC-1 multisubbasin model of the Rahway River basin, and simulate the storm of August 27. Use initial estimates of rainfall loss parameters for each of the subbasins based on the results of the parameter estimation analysis of this storm in Workshop Problem 5.2. Use the values of T_c and R adopted in Workshop Problem 6.1 for the five subbasins.

2. Estimate the rainfall loss function parameters of the subbasins using the automatic calibration option of HEC-1. Adjust the parameters as necessary to obtain a "best" fit between the computed hydrograph and the observed hydrograph.

References

1. Hydrologic Engineering Center, *HEC-1 Flood Hydrograph Package, Users Manual,* U.S. Army Corps of Engineers, Davis, Calif., 1982.
2. Hydrologic Engineering Center, *Hydrologic Analysis of Ungaged Watersheds Using HEC-1,* Training Document 15, U.S. Army Corps of Engineers, Davis, Calif., 1982.
3. Hydrologic Engineering Center, *Multiple Linear Regression,* generalized computer program, U.S. Army Corps of Engineers, Davis, Calif., September 1970 (rev. January 1983).
4. Hydrologic Engineering Center, *Statistical Methods in Hydrology,* U.S. Army Corps of Engineers, Davis, Calif., 1962.
5. Hydrologic Engineering Center, "Hydrologic Frequency Analysis," Engineer Manual EM 1110-2-1415, U.S. Army Corps of Engineers, Davis, Calif., April 1985. (Draft.)
6. Hydrologic Engineering Center, *Hydrologic-Hydraulic Simulation,* Special Projects Memo 469, U.S. Army Corps of Engineers, Davis, Calif., November 1976.
7. Hydrologic Engineering Center, *Regional Frequency Study, Upper Delaware and Hudson River Basins, New York District,* U.S. Army Corps of Engineers, Davis, Calif., November 1974.
8. Division of Water Resources, *Magnitude and Frequency of Floods in New Jersey with the Effects of Urbanization,* Special Report 38, State of New Jersey Department of Environmental Protection, 1974.

Frequency Analysis

Introduction

Flood risk

Flood control projects and other flood protection measures are designed to reduce the undesirable effects of floods, which occur randomly. A project that would eliminate all damage one year may not be large enough to provide complete protection the next year. Thus, projects are designed on the basis of the analysis of a wide range of floods and estimates of long-term impacts.

"Risk" may be defined as exposure to an undesirable event. "Probability" is a measure of risk, and one of the prime ways of estimating the probability of a flood is by analyzing historical flood-flow data. The reliability of the estimate depends on the quantity and quality of the historical data; a long record of dependable data permits a more reliable estimate.

Deterministic vs. stochastic models

Hydrologic phenomena such as floods are generally analyzed through the use of a model, whether the model is merely a mental conceptualization, an empirical relationship, a physical device, or a collection of mathematical and statistical equations. Most quantitative hydrologic models can be classified as deterministic models, stochastic models, or a combination of these. A deterministic model is one that is based on physical relationships and requires no experimental or sample data for its application. A stochastic, or "probabilistic," model is one involving random variables having distributions in probability and whose outputs are predictable only in a statistical sense. Repeated use of a given set of inputs in a stochastic model produces outputs that are not the same but that follow certain statistical patterns. A parametric model may be thought of as a combination of these basic types. It is deterministic in the sense that once model parameters are determined, the model always produces the same output from a given input. It is stochastic in the sense that parameter estimates depend on observed data and will change as the observed data changes.[1]

There are numerous applications of these models that could be used as examples. The use of the "probable maximum flood" in spillway design is an example of the deterministic approach. The storm size is based on hydrologic and meteorologic characteristics of the watershed, not on statistical analysis of historical events. A rainfall-runoff model such as HEC-1 might be considered a parametric model. However, most H&H flood analyses, including those involving HEC-1, consider a range of possible storm and runoff events and their likelihood of occurrence. This requires an understanding of the principles of probability and methods of frequency analysis.

Paired data vs. time-series data

Data to be analyzed statistically can be classified as paired data or time-series data. Paired data are related in terms of dependency: the values of one variable, the dependent variable, depend on the values of another variable, the independent variable. The value of a unit hydrograph parameter (the dependent variable) and the area of the drainage basin associated with the unit hydrograph (the independent variable) constitute an example of paired data. The regression techniques discussed in Chap. 6 are used to determine relationships for predicting values of the dependent variable on the basis of certain values of the independent variable, or variables.

Time-series data consists of observations taken at specified times, usually at equal intervals. Daily precipitation measurements and hourly temperatures are examples of time-series data. One of the purposes of analyzing time-series data is to predict the probability or frequency of occurrence of a specified value. For example, the purpose of an investigation might be to determine the probability that the flow in a river will exceed 100,000 ft^3/s or, stated in different terms, to determine the frequency with which flows can be expected to exceed 100,000 ft^3/s in the river.

Fundamentals of Frequency Analysis

Frequency concepts

Sampling theory.   Sampling and probability theory are the basis for analyzing chance events such as the occurrence of a flood of a certain size. Samples drawn from an entire group of events, termed the "population," are used to estimate unknown population parameters, such as the mean and standard deviation. These parameters characterize the population and provide a basis for predicting the occurrence of certain events. Stream discharge records spanning 50 years are but a small sample of the population of streamflow events. However, if this data is a reliable and representative time sample of random homogeneous events, it can be used in analyzing the frequency of floods of various sizes.

Probability theory. The relative frequency concept of probability defines the probability of a random event as the number of occurrences of the event divided by the number of trials, when the number of trials extends indefinitely. This fundamental relationship is expressed in the following equation:

$$P(X) = \lim_{n \to \infty} \frac{n_X}{n} \tag{7.1}$$

where n_X is the number of occurrences and n is the total number of trials.

This concept can be demonstrated with the familiar die-tossing experiment. Tossing any of the numbers 1 through 6, which appear on the six faces of the die, has an equal chance of occurring. The outcome of each toss has a finite probability, and the sum of the probabilities of all possible outcomes is 1. Thus, the probability of tossing a 1—$P(1)$— is $\frac{1}{6}$; the probability of tossing a 2—$P(2)$—is $\frac{1}{6}$; and so on. The outcomes also are "mutually exclusive" because if one of the numbers occurs, none of the other numbers can occur in the same toss.

In this simple example, computing the probabilities of the outcomes of tossing the die is easy. If it were necessary to estimate the probability by sampling techniques, a large number of tosses would be required to obtain a good estimate. The problem of estimating probability this way is depicted in Fig. 7.1. The relative frequency is plotted as a function of the number of tosses. In other words, the probability of tossing a 6 is estimated after each toss. If a 6 is obtained on the first toss, the probability is $P = \frac{1}{1} = 1$, not a very good estimate. If a 3 is obtained on the next toss, $P = \frac{1}{2} = .5$, still not a very good estimate. Several hundred tosses may be required to obtain a close estimate, as the figure suggests.

Two of the basic rules of probability governing the tossing of a die can be summarized as follows:

1. The probability of an event is equal to or greater than zero and equal to or less than 1.

$$0 \le P(X_i) \le 1$$

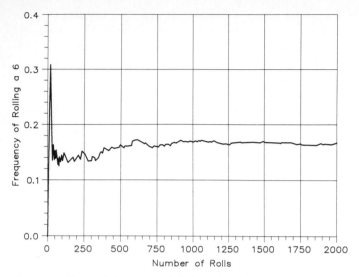

Figure 7.1   Determining the probability of "rolling a 6" by a relative frequency estimate.

2. The sum of probabilities of all possible outcomes in a single trial is 1.

$$\sum_{i=1}^{n} P(X_i) = 1$$

A probability distribution represents the probabilities attached to specific values that a random variable may have over the range of its occurrence. For example, consider the population, or entire group of events, consisting of all possible sums of numbers obtained by tossing two dice.

Sum (S)	2	3	4	5	6	7	8	9	10	11	12
$P(S)$	$\frac{1}{36}$	$\frac{2}{36}$	$\frac{3}{36}$	$\frac{4}{36}$	$\frac{5}{36}$	$\frac{6}{36}$	$\frac{5}{36}$	$\frac{4}{36}$	$\frac{3}{36}$	$\frac{2}{36}$	$\frac{1}{36}$

Since the variable S assumes a discrete set of values, this distribution of probabilities is called a "discrete probability distribution."

If a variable may assume a continuous set of values, the distribution of its probabilities is called a "continuous probability distribution." The curve defining this distribution is sometimes called the "probability density function," and the total area under this curve (Fig. 7.2) is equal to 1. The shaded area under the curve between points $X = a$ and $X = b$ gives the probability that X lies between a and b. The area under the curve to the right of b equals the probability that X is equal to or greater than b, and the area to the left of a is the probability that

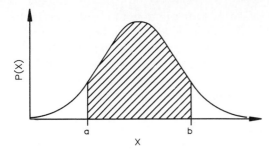

Figure 7.2 Probability density function.

X is equal to or less than a. These relationships have particular significance in frequency analysis, as will be explained.

Frequency analysis

Frequency distributions. In analyzing large amounts of data, such as flow data, it is useful to distribute the data into classes or categories and determine the number of individual events belonging to each class, called the "class frequency." A tabulation of data by classes together with the corresponding class frequencies constitutes a "frequency distribution." And a frequency distribution can be represented graphically by a "histogram" such as the one shown in Fig. 7.3. Mean annual flow data for 85 years from the same population is represented

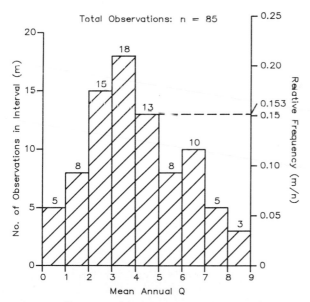

Figure 7.3 Frequency histogram.

in this histogram. The range of values for the mean annual flow is divided into nine class intervals of 1000 ft^3/s along the horizontal axis. The height of each bar equals the class frequency, or (in other words) the number of values of mean annual flow in a particular class. For example, there are five flows equal to or greater than zero and less than 1000 ft^3/s, eight flows equal to or greater than 1000 ft^3/s and less than 2000 ft^3/s, and so on.

The relative frequency of a class is the class frequency divided by the total frequency of all classes. For example, the relative frequency of flows in the class interval between 4000 and 5000 ft^3/s is $^{13}/_{85} = 0.153$. A relative frequency scale is shown on the right side of the histogram.

According to the definition of probability [Eq. (7.1)], the relative frequency as defined for the frequency distribution is an estimate of probability. Frequency distributions are distributions for samples drawn from populations, and probability distributions are distributions for populations. The frequency distribution approaches the probability distribution as the number of observations or trials is made very large.

A cumulative frequency distribution also can be determined for data which has been distributed into classes. The total frequency of all values less than the upper boundary of a given class interval is called the "cumulative frequency" up to and including that class interval. A relative cumulative frequency curve for the data shown in Fig. 7.3 is plotted in Fig. 7.4. The cumulative frequency up to and including a particular class interval is obtained by adding the class frequencies up to that point, beginning at the left side of the histogram. For the class interval between 4000 and 5000 ft^3/s, the cumulative frequency is

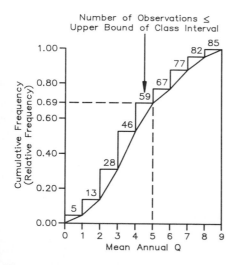

Figure 7.4 Cumulative frequency curve.

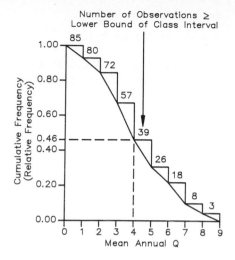

Figure 7.5 Exceedance frequency distribution.

$5 + 8 + 15 + 18 + 13 = 59$, indicating that 59 events had flows of less than 5000 ft^3/s. The relative cumulative frequency is $^{59}/_{85} = 0.69$.

In flood hydrology it usually is of greater interest to consider a cumulative frequency distribution of values greater than or equal to the lower boundary of each class interval. An "exceedance frequency distribution," which this is commonly called, is plotted in Fig. 7.5. This plot can be constructed by beginning it on the right side and adding the class frequencies from the histogram in reverse order, that is, from the highest class interval to the lowest. The cumulative frequency greater than or equal to the lower boundary of the class interval between 4000 and 5000 ft^3/s is $3 + 5 + 10 + 8 + 13 = 39$, indicating that 39 events had flows equal to or greater than 4000 ft^3/s. The relative exceedance frequency plotted on the curve is $^{39}/_{85} = 0.46$. As an estimate of probability, this indicates that there is a 46 percent chance that the flows will equal or exceed 4000 ft^3/s.

It is common practice in hydrology to rotate the axes of the exceedance frequency distribution 90° so that the discharge is shown on the vertical axis and the relative exceedance frequency on the horizontal axis. It is also common to plot the values on log-probability paper or other types of graph paper that tend to produce straight-line plots. The relative exceedance frequency is represented in a variety of ways and given a number of different labels, including the following:

Exceedance frequency = relative exceedance frequency × 100

Exceedance probability = relative exceedance frequency

Percent chance exceedance = relative exceedance frequency × 100

Return period = 1/relative exceedance frequency

Recurrence interval = 1/relative exceedance frequency

Exceedance Interval = 1/relative exceedance frequency

The "percent chance exceedance" is the terminology preferred by some as causing the least confusion regarding the underlying concept.

Distribution characteristics. The characteristics of probability distributions may be defined by the parameters of probability functions expressed in terms of moments. The principle characteristics are "central tendency," the grouping of observations about a central value; the "variability," the dispersion of observations; and "skewness," the degree of asymmetry of the distribution.[2] The distribution curves shown in Fig. 7.6 exhibit approximately the same grouping about a central value, but curve *b* has a greater variability than curve *a*, and curve *b* has a right skew, while curve *a* is symmetrical.

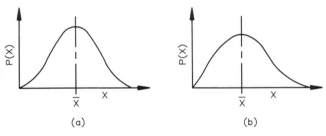

Figure 7.6   Probability distribution characteristics. (*a*) Symmetrical; (*b*) skewed.

The mean.   An average value for a set of data is a measure of central tendency. Of the several measures of central tendency, including the "mean," the "median," and the "mode," the most common measure is the mean μ. It is defined as the first moment about the origin, and the sample estimate of the mean $\overline{X}$ is expressed as

$$\overline{X} = \frac{1}{n}\sum_{i=1}^{n} X_i$$

The standard deviation.   The most statistically important measure of dispersion, or variability of data about the mean, is the mean squared deviation S^2—called the "variance." It is measured by the second moment about the mean.

$$S^2 = \frac{1}{n-1}\sum_{i=1}^{n} (X_i - \overline{X})^2$$

The square root of the variance, called the "standard deviation," has the same units as the mean and the data. The standard deviation S of sample data is an estimate of the population standard deviation.

$$S = \sqrt{\frac{1}{n-1} \sum_{i=1}^{n} (X_i - \overline{X})^2}$$

The coefficient of variation C_v is useful in studying the relative variability between different sets of data.

$$C_v = \frac{S}{\overline{X}}$$

Skew coefficient.   The third moment about the mean is used to measure skewness. A sample estimate of population skewness is computed by

$$a = \frac{n}{(n-1)(n-2)} \sum_{i=1}^{n} (X_i - \overline{X})^3$$

$$C_s = \frac{a}{S^3}$$

For right skewness (long tail to right side), $C_s > 0$, and for left skewness, $C_s < 0$.

Theoretical frequency distributions.   Most hydrologic variables are continuous, and several theoretical continuous frequency distributions have been derived to fit historical data. These do not exactly represent the physical processes involved, but they have been found to fit various types of hydrologic data. However, they do not all fit the same types of data equally well; some may fit one type, some another. Theoretical distributions may be fit to sample data with the statistical parameters of the distribution, such as the mean and the standard deviation. Sample estimates of the parameters are assumed to be population parameters and are used in the theoretical distributions to solve for probabilities of certain events. Among the most common continuous distributions are the uniform, exponential, normal, log-normal, gamma, and extreme value.

The normal distribution is defined by the equation

$$P(X) = \frac{1}{\sigma\sqrt{2\pi}} \exp\left[-\frac{1}{2}\left(\frac{X-\mu}{\sigma^2}\right)^2\right] \tag{7.2}$$

where μ is the mean and σ is the standard deviation.

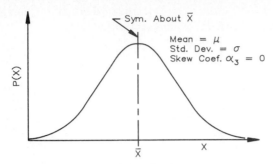

Figure 7.7   The normal distribution.

This bell-shaped frequency function (Fig. 7.7), also known as the gaussian distribution, describes many processes subject to random variation. Although it often does not fit hydrologic data directly, it has wide application with transformed hydrologic data and in estimating sample variability by virtue of the "central limit theorem." The central limit theorem asserts that means and standard deviations of a large number of samples drawn from the same population have a normal distribution regardless of the probability distribution of the population.

Many hydrologic variables have a positive skewness partly due to the influence of natural phenomena having values that are greater than zero or some other lower limit and that are theoretically unconstrained in their upper range. Their frequencies do not follow the normal distribution, but the frequencies of their logarithms do. The lognormal distribution that is obtained by replacing the variable X with its logarithm (log X) in the normal distribution is especially useful because the transformation opens the extensive body of theoretical and applied uses of the normal distribution.[2]

Application of the normal distribution or one of the other distributions in determining the frequency of an event is accomplished by making an estimate of the distribution's parameters from a sample or samples taken from the population. For the normal distribution, this means computing the mean and standard deviation and then using the resulting values to solve Eq. (7.2). Tabulated values for the standard normal distribution (Table 7.1) can be used to compute percent chance exceedance.

Data required for frequency analysis

Sources.   Sources of data for frequency analysis in gaged basins are records of gaged data, such as the streamflow records of the U.S. Geo-

TABLE 7.1   Normal Distribution

Percent Chance Exceedance for Given Normal Standard Deviate K

K				Increments to K in column 1					
	.00	.01	.02	.03	.04	.05	.06	.07	.08	.09
.0	50.00	49.60	49.20	48.80	48.40	48.01	47.61	47.21	46.81	46.41
.1	46.02	45.62	45.22	44.83	44.43	44.04	43.64	43.25	42.86	42.47
.2	42.07	41.68	41.29	40.90	40.52	40.13	39.74	39.36	38.97	38.59
.3	38.21	37.83	37.45	37.07	36.69	36.32	35.94	35.57	35.20	34.83
.4	34.46	34.09	33.72	33.36	33.00	32.64	32.28	31.92	31.56	31.21
.5	30.85	30.50	30.15	29.81	29.46	29.12	28.77	28.43	28.10	27.76
.6	27.43	27.09	26.76	26.43	26.11	25.78	25.46	25.14	24.83	24.51
.7	24.20	23.89	23.58	23.27	22.96	22.66	22.36	22.06	21.77	21.48
.8	21.19	20.90	20.61	20.33	20.05	19.77	19.49	19.22	18.94	18.67
.9	18.41	18.14	17.88	17.62	17.36	17.11	16.85	16.60	16.35	16.11
1.0	15.87	15.62	15.39	15.15	14.92	14.69	14.46	14.23	14.01	13.79
1.1	13.57	13.35	13.14	12.92	12.71	12.51	12.30	12.10	11.90	11.70
1.2	11.51	11.31	11.12	10.93	10.75	10.56	10.38	10.20	10.03	9.85
1.3	9.68	9.51	9.34	9.18	9.01	8.85	8.69	8.53	8.38	8.23
1.4	8.08	7.93	7.78	7.64	7.49	7.35	7.21	7.08	6.94	6.81
1.5	6.68	6.55	6.43	6.30	6.18	6.06	5.94	5.82	5.71	5.59
1.6	5.48	5.37	5.26	5.16	5.05	4.95	4.85	4.75	4.65	4.55
1.7	4.46	4.36	4.27	4.18	4.09	4.01	3.92	3.84	3.75	3.67
1.8	3.59	3.51	3.44	3.36	3.29	3.22	3.14	3.07	3.01	2.94
1.9	2.87	2.81	2.74	2.68	2.62	2.56	2.50	2.44	2.39	2.33
2.0	2.28	2.22	2.17	2.12	2.07	2.02	1.97	1.92	1.88	1.83
2.1	1.79	1.74	1.70	1.66	1.62	1.58	1.54	1.50	1.46	1.43
2.2	1.39	1.36	1.32	1.29	1.25	1.22	1.19	1.16	1.13	1.10
2.3	1.072	1.044	1.017	.990	.964	.939	.914	.889	.866	.842
2.4	.820	.798	.776	.755	.734	.714	.695	.676	.657	.639
2.5	.621	.604	.587	.570	.554	.539	.523	.508	.494	.480
2.6	.466	.453	.440	.427	.415	.402	.391	.379	.368	.357
2.7	.347	.336	.326	.317	.307	.298	.289	.280	.272	.264
2.8	.256	.248	.240	.233	.226	.219	.212	.205	.199	.193
2.9	.187	.181	.175	.169	.164	.159	.154	.149	.144	.139
3.0	.135	.131	.126	.122	.118	.114	.111	.107	.104	.100
3.1	.0968	.0936	.0904	.0874	.0845	.0816	.0789	.0762	.0736	.0711
3.2	.0687	.0664	.0641	.0619	.0598	.0577	.0557	.0538	.0519	.0501
3.3	.0483	.0467	.0450	.0434	.0419	.0404	.0390	.0376	.0362	.0350
3.4	.0337	.0325	.0313	.0302	.0291	.0280	.0270	.0260	.0251	.0242
3.5	.0233	.0224	.0216	.0208	.0200	.0193	.0185	.0179	.0172	.0165
3.6	.0159	.0153	.0147	.0142	.0136	.0131	.0126	.0121	.0117	.0112
3.7	.01078	.01037	.0099	.00958	.00920	.00884	.00850	.00817	.00784	.0075
3.8	.00724	.00695	.00667	.00641	.00615	.00591	.00567	.00544	.00522	.0050
3.9	.00481	.00462	.00443	.00425	.00408	.00391	.00375	.00360	.00345	.0033
4.0	.00317	.00304	.00291	.00279	.00267	.00256	.00245	.00235	.00225	.0021
5.0	.00002867									
6.0	.0000000987									

logical Survey (USGS) and other governmental agencies and the climatological records of the National Weather Service (NWS). These records ordinarily can be obtained from publications as well as electronic media.

Data adjustments.   Recorded flow data may require adjustment due to breaks in the record, incompleteness, zero flows, mixed populations, outliers, and shortness of record.

Breaks in the record.   It is not uncommon to find periods of time for which there is no data in time-series records (Fig. 7.8). In determining

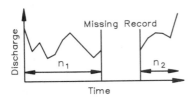

Figure 7.8   Missing record.

how to adjust for this condition, one needs to know why the data is missing. Was the gage inoperative due to lack of funding support, or was it washed out due to a flood? If it was washed out by a flood, it usually means that there was a significant flood peak during the missing period that needs to be accounted for. If the gage was out of operation because the agency responsible for it ran short of operating money or because of some other similar reason, it is common practice to eliminate the gap by joining n_1 and n_2 and treating the record as if it were continuous. An exception to this approach would be appropriate if some physical change in the watershed occurred between segments n_1 and n_2 which would make the total record nonhomogeneous.

Incomplete record.   A record may be incomplete because the flows were either too high or too low to be recorded by the gage. Missing high data and missing low data require different treatment. When one or more high annual peaks have not been recorded, there is usually information available from the data collection agency for estimating the peak discharges. In most instances, the data collection agency routinely provides such estimates. At crest gage sites where the bottom of the gage is not reached in some years, the use of a conditional probability adjustment is recommended.[3]

Zero-flow years.   Some streams in arid regions have no flow for an entire year, so the annual time-series record for such streams would have one or more zero values (Fig. 7.9). This precludes analysis of this

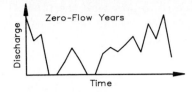

Figure 7.9 Zero-flow years.

data by the recommended Log Pearson Type III distribution (see a discussion of this method under the heading "Streamflow Frequency Analysis") because the logarithm of zero is minus infinity. A conditional probability adjustment can be used to determine frequency curves for records with zero-flow years.[3]

Mixed populations. Flooding sometimes is created by combinations of events such as rainstorms with snowmelt or intense tropical storms with general cyclonic storms. Hydrologic factors and other relationships are usually quite different for the phenomena which are combined, such as a winter rain and snowmelt. When it can be determined that there are two or more distinct and generally independent causes of a flood, it may be more reliable to segregate the flood data by cause, analyze each set separately, and combine the results.[4] If two or more populations cannot be identified and separated by a meaningful criterion, the record should be treated as coming from one population.[4]

Outliers. Outliers, which are data points that depart greatly from the trend of the other data (Fig. 7.10), require special treatment. The

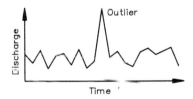

Figure 7.10 Outliers.

decision to retain such points or to delete or modify them can significantly affect the parameters computed from the data, especially for small samples. High and low outliers are treated differently. Procedures for treating outliers require both mathematical and hydrologic considerations.[4]

Short record. Dealing with a short record is always a problem to some degree in hydrologic analysis. Streamflow records, if available, are seldom longer than 30 to 40 years. Analysis of such short records involves considerable uncertainty with regard to how well the probability of the parameters of the population can be estimated. To compensate for a short record, the usual approach is to add data in some manner. This might be done by making adjustments with historical data, described in the next paragraph, if it is available; by transferring data from other locations through regional analysis; and by using synthetic data generated by a rainfall-runoff simulation model.

Historical data is considered to be flood information outside of the usual, or "systematic," record, including, for example, the knowledge of a large flood peak that was the largest event over a period longer than that of the systematic record. This information can often be used to weight the largest events, including high outliers, over a period longer than that of the systematic record.

A guide (based on length of record) to the appropriate analyses to include in determining flood magnitudes with 1 percent chance exceedance is as follows:[4]

	Length of record available, years		
Analyses to include	10 to 24	25 to 50	50 or more
Statistical analysis	X	X	X
Comparisons with similar watersheds	X	X	
Flood estimates from precipitation	X		

Streamflow Frequency Analysis

Annual series vs. partial-duration series

Flood events can be analyzed using either an annual series or a partial-duration series. An annual flood series consists of all the maximum annual flood peaks; i.e., only the maximum event for each year is included, irrespective of its magnitude. It is not unusual for the second greatest event in one year to exceed the annual maximum in some other year, but the analysis of annual flood peaks neglects such events.

The partial-duration series consists of all independent flood peaks equal to or greater than a predefined magnitude or threshold. These two types of series are contrasted in Fig. 7.11. While there are three annual peaks, one for each year, there are two partial-duration peaks in year 1, none in year 2, and one in year 3. Generally, the largest peaks in a record are included in both series, and the difference in the analysis of floods with recurrence intervals greater than 10 years is negligible.[2]

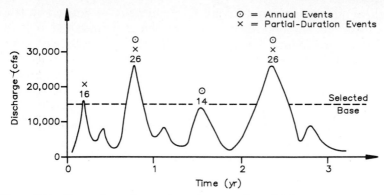

Figure 7.11 Comparison of annual maximum and partial-duration series.

In an annual series, peaks occurring within the same year are indexed to the maximum peak for the year, but in a partial-duration series, each event is considered independent, irrespective of other events in the same year. It should be noted that independence might be defined differently in flood damage analysis than it is in hydrology. In hydrology, two flood peaks are considered independent if all the runoff for the first has left the basin prior to the start of runoff for the second. On the other hand, if the first peak caused severe damage, the second may cause little, if any, because the damage has already been done. In an economic sense, the two events are definitely not independent. This is an important point to remember in flood damage analysis.

Factors affecting homogeneity of data

Data used in frequency analysis must be from a homogeneous and stationary population. In other words, it must be representative of a consistent set of conditions over time. Common causes of nonhomogeneous data include relocation of gages, diversions, construction of dams, land-use changes, and catastrophic natural influences such as hurricanes and earthquakes. Mixed populations resulting from the inclusion of rainfall, snowmelt, and combined rainfall-snowmelt events in a record present a similar problem. Thus, the assessment of the adequacy of the flood records is a necessary first step in frequency analysis. In some cases, it may be necessary to adjust parts of the record that have been subject to storage changes, urbanization, and other factors to obtain a homogeneous data set.

Storage changes. Large reservoirs and other smaller storage projects tend to change flood peaks from what they were prior to development.

The effects of storage on annual peak frequency curves depend on the amount of storage, the operating rules for the impoundment if it is regulated, and the amount of uncontrolled drainage area between the storage site and the location of interest. A frequency curve associated with regulated storage is usually nonlinear, the nonlinearity increasing with the degree of regulation.

Urbanization.   The increase in imperviousness associated with urbanization tends to cause greater volumes of runoff and higher peak flows for frequent to moderately frequent flood events. The frequency curve for a stream in a large urbanized area generally has a higher mean, a lower standard deviation, and a different skew coefficient than the frequency curve for the same stream prior to development.

Storm type and other factors.   There are many other factors that can affect the homogeneity of runoff data. The runoff may be generated by a variety of events, including frontal type storms, thunderstorms, hurricanes, and melting snow. It is desirable to separate these different types of events for frequency analysis and then combine the results. However, this may be impractical for some types of events. There may not be a large enough number of hurricane events to analyze, for example. Other developments or occurrences that can affect runoff characteristics of a basin and a frequency curve include logging, forest fires, channelization, levee construction, and diversions.

Frequency analysis techniques

There are two basic methods of frequency analysis: graphic and analytic. Each has certain advantages, and there are some applications where one is preferred over the other.

Graphic Method.   Graphic analysis is based on ordinal statistics in which the data is arranged in ascending or descending order according to magnitude. The data is assigned "plotting positions," which are, in effect, probability estimates, and these are plotted on probability graph paper to obtain the discharge-frequency curve. Annual flood peaks and several other hydrologic variables are usually plotted on log-probability paper. Annual river stages and a few other variables are plotted on arithmetic probability paper.

   Plotting positions.   Several formulas have been derived for calculating plotting positions; three of the most widely used of these are the "median," the Weibull, and the Hazen formulas.

$$P = \frac{100(m - 0.3)}{N + 0.4} \quad \text{(median)}$$

$$P = \frac{100m}{N - 1} \quad \text{(Weibull)}$$

$$P = \frac{100(m - 0.5)}{N} \quad \text{(Hazen)}$$

where P = percent chance exceedance
m = rank (order number) of event
N = number of years in sample

Median plotting positions are provided in Table 7.2.

Example of a graphic-frequency curve. An example of 24 annual peak flows for Fishkill Creek ranked according to Weibull plotting positions is shown in Fig. 7.12a. These points are plotted on a logarithmic normal grid in Fig. 7.12b. A smooth curve approximating a straight line is generally drawn through the plotted points. In this example, the pattern of the data points tends to curve upward, so a slightly curved line was drawn as a best fit line.

Advantages and disadvantages of graphic analysis. Graphic analysis generally is quick and easy to use and can serve several useful purposes. It can be used to analyze nonlinear variables such as river stage, reservoir elevation, and regulated flow that do not fit theoretical probability distributions. Very high or very low data points (outliers) can be arbitrarily weighted in a graphic procedure. Graphically derived plots can be used for comparing observed data with an analytically derived curve. And plots of observed data may provide some help in judging the validity of extending a frequency curve beyond the range of data to estimate the magnitude of rare events.

A disadvantage of graphic analysis is the lack of precision and consistency in "eyeballing" the curve. Two individuals would normally draw different curves through the same points. Also, the reliability of hand-drawn curves is difficult, if not impossible, to evaluate.

Analytic method. In the analytic method a theoretical probability distribution is selected for the population to be analyzed; the parameters of the selected distribution are estimated from observed data; several points are computed to plot the curve; and the analytically derived curve is compared with observed data plotted with the graphic method.

Selecting a theoretical probability distribution. The selection of a distribution to fit the data should be based on more than just one sample.

TABLE 7.2 Median Plotting Positions

Plotting Positions in Percent Chance Exceedance

Order m	Number of values in array N													Order m
	1	2	3	4	5	6	7	8	9	10	11	12	13	
1	50.00	29.29	20.63	15.91	12.94	10.91	9.43	8.30	7.41	6.70	6.11	5.61	5.19	1
		70.71	50.00	38.64	31.47	26.55	22.95	20.21	18.06	16.32	14.89	13.68	12.66	2
52	98.68		79.37	61.36	50.00	42.18	36.48	32.13	28.71	25.94	23.66	21.75	20.13	3
51	96.77	98.65		84.09	68.53	57.82	50.00	44.04	39.35	35.57	32.44	29.82	27.60	4
50	94.86	96.70	98.62		87.06	73.45	63.52	55.96	50.00	45.19	41.22	37.89	35.06	5
49	92.95	94.76	96.64	98.60		89.09	77.05	67.87	60.65	54.81	50.00	45.96	42.53	6
48	91.04	92.81	94.65	96.57	98.57		90.57	79.79	71.29	64.43	58.78	54.04	50.00	7
47	89.13	90.87	92.67	94.55	96.50	98.54		91.70	81.94	74.06	67.56	62.11	57.47	8
46	87.22	88.92	90.68	92.52	94.43	96.43	98.50		92.59	83.68	76.34	70.18	64.94	9
45	85.31	86.97	88.70	90.50	92.37	94.32	96.35	98.47		93.30	85.11	78.25	72.40	10
44	83.40	85.03	86.72	88.47	90.30	92.21	94.19	96.27	98.44		93.89	86.32	79.87	11
43	81.50	83.08	84.73	86.45	88.23	90.10	92.04	94.06	96.18	98.40		94.39	87.34	12
42	79.59	81.14	82.75	84.42	86.17	87.98	89.88	91.86	93.93	96.10	98.36		94.81	13
41	77.68	79.19	80.76	82.40	84.10	85.87	87.73	89.66	91.68	93.79	96.00	98.32		
40	75.77	77.24	78.78	80.37	82.03	83.76	85.57	87.46	89.43	91.49	93.64	95.91	98.28	40
39	73.86	75.30	76.79	78.35	79.97	81.65	83.41	85.25	87.17	89.18	91.29	93.49	95.81	39
38	71.95	73.35	74.81	76.32	77.90	79.54	81.26	83.05	84.92	86.88	88.93	91.08	93.33	38
37	70.04	71.41	72.82	74.30	75.83	77.43	79.10	80.85	82.67	84.57	86.57	88.66	90.85	37
36	68.13	69.46	70.84	72.27	73.77	75.32	76.95	78.64	80.41	82.27	84.21	86.24	88.38	36
35	66.23	67.51	68.85	70.25	71.70	73.21	74.79	76.44	78.16	79.96	81.85	83.83	85.90	35
34	64.32	65.57	66.87	68.22	69.63	71.10	72.64	74.24	75.91	77.66	79.49	81.41	83.43	34
33	62.41	63.62	64.88	66.20	67.57	68.99	70.48	72.03	73.66	75.35	77.13	78.99	80.95	33
32	60.50	61.68	62.90	64.17	65.50	66.88	68.32	69.83	71.40	73.05	74.77	76.58	78.47	32
31	58.59	59.73	60.92	62.15	63.43	64.77	66.17	67.63	69.15	70.74	72.41	74.16	76.00	31
30	56.68	57.78	58.93	60.12	61.37	62.66	64.01	65.42	66.90	68.44	70.05	71.75	73.52	30
29	54.77	55.84	56.95	58.10	59.30	60.55	61.86	63.22	64.64	66.13	67.69	69.33	71.05	29
28	52.86	53.89	54.96	56.07	57.23	58.44	59.70	61.02	62.39	63.83	65.33	66.91	68.57	28
27	50.95	51.95	52.98	54.05	55.17	56.33	57.55	58.81	60.14	61.52	62.98	64.50	66.09	27
26	49.05	50.00	50.99	52.02	53.10	54.22	55.39	56.61	57.89	59.22	60.62	62.08	63.62	26
25	47.14	48.05	49.01	50.00	51.03	52.11	53.23	54.41	55.63	56.91	58.26	59.66	61.14	25
24	45.23	46.11	47.02	47.98	48.97	50.00	51.08	52.20	53.38	54.61	55.90	57.25	58.67	24
23	43.32	44.16	45.04	45.95	46.90	47.89	48.92	50.00	51.13	52.30	53.54	54.83	56.19	23
22	41.41	42.22	43.05	43.93	44.83	45.78	46.77	47.80	48.87	50.00	51.18	52.42	53.71	22
21	39.50	40.27	41.07	41.90	42.77	43.67	44.61	45.59	46.62	47.70	48.82	50.00	51.24	21
20	37.59	38.32	39.08	39.88	40.70	41.56	42.45	43.39	44.37	45.39	46.46	47.58	48.76	20
19	35.68	36.38	37.10	37.85	38.63	39.45	40.30	41.19	42.11	43.09	44.10	45.17	46.29	19
18	33.77	34.43	35.12	35.83	36.57	37.34	38.14	38.98	39.86	40.78	41.74	42.75	43.81	18
17	31.87	32.49	33.13	33.80	34.50	35.23	35.99	36.78	37.61	38.48	39.38	40.34	41.33	17
16	29.96	30.54	31.15	31.78	32.43	33.12	33.83	34.58	35.36	36.17	37.02	37.92	38.86	16
15	28.05	28.59	29.16	29.75	30.37	31.01	31.68	32.37	33.10	33.87	34.67	35.50	36.38	15
14	26.14	26.65	27.18	27.73	28.30	28.90	29.52	30.17	30.85	31.56	32.31	33.09	33.91	14
13	24.23	24.70	25.19	25.70	26.23	26.79	27.36	27.97	28.60	29.26	29.95	30.67	31.43	13
12	22.32	22.76	23.21	23.68	24.17	24.68	25.21	25.76	26.34	26.95	27.59	28.25	28.95	12
11	20.41	20.81	21.22	21.65	22.10	22.57	23.05	23.56	24.09	24.65	25.23	25.84	26.48	11
10	18.50	18.86	19.24	19.63	20.03	20.46	20.90	21.36	21.84	22.34	22.87	23.42	24.00	10
9	16.59	16.92	17.25	17.60	17.97	18.35	18.74	19.15	19.59	20.04	20.51	21.01	21.53	9
8	14.69	14.97	15.27	15.58	15.90	16.24	16.59	16.95	17.33	17.73	18.15	18.59	19.05	8
7	12.78	13.03	13.28	13.55	13.83	14.13	14.43	14.75	15.08	15.43	15.79	16.17	16.57	7
6	10.87	11.08	11.30	11.53	11.77	12.02	12.27	12.54	12.83	13.12	13.43	13.76	14.10	6
5	8.96	9.13	9.32	9.50	9.70	9.91	10.12	10.34	10.57	10.82	11.07	11.34	11.62	5
4	7.05	7.19	7.33	7.48	7.63	7.79	7.96	8.14	8.32	8.51	8.71	8.92	9.15	4
3	5.14	5.24	5.35	5.45	5.57	5.68	5.81	5.94	6.07	6.21	6.36	6.51	6.67	3
2	3.23	3.30	3.36	3.43	3.50	3.57	3.65	3.73	3.82	3.90	4.00	4.09	4.19	2
1	1.32	1.35	1.38	1.40	1.43	1.46	1.50	1.53	1.56	1.60	1.64	1.68	1.72	1
m	52	51	50	49	48	47	46	45	44	43	42	41	40	m

Order m	14	15	16	17	18	19	20	21	22	23	24	25	26	Order m
1	4.83	4.52	4.24	4.00	3.78	3.58	3.41	3.25	3.10	2.97	2.85	2.73	2.63	1
2	11.78	11.01	10.34	9.75	9.22	8.74	8.31	7.92	7.57	7.24	6.95	6.67	6.42	2
3	18.73	17.51	16.44	15.50	14.65	13.90	13.22	12.60	12.03	11.52	11.05	10.61	10.21	3
4	25.68	24.01	22.54	21.25	20.09	19.05	18.12	17.27	16.50	15.80	15.15	14.55	14.00	4
5	32.63	30.51	28.65	27.00	25.53	24.21	23.02	21.95	20.97	20.07	19.25	18.49	17.79	5
6	39.58	37.00	34.75	32.75	30.97	29.37	27.93	26.62	25.43	24.35	23.35	22.43	21.58	6
7	46.53	43.50	40.85	38.50	36.41	34.53	32.83	31.30	29.90	28.62	27.45	26.37	25.37	7
8	53.47	50.00	46.95	44.25	41.84	39.68	37.74	35.97	34.37	32.90	31.55	30.31	29.16	8
9	60.42	56.50	53.05	50.00	47.28	44.84	42.64	40.65	38.83	37.17	35.65	34.24	32.95	9
10	67.37	63.00	59.15	55.75	52.72	50.00	47.55	45.32	43.30	41.45	39.75	38.18	36.74	10
11	74.32	69.49	65.25	61.50	58.16	55.16	52.45	50.00	47.77	45.72	43.85	42.12	40.53	11
12	81.27	75.99	71.35	67.25	63.59	60.32	57.36	54.68	52.23	50.00	47.95	46.06	44.32	12
13	88.22	82.49	77.46	73.00	69.03	65.47	62.26	59.35	56.70	54.28	52.05	50.00	48.11	13
14	95.17	88.99	83.56	78.75	74.47	70.63	67.17	64.03	61.17	58.55	56.15	53.94	51.89	14
		95.48	89.66	84.50	79.91	75.79	72.07	68.70	65.63	62.83	60.25	57.88	55.68	15
39	98.24		95.76	90.25	85.35	80.95	76.98	73.38	70.10	67.10	64.35	61.82	59.47	16
38	95.70	98.19		96.00	90.78	86.10	81.88	78.05	74.57	71.38	68.65	65.76	63.26	17
37	93.16	95.59	98.14		96.22	91.26	86.78	82.73	79.03	75.65	72.55	69.69	67.05	18
36	90.62	92.98	95.47	98.09		96.42	91.69	87.40	83.50	79.93	76.65	73.63	70.84	19
35	88.08	90.38	92.79	95.34	98.04		96.59	92.08	87.97	84.20	80.75	77.57	74.63	20
34	85.54	87.77	90.12	92.60	95.21	97.98		96.75	92.43	88.48	84.85	81.51	78.42	21
33	83.01	85.17	87.45	89.85	92.39	95.07	97.92		96.90	92.76	88.95	85.45	82.21	22
32	80.47	82.56	84.77	87.10	89.56	92.17	94.93	97.86		97.03	93.05	89.39	86.00	23
31	77.93	79.96	82.10	84.35	86.74	89.26	91.93	94.77	97.79		97.15	93.33	89.79	24
30	75.39	77.35	79.42	81.60	83.91	86.35	88.94	91.68	94.60	97.72		97.27	93.58	25
29	72.85	74.75	76.75	78.86	81.08	83.44	85.94	88.59	91.42	94.43	97.64		97.37	26
28	70.31	72.14	74.07	76.11	78.26	80.53	82.95	85.51	88.23	91.13	94.24	97.55		27
27	67.77	69.54	71.40	73.36	75.43	77.63	79.95	82.42	85.05	87.84	90.83	94.03	97.47	27
26	65.23	66.93	68.72	70.61	72.61	74.72	76.96	79.33	81.86	84.55	87.43	90.51	93.81	26
25	62.69	64.33	66.05	67.86	69.78	71.81	73.96	76.24	78.67	81.26	84.03	86.99	90.16	25
24	60.16	61.72	63.37	65.11	66.95	68.90	70.97	73.16	75.49	77.97	80.62	83.46	86.51	24
23	57.62	59.12	60.70	62.37	64.13	65.99	67.97	70.07	72.30	74.68	77.22	79.94	82.86	23
22	55.08	56.51	58.02	59.62	61.30	63.09	64.98	66.98	69.12	71.39	73.82	76.42	79.21	22
21	52.54	53.91	55.35	56.87	58.48	60.18	61.98	63.89	65.93	68.10	70.42	72.90	75.56	21
20	50.00	51.30	52.67	54.12	55.65	57.27	58.99	60.81	62.74	64.81	67.01	69.37	71.91	20
19	47.46	48.70	50.00	51.37	52.83	54.36	55.99	57.72	59.56	61.52	63.61	65.85	68.26	19
18	44.92	46.09	47.33	48.63	50.00	51.45	53.00	54.63	56.37	58.23	60.21	62.33	64.60	18
17	42.38	43.49	44.65	45.88	47.17	48.55	50.00	51.54	53.19	54.94	56.81	58.81	60.95	17
16	39.84	40.88	41.98	43.13	44.35	45.64	47.00	48.46	50.00	51.65	53.40	55.28	57.30	16
15	37.31	38.28	39.30	40.38	41.52	42.73	44.01	45.37	46.81	48.35	50.00	51.76	53.65	15
14	34.77	35.67	36.63	37.63	38.70	39.82	41.01	42.28	43.63	45.06	46.60	48.24	50.00	14
13	32.23	33.07	33.95	34.89	35.87	36.91	38.02	39.19	40.44	41.77	43.19	44.72	46.35	13
12	29.69	30.46	31.28	32.14	33.05	34.01	35.02	36.11	37.26	38.48	39.79	41.19	42.70	12
11	27.15	27.86	28.60	29.39	30.22	31.10	32.03	33.02	34.07	35.19	36.39	37.67	39.05	11
10	24.61	25.25	25.93	26.64	27.39	28.19	29.03	29.93	30.88	31.90	32.99	34.15	35.40	10
9	22.07	22.65	23.25	23.89	24.57	25.28	26.04	26.84	27.70	28.61	29.58	30.63	31.74	9
8	19.53	20.04	20.58	21.14	21.74	22.37	23.04	23.76	24.51	25.32	26.18	27.10	28.09	8
7	16.99	17.44	17.90	18.40	18.92	19.47	20.05	20.67	21.33	22.03	22.78	23.58	24.44	7
6	14.46	14.83	15.23	15.65	16.09	16.56	17.05	17.58	18.14	18.74	19.38	20.06	20.79	6
5	11.92	12.23	12.55	12.90	13.26	13.65	14.06	14.49	14.95	15.45	15.97	16.54	17.14	5
4	9.38	9.62	9.88	10.15	10.44	10.74	11.06	11.41	11.77	12.16	12.57	13.01	13.49	4
3	6.84	7.02	7.21	7.40	7.61	7.83	8.07	8.32	8.58	8.87	9.17	9.49	9.84	3
2	4.30	4.41	4.53	4.66	4.79	4.93	5.07	5.23	5.40	5.57	5.76	5.97	6.19	2
1	1.76	1.81	1.86	1.91	1.96	2.02	2.08	2.14	2.21	2.28	2.36	2.45	2.53	1
m	39	38	37	36	35	34	33	32	31	30	29	28	27	m

```
-PLOTTING POSITIONS- 01-3735   FISHKILL CREEK AT BEACON, NEW YORK
*******************************************************************
*.....EVENTS ANALYZED......*.........*.........ORDERED EVENTS.........*
*                          *         *                       WEIBULL  *
*                          *         *            WATER                *
* MON DAY YEAR FLOW,CFS    * RANK    * YEAR  FLOW,CFS  PLOT POS *
*                          *         *                         *
*  3   5  1945  2290.      *   1     * 1955   8800.    .0400    *
* 12  27  1945  1470.      *   2     * 1956   8280.    .0800    *
*  3  15  1947  2220.      *   3     * 1961   4340.    .1200    *
*  3  18  1948  2970.      *   4     * 1968   3630.    .1600    *
*  1   1  1949  3020.      *   5     * 1953   3220.    .2000    *
*  3   9  1950  1210.      *   6     * 1952   3170.    .2400    *
*  4   1  1951  2490.      *   7     * 1962   3060.    .2800    *
*  3  12  1952  3170.      *   8     * 1949   3020.    .3200    *
*  1  25  1953  3220.      *   9     * 1948   2970.    .3600    *
*  9  13  1954  1760.      *  10     * 1958   2500.    .4000    *
*  8  20  1955  8800.      *  11     * 1951   2490.    .4400    *
* 10  16  1955  8280.      *  12     * 1945   2290.    .4800    *
*  4  10  1957  1310.      *  13     * 1947   2220.    .5200    *
* 12  21  1957  2500.      *  14     * 1960   2140.    .5600    *
*  2  11  1959  1960.      *  15     * 1959   1960.    .6000    *
*  4   6  1960  2140.      *  16     * 1963   1780.    .6400    *
*  2  26  1961  4340.      *  17     * 1954   1760.    .6800    *
*  3  13  1962  3060.      *  18     * 1967   1580.    .7200    *
*  3  28  1963  1780.      *  19     * 1946   1470.    .7600    *
*  1  26  1964  1380.      *  20     * 1964   1380.    .8000    *
*  2   9  1965   980.      *  21     * 1957   1310.    .8400    *
*  2  15  1966  1040.      *  22     * 1950   1210.    .8800    *
*  3  30  1967  1580.      *  23     * 1966   1040.    .9200    *
*  3  19  1968  3630.      *  24     * 1965    980.    .9600    *
*******************************************************************
```

(a)

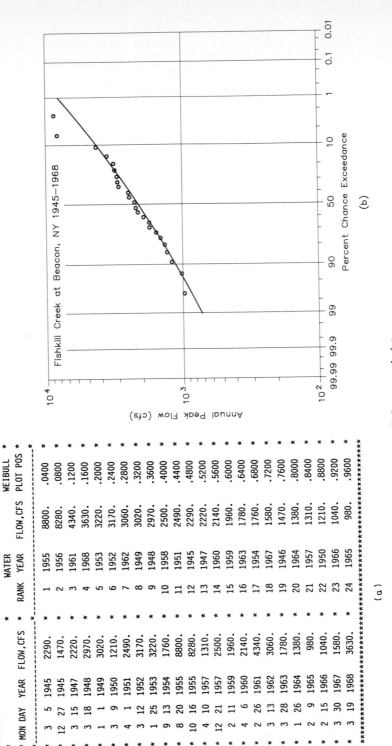

(b)

Figure 7.12 Graphic frequency analysis. (*a*) Plotting positions; (*b*) log-normal plot.

Knowledge and experience from previous studies regarding the variable should be utilized if available. Although a number of different theoretical distributions may be applicable, the U.S. government has set a standard for federal agencies: use of the Log Pearson Type III method for flood-flow frequency analysis.[4]. Other distributions can be used for other variables.

Log Pearson Type III distribution

General. The parameters of the Log Pearson Type III distribution are the mean, the standard deviation, and the skew coefficient. The cumulative frequency curve is computed with Eq (7.3).

$$\log Q = \overline{X} + KS \tag{7.3}$$

where log Q = logarithm of annual flood discharge
$\overline{X}$ = mean of the logarithms of the flood peaks
K = Pearson Type III deviate, which is a function of the skew coefficient and a selected exceedance probability (Table 7.3)
S = standard deviation of the logarithms

The Log Pearson Type III distribution is particularly useful for hydrologic analysis because of the skewness parameter, which enables it to fit samples that do not fit the normal distribution. When the skew is zero, this distribution becomes a two-parameter distribution identical with the log-normal. The Log Pearson Type III frequency curve is characterized by its three parameters. The mean represents the average ordinate, the standard deviation represents the slope of the curve, and the skew coefficient represents the degree of curvature.

Weighted skew coefficients. The skew coefficient computed from a small sample is not reliable as an estimate of the population parameter. This is evident from an examination of Fig. 7.13, which shows a high variation in coefficients sequentially computed after adding the annual peak each year. The extremely high peak flow recorded in 1955, for example, greatly affects the computed skew coefficient. The reliability of a sample skew coefficient can be increased by weighting it with a generalized coefficient estimated by pooling information from nearby sites.

Deriving generalized skew coefficients requires the use of at least 40 stations, or all stations within a 100-mi radius, each station having 25 or more years of record.[4] Generalized skew coefficients for a region can be estimated with one of three different methods: (1) using lines of

TABLE 7.3 Deviates (K values) for Pearson Type III Distribution

Skew coeffi-cient G	Percent chance exceedance												
	99.0	95.0	90.0	80.0	50.0	20.0	10.0	5.0	2.0	1.0	0.5	0.2	0.1
	Positive Skew												
0.0	-2.32635	-1.64485	-1.28155	-0.84162	0.00000	0.84162	1.28155	1.64485	2.05375	2.32635	2.57583	2.87816	3.09023
0.1	-2.25258	-1.61594	-1.27037	-0.84611	-0.01662	0.83639	1.29178	1.67279	2.10697	2.39961	2.66965	2.99978	3.23322
0.2	-2.17840	-1.58607	-1.25824	-0.84986	-0.03325	0.83044	1.30105	1.69971	2.15935	2.47226	2.76321	3.12169	3.37703
0.3	-2.10394	-1.55527	-1.24516	-0.85285	-0.04993	0.82377	1.30936	1.72562	2.21081	2.54421	2.85636	3.24371	3.52139
0.4	-2.02933	-1.52357	-1.23114	-0.85508	-0.06651	0.81638	1.31671	1.75048	2.26133	2.61539	2.94900	3.36566	3.66608
0.5	-1.95472	-1.49101	-1.21618	-0.85653	-0.08302	0.80829	1.32309	1.77428	2.31084	2.68572	3.04102	3.48737	3.81090
0.6	-1.88029	-1.45762	-1.20028	-0.85718	-0.09945	0.79950	1.32850	1.79701	2.35931	2.75514	3.13232	3.60872	3.95567
0.7	-1.80621	-1.42345	-1.18347	-0.85703	-0.11578	0.79002	1.33294	1.81864	2.40670	2.82359	3.22281	3.72957	4.10022
0.8	-1.73271	-1.38855	-1.16574	-0.85607	-0.13199	0.77986	1.33640	1.83916	2.45298	2.89101	3.31243	3.84981	4.24439
0.9	-1.66001	-1.35299	-1.14712	-0.85426	-0.14807	0.76902	1.33889	1.85856	2.49811	2.95735	3.40109	3.96932	4.38807
1.0	-1.58838	-1.31684	-1.12762	-0.85161	-0.16397	0.75752	1.34039	1.87683	2.54206	3.02256	3.48874	4.08802	4.53112
1.1	-1.51808	-1.28019	-1.10726	-0.84809	-0.17968	0.74537	1.34092	1.89395	2.58480	3.08660	3.57530	4.20582	4.67344
1.2	-1.44942	-1.24313	-1.08608	-0.84369	-0.19517	0.73257	1.34047	1.90992	2.62631	3.14944	3.66073	4.32263	4.81492
1.3	-1.38267	-1.20578	-1.06413	-0.83841	-0.21040	0.71915	1.33904	1.92472	2.66657	3.21103	3.74497	4.43839	4.95549
1.4	-1.31815	-1.16827	-1.04144	-0.83223	-0.22535	0.70512	1.33665	1.93836	2.70556	3.27134	3.82798	4.55304	5.09505
1.5	-1.25611	-1.13075	-1.01810	-0.82516	-0.23996	0.69050	1.33330	1.95083	2.74325	3.33035	3.90973	4.66651	5.23353
1.6	-1.19680	-1.09338	-0.99418	-0.81720	-0.25422	0.67532	1.32900	1.96213	2.77964	3.38804	3.99016	4.77875	5.37087
1.7	-1.14042	-1.05631	-0.96977	-0.80837	-0.26808	0.65959	1.32376	1.97227	2.81472	3.44438	4.06926	4.88971	5.50701
1.8	-1.08711	-1.01973	-0.94496	-0.79868	-0.28150	0.64335	1.31760	1.98124	2.84848	3.49935	4.14700	4.99937	5.64190
1.9	-1.03695	-0.98381	-0.91988	-0.78816	-0.29443	0.62662	1.31054	1.98906	2.88091	3.55295	4.22336	5.10768	5.77549
2.0	-0.98995	-0.94871	-0.89464	-0.77686	-0.30685	0.60944	1.30259	1.99573	2.91202	3.60517	4.29832	5.21461	5.90776
2.1	-0.94607	-0.91458	-0.86938	-0.76482	-0.31872	0.59183	1.29377	2.00178	2.94181	3.65600	4.37186	5.32014	6.03865
2.2	-0.90521	-0.88156	-0.84422	-0.75211	-0.32999	0.57383	1.28412	2.00570	2.97028	3.70543	4.44398	5.42426	6.16816
2.3	-0.86723	-0.84976	-0.81929	-0.73880	-0.34063	0.55549	1.27365	2.00903	2.99744	3.75347	4.51467	5.52694	6.29626
2.4	-0.83196	-0.81927	-0.79472	-0.72495	-0.35062	0.53683	1.26240	2.01128	3.02330	3.80013	4.58393	5.62818	6.42292
2.5	-0.79921	-0.79015	-0.77062	-0.71067	-0.35992	0.51789	1.25039	2.01247	3.04787	3.84540	4.65176	5.72796	6.54814
2.6	-0.76878	-0.76242	-0.74709	-0.69602	-0.36852	0.49872	1.23766	2.01263	3.07116	3.88930	4.71815	5.82629	6.67191
2.7	-0.74049	-0.73610	-0.72422	-0.68111	-0.37640	0.47934	1.22422	2.01177	3.09320	3.93183	4.78313	5.92316	6.79421
2.8	-0.71415	-0.71116	-0.70209	-0.66603	-0.38353	0.45980	1.21013	2.00992	3.11399	3.97301	4.84669	6.01858	6.91505
2.9	-0.68959	-0.68759	-0.68075	-0.65086	-0.38991	0.44015	1.19539	2.00710	3.13356	4.01286	4.90884	6.11254	7.03443
3.0	-0.66663	-0.66532	-0.66023	-0.63569	-0.39554	0.42040	1.18006	2.00335	3.15193	4.05138	4.96959	6.20506	7.15235

Negative Skew

0.0	3.09023	2.87816	2.57583	2.32635	2.05375	1.64485	1.28155	0.84162	0.00000	−0.84162	−1.28155	−1.64485	−2.32635
−0.1	2.94834	2.75706	2.48187	2.25258	1.99973	1.61594	1.27037	0.84611	0.01662	−0.83639	−1.29178	−1.67279	−2.39961
−0.2	2.80786	2.63672	2.38795	2.17840	1.94499	1.58607	1.25824	0.84986	0.03325	−0.83044	−1.30105	−1.69971	−2.47226
−0.3	2.66915	2.51741	2.29423	2.10394	1.88959	1.55527	1.24516	0.85285	0.04993	−0.82377	−1.30936	−1.72562	−2.54421
−0.4	2.53261	2.39942	2.20092	2.02933	1.83361	1.52357	1.23114	0.85508	0.06651	−0.81638	−1.31671	−1.75048	−2.61539
−0.5	2.39867	2.28311	2.10825	1.95472	1.77716	1.49101	1.21618	0.85653	0.08302	−0.80829	−1.32309	−1.77428	−2.68572
−0.6	2.26780	2.16884	2.01644	1.88029	1.72033	1.45762	1.20028	0.85718	0.09945	−0.79950	−1.32850	−1.79701	−2.75514
−0.7	2.14053	2.05701	1.92580	1.80621	1.66325	1.42345	1.18347	0.85703	0.11578	−0.79002	−1.33294	−1.81864	−2.82359
−0.8	2.01739	1.94806	1.83660	1.73271	1.60604	1.38855	1.16574	0.85607	0.13199	−0.77986	−1.33640	−1.83916	−2.89101
−0.9	1.89894	1.84244	1.74919	1.66001	1.54886	1.35299	1.14712	0.85426	0.14807	−0.76902	−1.33889	−1.85856	−2.95735
−1.0	1.78572	1.74062	1.66390	1.58838	1.49188	1.31684	1.12762	0.85161	0.16397	−0.75752	−1.34039	−1.87683	−3.02256
−1.1	1.67825	1.64305	1.58110	1.51808	1.43529	1.28019	1.10726	0.84809	0.17968	−0.74537	−1.34092	−1.89395	−3.08660
−1.2	1.57695	1.55016	1.50114	1.44942	1.37929	1.24313	1.08608	0.84369	0.19517	−0.73257	−1.34047	−1.90992	−3.14944
−1.3	1.48216	1.46232	1.42439	1.38267	1.32412	1.20578	1.06413	0.83841	0.21040	−0.71915	−1.33904	−1.92472	−3.21103
−1.4	1.39408	1.37981	1.35114	1.31815	1.26999	1.16827	1.04144	0.83223	0.22535	−0.70512	−1.33665	−1.93836	−3.27134
−1.5	1.31275	1.30279	1.28167	1.25611	1.21716	1.13075	1.01810	0.82516	0.23996	−0.69050	−1.33330	−1.95083	−3.33035
−1.6	1.23805	1.23132	1.21618	1.19680	1.16584	1.09338	0.99418	0.81720	0.25422	−0.67532	−1.32900	−1.96213	−3.38804
−1.7	1.16974	1.16534	1.15477	1.14042	1.11628	1.05631	0.96977	0.80837	0.26808	−0.65959	−1.32376	−1.97227	−3.44438
−1.8	1.10743	1.10465	1.09749	1.08711	1.06864	1.01973	0.94496	0.79868	0.28150	−0.64335	−1.31760	−1.98124	−3.49935
−1.9	1.05068	1.04898	1.04427	1.03695	1.02311	0.98381	0.91988	0.78816	0.29443	−0.62662	−1.31054	−1.98906	−3.55295
−2.0	0.99900	0.99800	0.99499	0.98995	0.97980	0.94871	0.89464	0.77686	0.30685	−0.60944	−1.30259	−1.99573	−3.60517
−2.1	0.95188	0.95131	0.94945	0.94607	0.93878	0.91458	0.86938	0.76482	0.31872	−0.59183	−1.29377	−2.00128	−3.65600
−2.2	0.90885	0.90854	0.90742	0.90521	0.90009	0.88156	0.84422	0.75211	0.32999	−0.57383	−1.28412	−2.00570	−3.70543
−2.3	0.86945	0.86929	0.86863	0.86723	0.86371	0.84976	0.81929	0.73880	0.34063	−0.55549	−1.27365	−2.00903	−3.75347
−2.4	0.83328	0.83320	0.83283	0.83196	0.82959	0.81927	0.79472	0.72495	0.35062	−0.53683	−1.26240	−2.01128	−3.80013
−2.5	0.79998	0.79994	0.79973	0.79921	0.79765	0.79015	0.77062	0.71067	0.35992	−0.51789	−1.25039	−2.01247	−3.84540
−2.6	0.76922	0.76920	0.76909	0.76878	0.76779	0.76242	0.74709	0.69602	0.36852	−0.49872	−1.23766	−2.01263	−3.88930
−2.7	0.74074	0.74073	0.74067	0.74049	0.73987	0.73610	0.72422	0.68111	0.37640	−0.47934	−1.22422	−2.01177	−3.93183
−2.8	0.71428	0.71428	0.71425	0.71415	0.71377	0.71116	0.70209	0.66603	0.38353	−0.45980	−1.21013	−2.00992	−3.97301
−2.9	0.68965	0.68965	0.68964	0.68959	0.68935	0.68759	0.68075	0.65086	0.38991	−0.44015	−1.19539	−2.00710	−4.01286
−3.0	0.66667	0.66667	0.66666	0.66663	0.66649	0.66532	0.66023	0.63569	0.39554	−0.42040	−1.13006	−2.00335	−4.05138

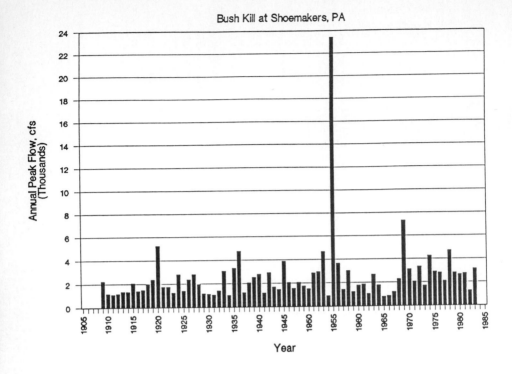

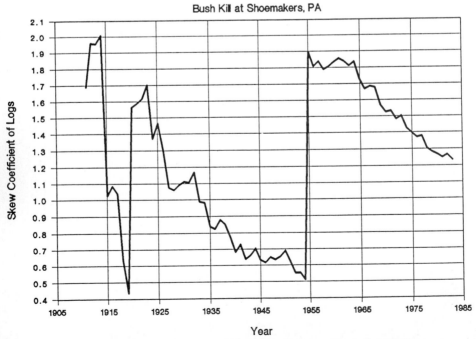

Figure 7.13 Variability of the skew coefficient caused by 1955 high peak flow.

equal skewness drawn on a map, (2) using a skew prediction equation, and (3) using the mean of station skew values. The method that provides the most accurate estimate in a given situation is the one that should be used. Using a generalized skew coefficient map of the United States (Fig. 7.14) is an alternative to the regional methods just described, but the accuracy is generally not as good.

It is assumed that a generalized skew coefficient is not biased by and is independent of a station skew coefficient and that the two can be combined to obtain a better estimate of skew for a basin. A combined, or weighted, station coefficient G_w is obtained by weighting the station and generalized skew coefficients in inverse proportion to their individual mean square errors [Eq. (7.4)].

$$G_w = \frac{\text{MSE}_{\overline{G}}(G) + \text{MSE}_G(\overline{G})}{\text{MSE}_{\overline{G}} + \text{MSE}_G} \tag{7.4}$$

where G_w = weighted skew coefficient
$\quad\quad G$ = station skew
$\quad\quad \overline{G}$ = generalized skew
$\quad \text{MSE}_{\overline{G}}$ = mean square error of generalized skew
$\quad \text{MSE}_G$ = mean square error of station skew

When a generalized skew coefficient is taken from Fig. 7.14, the value of 0.302 should be used for the mean square error of generalized skew MSE_G. The mean square error of station skew MSE_G can be approximated with Eq. (7.5).

$$\text{MSE}_G \simeq 10^{A - B\,[\log_{10}(N/10)]} \tag{7.5}$$

where $A = \; -0.33 + 0.08|G| \quad$ if $|G| \le 0.90$
$\quad\quad\quad\quad -0.52 + 0.30|G| \quad$ if $|G| > 0.90$
$\quad\; B - 0.94 - 0.26|G| \quad\;$ if $|G| \le 1.50$
$\quad\quad\quad\; 0.55 \quad\quad\quad\quad\quad$ if $|G| > 1.50$

If the frequency data has been adjusted with historical information, the historically adjusted skew and the historical period of record should be used for G and N, respectively, in Eq. (7.5).

Outliers. The procedure for testing for outliers varies, depending on the skew coefficient. If the station skew is greater than $+0.04$, high outliers are tested first and appropriate adjustments made. If it is less than -0.04, low outliers are tested first and adjustments made. If the station skew is between these numbers, tests for both are applied,

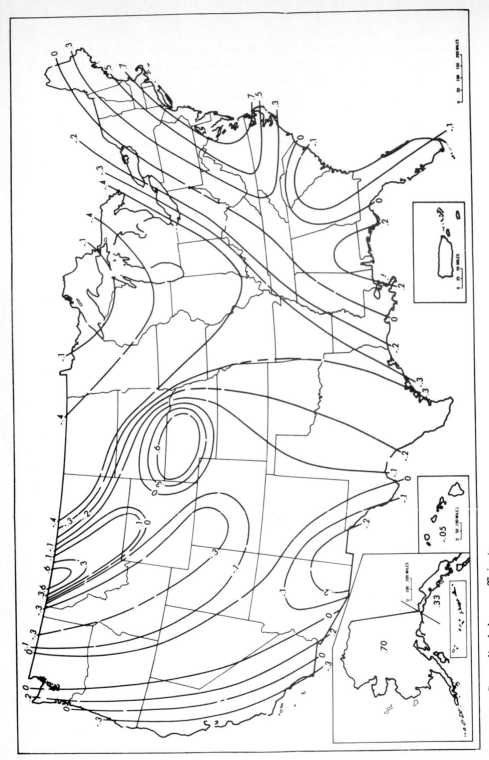

Figure 7.14 Generalized skew coefficients.

198

without regard to the order, before any adjustments are made. The equation used to screen for outliers is as follows:

$$X_o = \overline{X} + K_N S \qquad (7.6)$$

where X_o = outlier threshold in log units
$\overline{X}$ = mean logarithm (may have been adjusted for high or low outliers and historical information)
K_N = K value from Table 7.4 [use with + sign in Eq. (7.6) for high outlier; use with − sign for low outlier]
N = sample size (may be historical period if sample has been adjusted for historical information)
S = standard deviation (also may have been adjusted, as described for the mean)

High outliers may be treated as historical events if historical information is available. Low outliers are eliminated from the sample, and the frequency curve is adjusted for conditional probability.[4,5]

Adjustment for expected probability. A frequency curve computed with Eq. (7.3) is biased in relation to future expectation because of uncertainty in the sample estimates of the population mean and standard deviation. The effect of this bias can be eliminated by an "expected probability" adjustment based on the normal distribution of sampling error.[6,7] An adjusted frequency curve can be obtained with "percent chance exceedance" values taken from the Table 7.5. For selected values greater than 50 percent, the appropriate value equals 100 minus the value in the table.

Confidence limits. How well a frequency curve approximates the true population curve depends on sample size and sample accuracy, assuming that the underlying theoretical distribution is known. Confidence limits are constructed to gauge sampling accuracy. Upper and lower confidence limits define an interval that contains the population value for a given frequency with some specified degree of probability. For example, there is a 90 percent chance that the interval between a 0.05 confidence limit and a 0.95 confidence limit contains the population value.

The general form of the confidence limits for the Log Pearson Type III distribution is expressed as

$$U_{P,c}(X) = \overline{X} + S(K^U_{P,c}) \qquad (7.7)$$

$$L_{P,c}(X) = \overline{X} + S(K^L_{P,c}) \qquad (7.8)$$

TABLE 7.4   Outlier Test *K* Values

(10 Percent Significance Level)

Sample size	Outlier K value	Sample size	Outlier K value	Sample size	Outlier K value	Sample size	Outlier K value
10	2.036	45	2.727	80	2.940	115	3.064
11	2.088	46	2.736	81	2.945	116	3.067
12	2.134	47	2.744	82	2.949	117	3.070
13	2.175	48	2.753	83	2.953	118	3.073
14	2.213	49	2.760	84	2.957	119	3.075
15	2.247	50	2.768	85	2.961	120	3.078
16	2.279	51	2.775	86	2.966	121	3.081
17	2.309	52	2.783	87	2.970	122	3.083
18	2.335	53	2.790	88	2.973	123	3.086
19	2.361	54	2.798	89	2.977	124	3.089
20	2.385	55	2.804	90	2.981	125	3.092
21	2.408	56	2.811	91	2.984	126	3.095
22	2.429	57	2.818	92	2.989	127	3.097
23	2.448	58	2.824	93	2.993	128	3.100
24	2.467	59	2.831	94	2.996	129	3.102
25	2.486	60	2.837	95	3.000	130	3.104
26	2.502	61	2.842	96	3.003	131	3.107
27	2.519	62	2.849	97	3.006	132	3.109
28	2.534	63	2.854	98	3.011	133	3.112
29	2.549	64	2.860	99	3.014	134	3.114
30	2.563	65	2.866	100	3.017	135	3.116
31	2.577	66	2.871	101	3.021	136	3.119
32	2.591	67	2.877	102	3.024	137	3.122
33	2.604	68	2.883	103	3.027	138	3.124
34	2.616	69	2.888	104	3.030	139	3.126
35	2.628	70	2.893	105	3.033	140	3.129
36	2.639	71	2.897	106	3.037	141	3.131
37	2.650	72	2.903	107	3.040	142	3.133
38	2.661	73	2.908	108	3.043	143	3.135
39	2.671	74	2.912	109	3.046	144	3.138
40	2.682	75	2.917	110	3.049	145	3.140
41	2.692	76	2.922	111	3.052	146	3.142
42	2.700	77	2.927	112	3.055	147	3.144
43	2.710	78	2.931	113	3.058	148	3.146
44	2.719	79	2.935	114	3.061	149	3.148

NOTE: Table contains one-sided 10 percent significance level deviates for the normal distribution.

where $\overline{X}$ = logarithmic mean of frequency curve
S = standard deviation of frequency curve
$K_{P,c}^{U}$ = upper confidence coefficient
$K_{P,c}^{L}$ = lower confidence coefficient

and
$$K_{P,c}^{U} = \frac{K_{G_w,P} + \sqrt{K_{G_w,P}^2 - ab}}{a}$$
(7.9)

TABLE 7.5 Percentages for the Expected Probability Adjustment

N						Percent chance exceedance								
	40.0	30.0	20.0	10.0	5.0	4.0	2.0	1.0	0.5	0.2	0.1	0.05	0.02	0.01
2	43.5	37.1	30.8	24.3	20.37	19.43	17.12	15.43	14.13	12.81	12.011	11.342	10.602	10.126
3	42.3	34.7	27.1	19.1	14.52	13.44	10.86	9.08	7.77	6.51	5.793	5.212	4.598	4.219
4	41.8	33.6	25.3	16.7	11.88	10.77	8.18	6.45	5.23	4.11	3.496	3.018	2.531	2.242
5	41.4	32.9	24.3	15.3	10.38	9.26	6.70	5.05	3.92	2.92	2.389	1.990	1.596	1.370
6	41.2	32.4	23.6	14.4	9.41	8.30	5.78	4.19	3.14	2.23	1.768	1.427	1.101	.919
7	41.0	32.1	23.1	13.8	8.74	7.63	5.16	3.62	2.63	1.80	1.384	1.086	.809	.658
8	40.9	31.8	22.7	13.3	8.24	7.14	4.70	3.22	2.28	1.50	1.127	.863	.623	.495
9	40.8	31.6	22.4	12.9	7.86	6.77	4.36	2.92	2.02	1.29	.947	.710	.498	.388
10	40.7	31.5	22.1	12.6	7.56	6.47	4.09	2.69	1.82	1.13	.816	.599	.409	.313
11	40.7	31.3	22.0	12.4	7.32	6.23	3.83	2.50	1.67	1.01	.716	.516	.345	.259
12	40.6	31.2	21.8	12.2	7.12	6.04	3.71	2.36	1.54	.92	.638	.453	.296	.219
13	40.6	31.1	21.7	12.0	6.95	5.87	3.56	2.23	1.44	.84	.577	.403	.258	.188
14	40.5	31.0	21.5	11.9	6.80	5.73	3.44	2.13	1.36	.78	.526	.363	.228	.164
15	40.5	31.0	21.4	11.8	6.68	5.61	3.33	2.04	1.29	.73	.485	.330	.204	.145
16	40.5	30.9	21.3	11.6	6.57	5.50	3.24	1.97	1.23	.68	.450	.303	.184	.129
17	40.4	30.9	21.3	11.5	6.47	5.41	3.16	1.90	1.18	.65	.421	.280	.168	.116
18	40.4	30.8	21.2	11.5	6.39	5.33	3.09	1.85	1.13	.61	.396	.261	.154	.106
19	40.4	30.8	21.1	11.4	6.31	5.26	3.03	1.80	1.09	.59	.374	.244	.143	.097
20	40.4	30.7	21.1	11.3	6.25	5.19	2.98	1.75	1.06	.56	.355	.230	.133	.089
21	40.4	30.7	21.0	11.2	6.19	5.13	2.93	1.71	1.02	.54	.339	.217	.124	.083
22	40.3	30.7	21.0	11.2	6.13	5.08	2.88	1.67	1.00	.52	.324	.206	.117	.077
23	40.3	30.6	20.9	11.1	6.08	5.03	2.84	1.64	.97	.50	.311	.197	.110	.072
24	40.3	30.6	20.9	11.1	6.03	4.99	2.80	1.61	.95	.49	.299	.188	.104	.068
25	40.3	30.6	20.9	11.0	5.99	4.95	2.77	1.59	.93	.47	.289	.180	.099	.064
26	40.3	30.6	20.8	11.0	5.95	4.91	2.74	1.56	.91	.46	.279	.173	.094	.061
27	40.3	30.5	20.8	11.0	5.92	4.87	2.71	1.54	.89	.45	.271	.167	.090	.058
28	40.3	30.5	20.8	10.9	5.88	4.84	2.58	1.52	.88	.44	.263	.161	.086	.055
29	40.3	30.5	20.7	10.9	5.85	4.81	2.66	1.50	.86	.43	.255	.156	.083	.052
30	40.2	30.5	20.7	10.9	5.82	4.78	2.63	1.48	.85	.42	.249	.151	.080	.050

TABLE 7.5 Percentages for the Expected Probability Adjustment (Continued)

						Percent chance exceedance							
N	40.0	30.0	20.0	10.0	5.0	4.0	2.0	1.0	0.5	0.2	0.1	0.05	0.02	0.01
40	40.2	30.4	20.5	10.7	5.61	4.58	2.47	1.35	.75	.35	.204	.119	.060	.036
50	40.1	30.3	20.4	10.5	5.49	4.47	2.37	1.28	.70	.32	.179	.102	.049	.029
60	40.1	30.2	20.4	10.4	5.41	4.39	2.31	1.23	.66	.30	.164	.092	.043	.025
70	40.1	30.2	20.3	10.4	5.35	4.33	2.26	1.19	.64	.28	.154	.085	.039	.022
80	40.1	30.2	20.3	10.3	5.30	4.29	2.23	1.17	.62	.27	.146	.080	.036	.020
90	40.1	30.2	20.2	10.3	5.27	4.26	2.20	1.15	.61	.26	.141	.076	.034	.019
100	40.1	30.1	20.2	10.3	5.24	4.23	2.18	1.13	.59	.26	.136	.073	.032	.018
110	40.1	30.1	20.2	10.2	5.22	4.21	2.17	1.12	.59	.25	.132	.071	.031	.017
120	40.1	30.1	20.2	10.2	5.20	4.19	2.15	1.11	.58	.25	.130	.069	.030	.016
INF	40.0	30.0	20.0	10.0	5.00	4.00	2.00	1.00	.50	.20	.100	.050	.020	.010

NOTE: Values have been generated by use of computer routines for the inverse normal and inverse t distributions. This table is based on samples drawn from a normal distribution. The above values may be used as approximate adjustments to Pearson Type III distributions having small skew coefficients.

$$K_{P,c}^{L} = \frac{K_{G_w,P} - \sqrt{K_{G_w,P}^2 - ab}}{a} \tag{7.10}$$

in which
$$a = 1 - \frac{z_c^2}{2(N-1)}$$

$$b = K_{G_w,P}^2 - \frac{z_c^2}{N}$$

An example computation of the Log Pearson Type III frequency curve for Fishkill Creek is presented in Bulletin 17B.[4]

Advantages and disadvantages of the analytic method

To some extent the advantages of the analytic method are the reverse of the disadvantages of the graphic method. More uniform and consistent estimates of population parameters are possible, statistical measures of reliability can be computed, and there is a theoretical basis for analyzing rare events. The disadvantages of the analytic method include the necessity of selecting a theoretical probability distribution that may not fit the data exactly and the greater potential for a false sense of accuracy, particularly in the estimation of rare events.

Frequency Analysis by Computer

Computer program HECWRC

The Flood Flow Frequency Analysis program, HECWRC,[5] computes flood frequencies with the Log Pearson Type III distribution, using annual maximum streamflows or other large events. The computations in the program are based on the guidelines presented in Bulletin 17B.[4]

Computational capabilities. The program computes and tabulates data for plotting with the graphic method and performs all the computations required for applying the Log Pearson Type III theoretical distribution. For the graphic method, it orders the data and assigns plotting positions by the Weibull, median, or Hazen formulas. For the theoretical distribution, it computes the frequency curve; automatically adjusts for a broken record, an incomplete record, and zero flood years; weights the station skew with a generalized skew coefficient;

tests and adjusts for outliers; incorporates historical events; adjusts for expected probability; and computes confidence limits.

Data input. Data input for the program generally follows the standard HEC format of eight-column fields in each line of input. Default values are provided for all decision variables, so only a minimum number of lines of input are required to provide the job title and annual flood peaks, and signal the end of data. Any options or nonstandard values are entered in optional lines of input such as lines with the identifiers J1 and J2. The Flood Flow Frequency Analysis program user's manual[5] should be used as a reference for input details.

Example problem: frequency analysis by computer. The data for 24 years of annual peaks on Fishkill Creek is used again here to demonstrate the use of the computer program HECWRC. The input file for the program is shown in Fig. 7.15. After the title lines and the job identification line, the

```
TT TEST NO. 1 FLOOD FLOW FREQUENCY ANALYSIS PROGRAM
TT WRC APPENDIX 12, EXAMPLE 1 - FITTING THE LOG PEARSON TYPE III DIST
TT FISHKILL CREEK AT BEACON, NY
ID 01-3735  FISHKILL CREEK AT BEACON, NEW YORK  DA=190 SQ MI           1945-68
GS  3735              .6
QR  373503051945    2290
QR  373512271945    1470
QR  373503151947    2220
QR  373503181948    2970
QR  373501011949    3020
QR  373503091950    1210
QR  373504011951    2490
QR  373503121952    3170
QR  373501251953    3220
QR  373509131954    1760
QR  373508201955    8800
QR  373510161955    8280
QR  373504101957    1310
QR  373512211957    2500
QR  373502111959    1960
QR  373504061960    2140
QR  373502261961    4340
QR  373503131962    3060
QR  373503281963    1780
QR  373501261964    1380
QR  373502091965     980
QR  373502151966    1040
QR  373503301967    1580
QR  373503191968    3630
ED
```

Figure 7.15 Input file for computer program HECWRC.

GS line indicates the station number and the generalized skew coefficient of .6 to be used in weighting the station skew. Each QR line specifies the following: the station number in field 1; the month, day, and year in field 2. Spaces for indicating the month and day may be left blank. The recorded flood peak is entered in field 3.

Since several of the optional lines of input, such as the J1 line, are omitted, default values for the variables are used.

Output for this example computation without the data input listing is tabulated in Fig. 7.16 and plotted in Figs. 7.17 and 7.18.

Flood volume frequency analysis with the computer program STATS

Flood volumes can be analyzed with the computer program STATS[8] to estimate the amount of storage required to control a flood of a given percent chance exceedance. STATS will compute volumes (mean flow rates) from daily flows for durations ranging from 1 to 183 days.

Frequency Analysis of Ungaged Basins

General

Many times, frequency analysis must be performed for a basin where little, if any, observed precipitation data and no discharge data are available. Techniques used for ungaged-basin analysis may be divided into four general categories: (1) statistical methods (regional analysis), (2) data transfer methods, (3) empirical equations, and (4) precipitation-runoff simulation. A more detailed breakdown of such techniques made by a task force of the U.S. Water Resources Council (WRC) consists of the following eight categories:[9]

1. Statistical estimation of peak flow for a given exceedance probability

2. Statistical estimation of moments (mean, standard deviation, and skew)

3. Index flood estimation

4. Transfer methods

5. Empirical equations

6. Single-event simulation

7. Multiple discrete events simulation

8. Continuous simulation for period of record

```
-OUTLIER TESTS  -
----------------------------------------------------------------
LOW OUTLIER TEST
----------------

BASED ON  24 EVENTS, 10 PERCENT OUTLIER TEST VALUE K(N) = 2.467

        0 LOW OUTLIER(S) IDENTIFIED BELOW TEST VALUE OF    578.7


------------------
HIGH OUTLIER TEST
------------------

BASED ON  24 EVENTS, 10 PERCENT OUTLIER TEST VALUE K(N) = 2.467

        0 HIGH OUTLIER(S) IDENTIFIED ABOVE TEST VALUE OF    9425.


----------------------------------------------------------------


-SKEW WEIGHTING -
----------------------------------------------------------------
BASED ON  24 EVENTS, MEAN SQUARE ERROR OF STATION SKEW =   .277
DEFAULT OR INPUT MEAN SQUARE ERROR OF GENERALIZED SKEW =   .302
----------------------------------------------------------------



   FINAL RESULTS
   -FREQUENCY CURVE- 01-3735  FISHKILL CREEK AT BEACON, NEW YORK
   ****************************************************************
   *........FLOW,CFS........*          *...CONFIDENCE LIMITS...*
   *              EXPECTED   * EXCEEDANCE *                      *
   * COMPUTED  PROBABILITY * PROBABILITY *  .05 LIMIT  .95 LIMIT *
   *---------------------*-------------*----------------------*
   *   19200.    28300.   *    .002   *    39100.    12300.  *
   *   14500.    19000.   *    .005   *    26900.     9740.  *
   *   11500.    14100.   *    .010   *    20100.     8080.  *
   *    9110.    10500.   *    .020   *    14800.     6640.  *
   *    6530.     7090.   *    .050   *     9680.     5010.  *
   *    4960.     5210.   *    .100   *     6850.     3950.  *
   *    3650.     3740.   *    .200   *     4710.     2990.  *
   *    2190.     2190.   *    .500   *     2650.     1790.  *
   *    1440.     1420.   *    .800   *     1760.     1110.  *
   *    1200.     1170.   *    .900   *     1490.      884.  *
   *    1040.     1010.   *    .950   *     1320.      746.  *
   *     841.      791.   *    .990   *     1100.      568.  *
   *++++++++++++++++++++++++++++++++++++++++++++++++++++++++++++*
   *   FREQUENCY CURVE STATISTICS   *    STATISTICS BASED ON     *
   *-----------------------------*----------------------------*
   *  MEAN LOGARITHM      3.3684  * HISTORIC EVENTS        0 *
   *  STANDARD DEVIATION   .2456  * HIGH OUTLIERS        0   *
   *  COMPUTED SKEW        .7300  * LOW OUTLIERS         0   *
   *  GENERALIZED SKEW     .6000  * ZERO OR MISSING      0   *
   *  ADOPTED SKEW         .7000  * SYSTEMATIC EVENTS      24 *
   ****************************************************************
```

Figure 7.16 Example output.

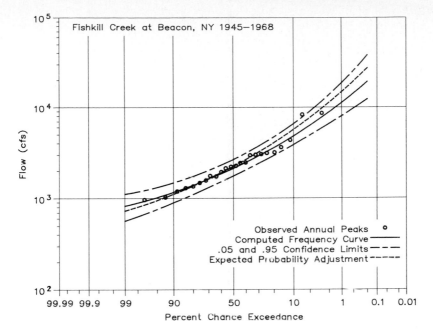

Figure 7.17   Example of frequency plots.

Selection of a technique

The selection of a particular technique depends on several factors: the information required, the data available, the location and hydrologic characteristics of the basin relative to gaged basins, the financial and other resources available for the study, and the familiarity of the investigators with particular techniques.

Although there may be no observed discharge data for the basin itself, the availability of precipitation data in the region and discharge data in nearby basins will determine if direct transfer of information from other basins or regional analysis is feasible. Of course, the similarity of hydrologic characteristics is also an important determinant. Other factors, such as the scope of the study, including the allocation of money and human resources, may dictate whether a complex modeling technique is employed.

Statistical methods (regional analysis). The first three of the eight WRC categories fall into the more general "statistical methods" category, frequently referred to as "regional analysis." All are based on statistically derived flood-frequency relationships for a region. The first uses equations developed by regression analysis for direct estimation of peak flows using watershed and climatic characteristics and

FINAL RESULTS
-FREQUENCY PLOT - 01-3735 FISHKILL CREEK AT BEACON, NEW YORK DA=190 SQ MI   1945-68
BASED ON EXPECTED PROBABILITY ADJUSTMENT, FLOW IN CUBIC FEET PER SECOND

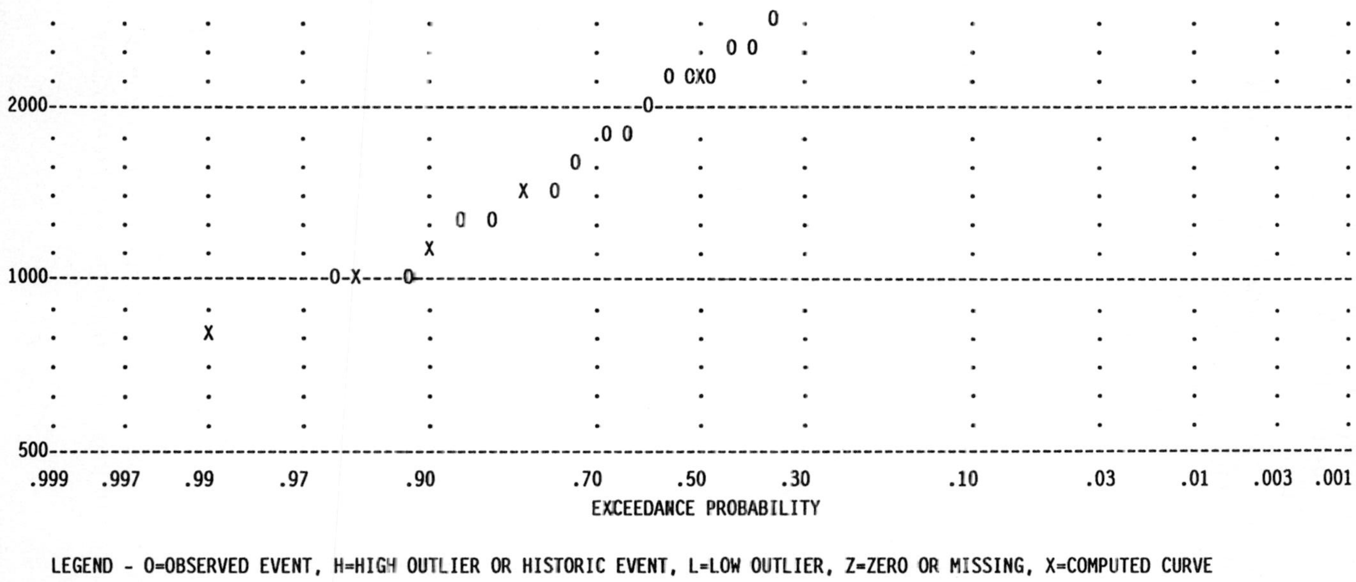

2000

1000

500

EXCEEDANCE PROBABILITY

.999 .997 .99 .97 .90 .70 .50 .30 .10 .03 .01 .003 .001

LEGEND - O=OBSERVED EVENT, H=HIGH OUTLIER OR HISTORIC EVENT, L=LOW OUTLIER, Z=ZERO OR MISSING, X=COMPUTED CURVE

Figure 7.18 Example line printer plot of frequency curve and observed events.

peak flow frequency values from station data. The second uses equations relating the moments of a probability distribution of peak flows to watershed and climatic characteristics to predict flood flow frequency at ungaged locations. The third applies statistically determined ratios to an index flood, such as the mean annual flood, to determine selected frequency events.

Regional analysis as applied to unit hydrograph and loss rate parameters in modeling is discussed in Chap. 6. Similar procedures apply to the determination of flood flows of given frequencies for ungaged basins. However, the variables of peak discharge, statistical moments, and flood index ratios are correlated to basin and climatic characteristics rather than runoff parameters.[3,10]

Data transfer methods. The direct transfer of data from a gaged basin to an ungaged basin is considered valid if the basins are hydrologically similar. Criteria for determining similarity relate to the meteorologic regime as well as basin characteristics. Storm patterns and rainfall intensities must be similar, and average annual rainfall must be approximately the same. Physical characteristics of the basins—including size, land use, drainage density, channel slopes, soil conditions, and topographic relief—also must be similar. Under proper conditions, data for peak flows, data for single-storm events, or the entire record of historical data can be transferred.

Empirical methods. The use of simplified empirical methods of estimating runoff is generally limited to small homogeneous watersheds. Two well-known methods are the rational formula[11] and the SCS TR55 method.[12]

Precipitation-runoff simulation

Single-event simulation. Design storms with specified frequencies may be used in a single-event model, such as HEC-1, to develop frequency curves at prescribed locations. If the model has been calibrated to available information so that the computed runoff frequency approximates the rainfall frequency, the application of design storms produces hydrographs of corresponding frequencies at locations of interest.

Multiple discrete events simulation. A single-event model also may be used to simulate runoff from selected discrete rainfall events for each year of the period of record instead of design storms. The results yield maximum annual peak hydrographs for each year of the period of record, and these are used to develop corresponding frequency curves at desired locations. Long-record precipitation data must be available for this technique.

Continuous simulation for the period of record. A continuous watershed model[13] can be used to develop a continuous-flow hydrograph at a location of interest from a continuous record of precipitation and other climatic factors. A frequency curve is developed from the maximum annual flood peaks.

Field reconnaissance at ungaged locations

Field reconnaissance, including interviews with local agency personnel and residents, visual inspection of the study area, and review of local documents, is important in obtaining reliable flood-frequency estimates at ungaged locations. Examples of information that can be obtained are as follows:[14]

1. Meteorologic and physical basin characteristics
2. Historical high-water marks
3. General knowledge of direction of flood flows, debris blockage, frequency of bridge overtopping, etc.
4. Design discharges for highway bridges, storm sewers, culverts, channels, etc.
5. Knowledge of anticipated future development, including flood control works, subdivisions, and shopping centers
6. Descriptions of historical events from newspapers and other documents
7. Photographs of historical events

The information obtained from field reconnaissance should be used in formulating strategies for analyzing flood frequency at ungaged locations and in judging the results of simulation model computations and other procedures.

Workshop Problem 7.1: Frequency Analysis of Streamflow Record for the Gage Near Springfield (3945)

Purpose

The purpose of this workshop problem is to provide an understanding of graphic and analytic methods of frequency analysis. Bulletin 17B procedures are demonstrated with manual computations as well as with the computer program HECWRC.

Problem statement

A systematic record of peak flows at the Springfield gage (3945) from 1938 to 1986 is shown in Table P7.1. A discharge-frequency curve is to be developed from this data.

TABLE P7.1   Record of Peak Flows at the Springfield Gage

Water year	Date	Peak discharge, ft^3/s	Water year	Date	Peak discharge, ft^3/s
1938	07/23/38	2050		11/08/72	2870
	09/21/38	1590		11/14/72	1010
1939	02/03/39	699		02/02/73	1840
1940	05/31/40	1140		04/04/73	1210
1941	02/07/41	885		06/29/73	1760
1942	08/09/42	1320	1974	12/21/73	1870
1943	12/30/42	663		10/30/73	1770
1944	03/13/44	815		08/17/74	1650
1945	09/19/45	1370		09/03/74	1410
1946	06/02/46	975	1975	07/14/75	3110
1947	04/05/47	646		07/25/75	1290
1948	11/08/47	1280		09/26/75	1490
1949	01/06/49	834	1976	08/10/76	960
1950	03/23/50	501	1977	03/22/77	1950
1951	03/30/51	954		02/25/77	1260
1952	06/01/52	1280		04/05/77	1140
1953	03/13/53	1330	1978	11/08/77	2180
1954	09/11/54	947		01/26/78	1280
1955	08/13/55	1270		03/27/78	1660
	08/19/55	1060		05/14/78	1060
1956	10/14/55	643		08/06/78	1060
1957	04/05/57	538	1979	01/24/79	1540
1958	02/28/58	844		01/21/79	1510
1959	08/09/59	885		02/26/79	1050
1960	09/12/60	911		05/24/79	1120
1961	04/16/61	708		09/06/79	1300
1962	03/12/62	1530	1980	03/21/80	1250
1963	03/06/63	675	1981	05/11/81	926
1964	11/07/63	748	1982	01/04/82	1650
1965	02/08/65	838	1983	04/10/83	1360
1966	09/22/66	1520		03/19/83	1340
1967	03/07/67	1170		03/21/83	1140
1968	05/29/68	3370		04/03/83	1090
1969	07/29/69	1510		04/16/83	1290
	03/25/69	1390		06/19/83	1050
	09/04/69	1050	1984	04/05/84	1660
1970	07/31/70	1170		12/13/83	1530
1971	08/28/71	3430		12/22/83	1210
	09/12/71	1530		05/30/84	1430
1972	06/22/72	1160		07/07/84	1310
1973	08/02/73	5430	1985	09/27/85	1410
	10/07/72	1050	1986	11/17/85	1210
				04/17/86	1070

Tasks

1. Determine the graphic plotting positions for the annual peak discharges using the Weibull formula, and plot the points on log-probability paper. Draw in a smooth frequency curve to fit the points.
2. Compute the mean, the standard deviation, and the skew coefficient for this data.
3. Using a generalized skew of zero, calculate the weighted "adopted" skew coefficient G_w.
4. Using the adopted skew coefficient from task 3, rounded to the nearest tenth, calculate the analytic frequency curve. Plot this curve on the same graph as used in task 1.
5. Prepare an input file of annual flood peaks for the Flood Flow Frequency Analysis program, HECWRC, using the record of flows from Table P7.1. The record contains peak flows above a base discharge of 1000 ft^3/s. Run the program to compute a frequency curve. Obtain tabulated output and a line printer plot of the curve.
6. Compare the results of the computed output with the results of the graphic and analytic hand calculations.

References

1. Charles T. Haan, *Statistical Methods in Hydrology*, 3d ed., Iowa State University Press, Ames, 1982.
2. Warren Viessman, Jr., John W. Knapp, Gary L. Lewis, and Terence E. Harbaugh, *Introduction to Hydrology*, 2d ed., Harper & Row, New York, 1977.
3. "Hydrologic Frequency Analysis," engineer manual, U.S. Army Corps of Engineers, Apr. 1, 1985. (Draft.)
4. Interagency Advisory Committee on Water Data, *Guidelines for Determining Flood Flow Frequency*, Bulletin 17B, U.S. Geological Survey, Reston, Va. 1982.
5. Hydrologic Engineering Center, *Flood Flow Frequency Analysis, Users Manual*, U.S. Army Corps of Engineers, Davis, Calif. February 1982.
6. L. R. Beard, "Probability Estimates Based on Small Normal-Distribution Samples," *Journal of Geophysical Research*, July 1960.
7. L. R. Beard, *Statistical Methods in Hydrology*, Civil Works Investigation Project CW-151, U.S. Army Corps of Engineers, Sacramento, Calif. January 1962.
8. Hydrologic Engineering Center, "Statistical Analysis of Time Series Data," U.S. Army Corps of Engineers, Davis, Calif. 1986. (Draft input description for computer program STATS.)
9. *Estimating Peak Flow Frequencies for Natural Ungaged Watersheds*, U.S. Water Resources Council, Washington, D.C., 1981.
10. Hydrologic Engineering Center, *Hydrologic Analysis of Ungaged Watersheds Using HEC-1*, Training Document 15, U.S. Army Corps of Engineers, Davis, Calif., 1982.
11. Philip B. Bedient and Wayne C. Huber, *Hydrology and Floodplain Analysis*, Addison-Wesley, New York, 1988.
12. Soil Conservation Service, *Urban Hydrology for Small Watersheds*, Technical Release 55, U.S. Department of Agriculture, January 1975 (rev. June 1986).
13. R. C. Johanson, J. C. Imhoff, and H. H. Davis, *User's Manual for Hydrological Simulation Program—FORTRAN (HSPF)*, EPA-600/9-80-015, U.S. Environmental Protection Agency, Athens, Ga., 1980.
14. Hydrologic Engineering Center, *Adoption of Flood Flow Frequency Estimates at Ungaged Locations*, Training Document 11, U.S. Army Corps of Engineers, Davis, Calif., 1980.

Use of Design Storms in Rainfall-Runoff Simulation and Frequency Analysis

Introduction

Design storms, sometimes called hypothetical or synthetic storms, are often used for planning and design studies. As the names imply, these storms are synthesized from generalized rainfall data and regression relationships. They are derived in different ways and have a variety of applications.

Design storms are used in discharge-frequency analysis when gaged data is not available or the records are too short to be used to develop rainfall-frequency relationships. They are also used to evaluate the effect of development projects on flood levels at downstream locations and as a basis for the design of projects when substantial loss of property or the loss of life is a primary factor.

Three types of design storms can be generated in HEC-1: the standard project storm, the probable maximum storm, and design storms of different durations derived from depth-area data. Procedures for deriving standard project storms, used in the design of Corps of Engineers's projects, are specified in Corps criteria.[1] Storms based on depth-area data and "probable maximum precipitation" are discussed in the remainder of this chapter.

Design storm data

Sources. The primary sources of design storm information for the United States are various technical publications (TP and HMR series) of the National Weather Service (NWS) and its parent agency, the National Oceanic and Atmospheric Administration (NOAA). The development of design storms is based on generalized rainfall maps and regression equations in these publications.

The contiguous 48 states are divided into two geographic areas on the basis of precipitation characteristics. The 35 states east of the Rocky Mountains—being relatively free of orographic effects—are included in one region. The remaining 13 western states are in the other region, for which more detailed information on a state-by-state and site-specific basis is available. Similar detailed information is available for Alaska and Hawaii.

Key publications used for developing design storms include the following:

1. NWS TP-40 contains maps for precipitation depths with percent chance exceedance of 1, 2, 4, 10, 20, 50, and 100 for durations ranging from 30 min to 24 h for the eastern United States.[2]

2. NWS TP-49 contains maps for precipitation depths with 1 to 50 percent chance exceedance and durations ranging from 2 to 10 days for the eastern United States.[3]

3. NOAA Technical Memorandum NWS HYDRO-35 contains precipitation estimates for durations of 5 to 60 min for the eastern United States. The 30-min and 1-h maps in TP-40 have been superseded by data in HYDRO-35.[4]

4. NOAA *Atlas 2* contains precipitation depths with percent chance exceedance ranging from 1 to 50 for durations of 6 and 24 h for the western United States. Methods for estimating values for other durations are also included.[5]

Description of design storm data. National Weather Service depth-duration data is based on statistical analysis of long-term rain gage records in a region. Isopluvial maps are prepared showing lines of equal rainfall depth for a given storm duration and frequency. An isopluvial map for a 1 percent chance exceedance (100-year return period), 2-h rainfall is shown in Fig. 8.1. Maps such as this one can be used to determine point precipitation values for areas up to 10 mi^2 in size. For larger areas, the average value for an area is less than the maximum value at a point, so adjustments are required. Curves for making these adjustments are shown in Fig 8.2.[2]

The precipitation depths on NWS isopluvial maps are based on data from partial-duration series. To obtain annual series depths, multiply the values taken from the maps by the following adjustment factors:[2]

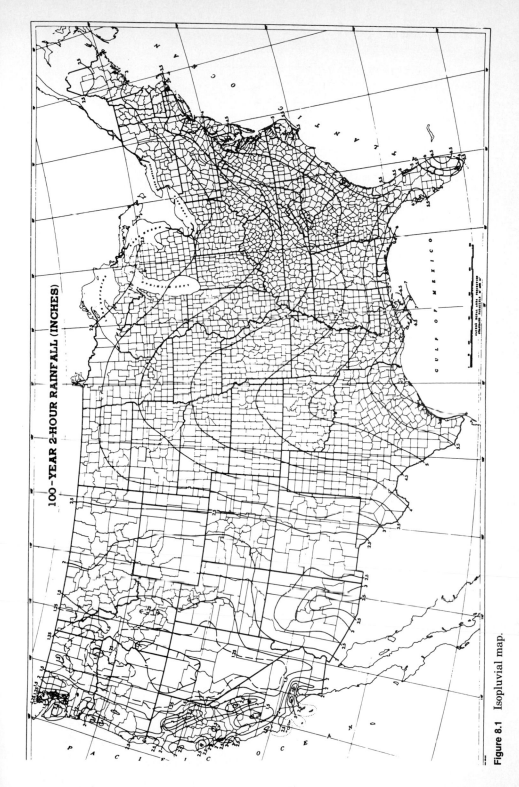

Figure 8.1 Isopluvial map.

217

Percent chance exceedance	Adjustment factor
50	0.88
20	0.96
10	0.99

For depths with percent chance exceedance smaller than 10, no adjustment is required because depths for the two series coincide beyond that point.

A design storm developed from NWS data is sometimes referred to as a "balanced" storm because its incremental depths may be arranged in a consistent depth-frequency relationship for each duration interval of the total storm. In other words, different duration intervals—1 h, 6 h, etc.—produce rainfall depths with the same percent chance exceedance. This consistent relationship, though not representative of natural storms because of their random nature, is used to ensure an appropriate depth of rainfall for a given frequency regardless of the time-response characteristics of a particular river basin. However, other arrangements of design storm data also may be used.

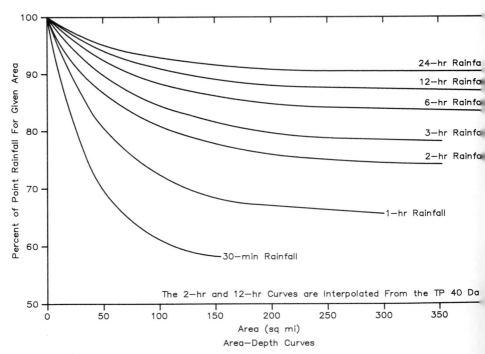

Figure 8.2 Precipitation depth adjustment curves.

Before a design storm can be developed from NWS data, two storm parameters must be estimated: the duration of the storm and the time interval for each rainfall increment. The duration needed is directly related to the time of concentration of the basin, as determined from an analysis of recorded data or by computation. The duration should be at least as long as, but preferably longer than, the time of concentration. A duration less than that would not allow all parts of the basin to contribute runoff simultaneously at the outlet during the course of the storm. Runoff from the lower parts of the basin would have left the basin before runoff from the upper parts had reached the outlet, and the estimated peak discharge would be too low. If total runoff volume as well as peak discharge is of importance, or if the basin has large amounts of floodplain storage, the storm duration should be well in excess of the time of concentration. A long-duration storm is required to capture the attenuation effects of large natural storage areas.

The time interval of storm increments should be small enough to define accurately the flood hydrograph. Normally, the interval is set according to the fastest-peaking subarea of the basin model. A time interval that produces at least four or five points on the rising limb of the unit hydrograph for this subarea is accurate enough for most applications.

Procedures for developing a design storm

The following steps, consistent with the procedure programmed in HEC-1, are used to develop a balanced design storm:[6]

1. Determine the duration of the storm on the basis of the time-response characteristics of the basin.

2. Determine precipitation depths from the appropriate NWS publication for the desired percent chance exceedance (recurrence interval) and durations.

3. Adjust the depths for storm area size using the curves in Fig. 8.2 if the area is larger than 10 mi^2.

4. Adjust depths from partial- to annual-duration series, if desired.

5. Plot a curve of adjusted depth vs. duration on log-log graph paper, and determine a complete set of incremental depths.

6. Arrange depths in descending order of magnitude.

7. If a balanced storm is desired, arrange the incremental depths with the largest value in the center of the storm, the second largest just ahead of the largest, the third largest just after the largest, and so

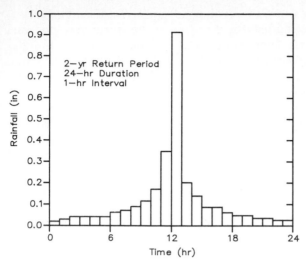

Figure 8.3 Design storm with a "balanced" configuration.

forth. In other words, alternate the locations of the incremental amounts from one side to the other side of the largest value to obtain a triangular distribution (Fig. 8.3).

Example problem: development of a design storm for the northeast corner of the Texas panhandle. A hydrologic analysis of a 100-mi^2 basin located in the northeast corner of the Texas panhandle[7] requires the development of a design storm with a 1 percent chance exceedance. The total storm duration has been determined to be 12 h with a time interval of 30 min.

Two NWS publications, TP-40 and HYDRO-35, are used to determine the characteristics of this storm. Rainfall depths for 2-, 3-, 6-, and 12-h durations with a 100-year recurrence interval are obtained from isopluvial maps in TP-40. Location of the 2-h depth of 4.35 in is shown, for example, in Fig. 8.4. Depths for 15-min and 60-min durations are obtained from maps in HYDRO-35. The values are as follows:

Duration	15-min	1-h	2-h	3-h	6-h	12-h
Depths, in	1.82	3.46	4.35	4.65	5.30	6.00

The 30-min depth is computed from Eq. 7 in HYDRO-35:

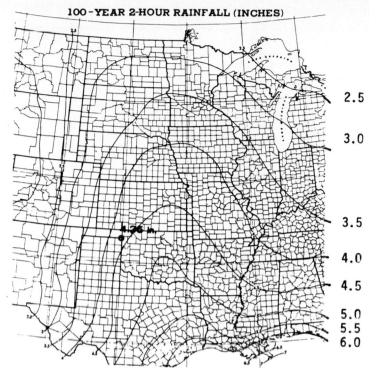

100-YEAR 2-HOUR RAINFALL (INCHES)

2.5

3.0

3.5

4.0

4.5

5.0
5.5
6.0

Figure 8.4 Location of 2-h rainfall depth for the northeast corner of the Texas panhandle.

$$30\text{-min depth} = 0.49(60\text{-min depth}) + 0.51(15\text{-min depth})$$

$$= 0.49(3.46) + 0.51(1.82)$$

$$= 2.62 \text{ in}$$

These values of accumulated point rainfall for various durations are shown in column (2) of Table 8.1. Area adjustment factors taken from Fig. 8.2 are shown in column (3). Adjusted depths are entered in column (4) and are plotted against corresponding durations in Fig. 8.5. Rainfall depths for intervening intervals (1.5 h, 2.5 h, etc.) are determined from the curve in Fig. 8.5 and entered in column (4). Since a 1 percent chance exceedance storm is being developed—i.e., percent chance exceedance < 10—no adjustment to annual series values is necessary. After the accumulated depths are transformed into increments, entered in column (5), and reordered into a "balanced" storm in column (6), the design storm is complete.

TABLE 8.1   Computation of a Design Storm for the Northeast Corner of the Texas Panhandle

Period, h (1)	Accumulated point rainfall, in (2)	Point rainfall factor (100 mi²) (3)	Accumulated depth, in (4)	Incremental depth, in (5)	Arranged incremental depth, in (6)
0.5	2.62	0.615	1.61	1.61	0.06
1.0	3.46	0.723	2.50	0.89	0.06
1.5			3.10	0.60	0.06
2.0	4.35	0.810	3.52	0.42	0.06
2.5			3.75	0.23	0.06
3.0	4.65	0.845	3.93	0.18	0.07
3.5			4.10	0.17	0.10
4.0			4.25	0.15	0.11
4.5			4.40	0.15	0.15
5.0			4.50	0.10	0.18
5.5			4.60	0.10	0.42
6.0	5.30	0.888	4.71	0.11	0.89
6.5			4.77	0.06	1.61
7.0			4.84	0.07	0.60
7.5			4.90	0.06	0.23
8.0			4.96	0.06	0.17
8.5			5.02	0.06	0.15
9.0			5.09	0.07	0.10
9.5			5.15	0.06	0.07
10.0			5.21	0.06	0.07
10.5			5.27	0.06	0.06
11.0			5.34	0.07	0.06
11.5			5.40	0.06	0.06
12.0	6.00	0.910	5.46	0.06	0.06

Example problem: development of a design storm for the western United States.   A design storm with a 50 percent chance exceedance (2-year return interval) is needed for a 2.6-mi² basin in Davis, California.[7] Annual series rainfall at a 10-min interval is required for a storm with a 3-h duration. The 2-year-return-period rainfall depth of 1.30 in for a 6-h duration is read from an isopluvial map in NOAA *Atlas 2*, Vol. 11: *California*.[5] See Fig. 8.6. The depth for a 1-h duration is determined with an equation provided for Region 4, in which Davis is located:

$$Y_2 = 0.107 + 0.315X_1 \qquad (8.1)$$

where Y_2 = 2-year, 1-h depth and X_1 = 2-year, 6-h depth. Substituting 1.30 in for X_1 in Eq.(8.1) yields a 2-year, 1-h depth of 0.52 in. Depths for 2-h and 3-h durations also may be determined from equations:[5]

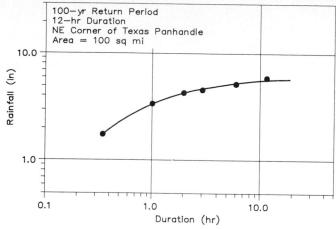

Figure 8.5 Plot of rainfall depth for the northeast corner of the
Texas panhandle.

$$2\text{-h depth} = 0.240(6\text{-h depth}) + 0.760(1\text{-h depth})$$

$$= 0.240(1.30) + 0.760(0.52)$$

$$= 0.71 \text{ in}$$

$$3\text{-h depth} = 0.468(1.30) + 0.532(0.52)$$

$$= 0.89 \text{ in}$$

Depths for durations of less than 1 h are found by applying values
from Table 12 in NOAA *Atlas 2*[5] to the 1-h depth:

$$10\text{-min depth} = 0.45(0.52) = 0.23 \text{ in}$$

$$15\text{-min depth} = 0.57(0.52) = 0.30 \text{ in}$$

$$30\text{-min depth} = 0.79(0.52) = 0.41 \text{ in}$$

For a drainage area as small as 2.6 mi^2 the adjustment for area is
not made, so the point values of rainfall are used for basin averages.
The adjustment factor of 0.88 for partial-duration series to annual series is taken from Table 2 in NOAA *Atlas 2*[5] for a 2-year return period. Applying this factor to the depths computed above yields the adjusted depths listed in Table 8.2 and plotted in Fig. 8.7. Depths for 10-
min intervals are read from Fig. 8.7, and these are transformed into
increments and arranged as shown in Table 8.3. The last column

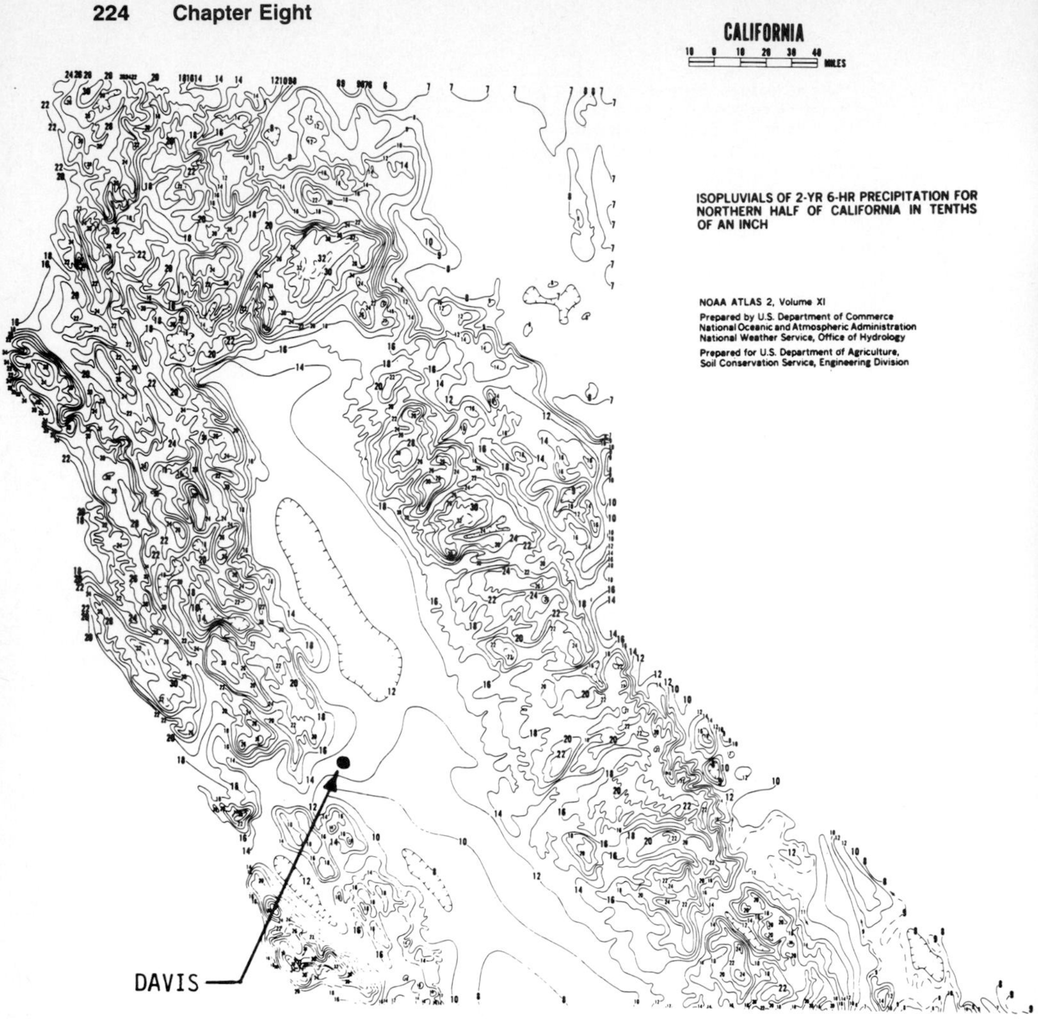

CALIFORNIA

ISOPLUVIALS OF 2-YR 6-HR PRECIPITATION FOR NORTHERN HALF OF CALIFORNIA IN TENTHS OF AN INCH

NOAA ATLAS 2, Volume XI

Prepared by U.S. Department of Commerce
National Oceanic and Atmospheric Administration
National Weather Service, Office of Hydrology

Prepared for U.S. Department of Agriculture,
Soil Conservation Service, Engineering Division

DAVIS

Figure 8.6   Location of 2-year, 6-h rainfall depth for Davis, California.

TABLE 8.2   Hypothetical 2-Year Rainfall at Davis, California

Period	Accumulated point rainfall depth, in	Areal adjustment	Annual series adjustment	Adjusted accumulated depth, in
10 min	0.23	None	0.88	0.20
15 min	0.30	None	0.88	0.26
30 min	0.41	None	0.88	0.36
60 min	0.52	None	0.88	0.46
2 h	0.71	None	0.88	0.62
3 h	0.89	None	0.88	0.78

shows the desired design storm which can be input into a basin rainfall-runoff simulation model.

Generation of a design storm with HEC-1.   The program HEC-1 has the capability to compute automatically a balanced design storm according to specified depth-duration data. Specified data, including storm frequency, storm area to be used in computing the reduction in point rainfall depths, and rainfall depths from appropriate NWS publications such as TP-40, is entered on a PH line in the input file. A storm of any duration from 5 minutes to 10 days can be generated.

Depth-area option of HEC-1

Concept.   Precipitation amounts associated with a design storm are a function of the size of the tributary area affected. Each size of storm

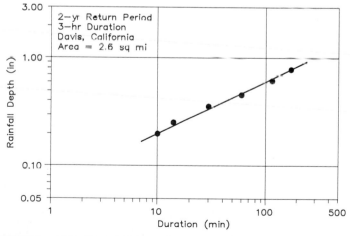

Figure 8.7   Plot of rainfall depth for Davis, California.

TABLE 8.3   Design Storm at Davis, California

(3-h Duration, 2-Year Return Period, 10-min Interval)

Period, min	Accumulated rain depth, in	Incremental depth, in	Arranged storm, in
10	0.20	0.20	0.02
20	0.29	0.09	0.02
30	0.35	0.06	0.02
40	0.40	0.05	0.02
50	0.45	0.05	0.03
60	0.49	0.04	0.03
70	0.52	0.03	0.04
80	0.55	0.03	0.05
90	0.58	0.03	0.09
100	0.61	0.03	0.20
110	0.63	0.02	0.06
120	0.65	0.02	0.05
130	0.67	0.02	0.03
140	0.70	0.03	0.03
150	0.72	0.02	0.03
160	0.74	0.02	0.02
170	0.76	0.02	0.02
180	0.78	0.02	0.02

has a unique depth and temporal distribution. Because of this, modeling several different design storms would ordinarily be required to generate a consistent set of discharge hydrographs for several different locations, each with a different drainage area.

The depth-area option[6] in HEC-1 provides an efficient means of dealing with this problem by using "index" hydrographs with which interpolation can be performed to obtain a consistent hydrograph at any location. Index hydrographs are computed at each computational point for a fixed set of hyetographs representing storm areas that encompass the full range of drainage areas in the basin.

Computational procedure. The precipitation depth–area computational procedure for a complex river basin is illustrated in Fig. 8.8. Four index hydrographs are computed for each subbasin with precipitation quantities of 15, 13, 10, and 8 in. The consistent hydrograph for a subbasin—i.e., the hydrograph that corresponds to the appropriate precipitation depth for the subbasin's drainage area—is depicted with a dashed line. This hydrograph is determined by interpolation

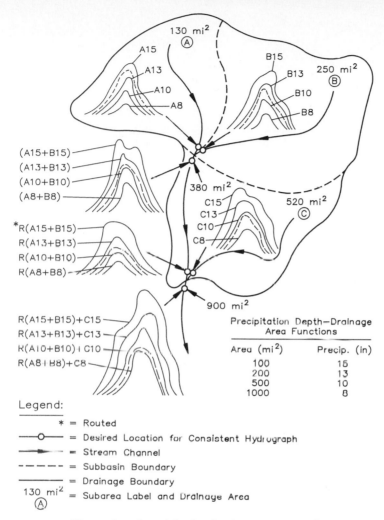

Figure 8.8   Illustration of precipitation depth–area computation.

between the two index hydrographs bracketing the subbasin's drainage area. Index hydrographs, not consistent hydrographs, are routed and combined to obtain index hydrographs for downstream points. A consistent hydrograph for any location is obtained by interpolation between appropriate index hydrographs based on the total contributing area for that location. The example problem that follows will clarify this procedure.

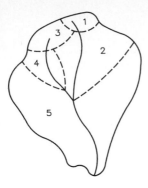

Figure 8.9 Example application of depth-area option, Never Dry River basin.

Example problem: application of the depth-area option. Discharge hydrographs for a 1 percent chance design storm are required at the outlet of each of the five subbasins shown in Fig. 8.9. The areas for the subbasins are as follows:

Subbasin	Area, mi^2
1	5.6
2	27.0
3	9.2
4	12.0
5	49.0
	102.8

The depth-area option in HEC-1 is used to obtain these hydrographs: appropriate depth-duration data is entered on JD and PH lines in the input file (Fig. 8.10). The remaining lines in the input file (not shown) following the last JD line are typical of any other HEC-1 model for computing subbasin runoff, routing, and combining. Up to nine index areas may be specified on JD lines; these should cover the range of areas in the basin. In this example, areas of 5, 10, 20, 50, and 105 mi^2 are specified, the last one (105) large enough to cover the total area at the basin outlet of 102.8 mi^2. The PH line, which must follow the first JD line, contains the percent chance exceedance for the design storm in the first field and the rainfall depths for various durations in fields 4 through 10. These depths are taken from an appropriate NWS publication, such as TP-40.

```
ID.......1.......2.......3.......4.......5.......6.......7.......8.......9......10

ID        NEVERDRY RIVER BASIN
ID        1% CHANCE HYPOTHETICAL STORM
ID        DEPTH-AREA OPTION
IT    30            100
JD            5
PH    1                 1.81   3.45   3.90   4.26   5.10   6.00   6.98
JD           10
JD           20
JD           50
JD          105
*
```

Figure 8.10   Depth-area data specified on JD and PH lines of input file.

Excerpts from the output for subbasin 1 in this example show the types of information provided. Precipitation patterns for each index storm are computed, such as the one shown for index storm no. 2 in Fig. 8.11. Hydrographs with various transposition areas are computed for each subbasin, such as the one for an area of 20 mi^2 for subbasin 1 shown in Fig. 8.12. The interpolated hydrograph for subbasin 1 is shown in Fig. 8.13.

This same information, though not shown here, can be obtained in output for each of the subbasins in the model. The index hydrographs are routed and combined through the entire basin, and interpolated hydrographs are computed at each location of interest. Tables and plots of index and interpolated hydrographs are obtained at any location desired by using appropriate output specifications on IO and KO lines in the input file. The peak flows shown in the runoff summary (Fig. 8.14) are from the interpolated hydrographs at each station indicated.

```
INDEX STORM NO. 2
      STRM      6.89  PRECIPITATION DEPTH
      TRDA     10.00  TRANSPOSITION DRAINAGE AREA

PRECIPITATION PATTERN
      .03       .03       .03       .04       .04       .04       .04       .04       .04       .05
      .05       .05       .06       .07       .07       .08       .08       .09       .12       .13
      .16       .17       .21       .84      2.44       .27       .20       .18       .14       .12
      .10       .09       .08       .07       .07       .06       .05       .05       .05       .05
      .04       .04       .04       .04       .04       .04       .03       .03
```

Figure 8.11   Computed precipitation pattern for storm no. 2.

```
******************************************************************************
```

HYDROGRAPH AT   SUB1
TRANSPOSITION AREA   20.0 SQ MI

```
******************************************************************************
```

DA	MON	HRMN	ORD	RAIN	LOSS	EXCESS	COMP Q		DA	MON	HRMN	ORD	RAIN	LOSS	EXCESS	COMP Q
1		0100	1	.00	.00	.00	1.	*	2		0230	52	.00	.00	.00	470.
1		0130	2	.03	.03	.00	1.	*	2		0300	53	.00	.00	.00	469.
1		0200	3	.03	.03	.00	1.	*	2		0330	54	.00	.00	.00	468.
1		0230	4	.04	.04	.00	1.	*	2		0400	55	.00	.00	.00	467.
1		0300	5	.04	.04	.00	1.	*	2		0430	56	.00	.00	.00	466.
1		0330	6	.04	.04	.00	1.	*	2		0500	57	.00	.00	.00	465.
1		0400	7	.04	.04	.00	1.	*	2		0530	58	.00	.00	.00	463.
1		0430	8	.04	.04	.00	1.	*	2		0600	59	.00	.00	.00	462.
1		0500	9	.04	.04	.00	1.	*	2		0630	60	.00	.00	.00	461.
1		0530	10	.05	.05	.00	1.	*	2		0700	61	.00	.00	.00	460.
1		0600	11	.05	.05	.00	1.	*	2		0730	62	.00	.00	.00	459.
1		0630	12	.05	.05	.00	1.	*	2		0800	63	.00	.00	.00	458.
1		0700	13	.05	.06	.00	1.	*	2		0830	64	.00	.00	.00	456.
1		0730	14	.06	.06	.00	1.	*	2		0900	65	.00	.00	.00	455.
1		0800	15	.07	.07	.00	1.	*	2		0930	66	.00	.00	.00	454.
1		0830	16	.07	.07	.00	1.	*	2		1000	67	.00	.00	.00	453.
1		0900	17	.08	.08	.00	1.	*	2		1030	68	.00	.00	.00	452.

1	0930	18	.08	.03	.00	1.	*	2	1100	69	.00	.00	.00	451.
1	1000	19	.09	.09	.00	1.	*	2	1130	70	.00	.00	.00	450.
1	1030	20	.12	.10	.02	4.	*	2	1200	71	.00	.00	.00	449.
1	1100	21	.13	.10	.03	20.	*	2	1230	72	.00	.00	.00	447.
1	1130	22	.16	.10	.06	59.	*	2	1300	73	.00	.00	.00	446.
1	1200	23	.18	.10	.08	128.	*	2	1330	74	.00	.00	.00	445.
1	1230	24	.22	.10	.12	230.	*	2	1400	75	.00	.00	.00	444.
1	1300	25	.85	.10	.75	493.	*	2	1430	76	.00	.00	.00	443.
1	1330	26	2.29	.10	2.19	1385.	*	2	1500	77	.00	.00	.00	442.
1	1400	27	.28	.10	.18	2897.	*	2	1530	78	.00	.00	.00	441.
1	1430	28	.20	.10	.10	4308.	*	2	1600	79	.00	.00	.00	440.

****

TOTAL RAINFALL = 6.82, TOTAL LOSS = 3.14, TOTAL EXCESS = 3.68

PEAK FLOW	TIME		MAXIMUM AVERAGE FLOW			
(CFS)	(HR)		6-HR	24-HR	72-HR	50.00-HR
4911.	14.00	(CFS)	2187.	897.	558.	558.
		(INCHES)	3.632	5.958	7.725	7.725
		(AC-FT)	1085.	1779.	2307.	2307.

CUMULATIVE AREA = 5.60 SQ MI

Figure 8.12 Program output: hydrograph for 20-mi² transposition area at subbasin 1.

INTERPOLATED HYDROGRAPH AT   SUB1

DA	MON	HRMN	ORD	FLOW		DA	MON	HRMN	ORD	FLOW		DA	MON	HRMN	ORD	FLOW
1		0100	1	1.		1		1400	27	3003.		2		0300	53	496.
1		0130	2	1.		1		1430	28	4525.		2		0330	54	495.
1		0200	3	1.		1		1500	29	5190.		2		0400	55	493.
1		0230	4	1.		1		1530	30	4667.		2		0430	56	492.
1		0300	5	1.		1		1600	31	3295.		2		0500	57	491.
1		0330	6	1.		1		1630	32	1979.		2		0530	58	490.
1		0400	7	1.		1		1700	33	1185.		2		0600	59	488.
1		0430	8	1.		1		1730	34	698.		2		0630	60	487.
1		0500	9	1.		1		1800	35	519.		2		0700	61	486.
1		0530	10	1.		1		1830	36	517.		2		0730	62	485.
1		0600	11	1.		1		1900	37	516.		2		0800	63	484.
1		0630	12	1.		1		1930	38	515.		2		0830	64	482.
1		0700	13	1.		1		2000	39	513.		2		0900	65	481.

DA	MON	HRMN	ORD	FLOW
2		1600	79	465.
2		1630	80	463.
2		1700	81	462.
2		1730	82	461.
2		1800	83	460.
2		1830	84	459.
2		1900	85	458.
2		1930	86	457.
2		2000	87	455.
2		2030	88	454.
2		2100	89	453.
2		2130	90	452.
2		2200	91	451.

PEAK FLOW	TIME
(CFS)	(HR)
5190.	14.00

MAXIMUM AVERAGE FLOW

	6-HR	24-HR	72-HR	50.00-HR
(CFS)	2288.	942.	586.	586.
(INCHES)	3.799	6.257	8.111	8.111
(AC-FT)	1135.	1869.	2423.	2423.

CUMULATIVE AREA =   5.60 SQ MI

Figure 8.12 Program output: interpolated hydrograph at subbasin 1.

RUNOFF SUMMARY
FLOW IN CUBIC FEET PER SECOND
TIME IN HOURS, AREA IN SQUARE MILES

OPERATION	STATION	PEAK FLOW	TIME OF PEAK	AVERAGE FLOW FOR MAXIMUM PERIOD			BASIN AREA	MAXIMUM STAGE	TIME OF MAX STAGE
				6-HOUR	24-HOUR	72-HOUR			
HYDROGRAPH AT	SUB1	5190.	14.00	2288.	942.	586.	5.60		
ROUTED TO	RT1	4428.	16.00	2255.	936.	569.	5.60		
HYDROGRAPH AT	SUB2	11366.	16.50	8539.	3156.	1807.	27.00		
2 COMBINED AT	C1&2	15097.	16.00	10498.	3959.	2301.	32.60		
HYDROGRAPH AT	SUB3	7782.	14.00	3668.	1473.	908.	9.20		
ROUTED TO	RT4	7242.	15.00	3641.	1466.	895.	9.20		
HYDROGRAPH AT	SUB4	5291.	16.50	3966.	1465.	838.	12.00		
2 COMBINED AT	C3&4	11489.	15.50	7321.	2838.	1677.	21.20		
2 COMBINED AT	C2&4	24152.	16.00	16861.	6433.	3767.	53.80		
ROUTED TO	RT5	21516.	19.00	16066.	6383.	3627.	53.80		
HYDROGRAPH AT	SUB5	15651.	17.00	13484.	5196.	2922.	49.00		
2 COMBINED AT	TOT	33529.	18.50	26314.	10682.	6124.	102.80		

Figure 8.14 Program output: runoff summary.

Development of Discharge-Frequency Curves Using Design Storms

Design storm frequency vs. design flood frequency

In the development of flood-frequency curves, it is sometimes assumed that the frequency of the flood generated by the rainfall-runoff model is the same as that of the design storm used as input. In the absence of other data, this may be the only approach available. However, antecedent moisture conditions and other factors can significantly alter the amount of runoff and, assuming that design storm frequency and design flood frequency are the same, may produce erroneous results. There are several ways to define a design storm of a specified frequency, and the methods of constructing and applying design storms is controversial.[8]

Frequency curves developed from the statistical analysis of historical data can be used to verify and improve frequency curves developed from design storms. They provide a basis for adjusting rainfall loss function parameters used in design storm simulation and for assigning frequencies to design storm flows.

Adjusting rainfall loss parameters with a frequency curve. Rainfall loss parameters in a rainfall-runoff model may be adjusted in a cut-and-try procedure to achieve an approximate match between the peak discharge computed with a design storm of a given frequency, say a 25-year storm, and the corresponding point on the frequency curve. Using this procedure with several design storms of different frequencies may result in several different values for the parameters. However, the variation should be consistent. Low loss rates are generally associated with the most infrequent or rarest events and higher loss rates with the more frequent events.

An alternative to adjusting the loss parameters to achieve consistency between the design storm frequency curve and the statistical frequency curve is to adjust the assigned frequencies associated with the design storms.

Using frequency curves at gaged locations to determine assigned frequencies at ungaged locations. The assigned-frequency curve developed at a gaged location from routing and combining design storm flows from upstream subbasins can be compared with the statistically derived frequency curve for that location. The assigned frequencies can be adjusted so that they are consistent with the statistical frequencies, and proportional adjustments can be made to assigned frequencies at ungaged locations upstream. Since the average depth of rainfall over

235

Design Storms

235

an area tends to decrease with the size of the contributing area, this factor should be taken into account in making these adjustments. Upstream tributaries, of course, will have smaller contributing areas. If the depth-area option in HEC-1, discussed earlier in this chapter, is used for developing consistent hydrographs throughout the basin, this factor will automatically be accounted for.

Using frequency curves as a basis for developing modified frequency curves for changed basin conditions. Frequency curves developed at gaged locations can be modified to reflect changed basin conditions through the use of design storms with the multiflood and multiplan options in HEC-1. This procedure as related to urbanization is described in Chap. 9.

Probable Maximum Storm

Concept

A probable maximum storm (PMS) is used in a rainfall-runoff model to generate a probable maximum flood (PMF). A PMF is an estimate of the flood that can be expected from the severest combination of conditions that are "reasonably possible" in a region. It is used in the design of projects or project features such as spillways for which virtually complete security from flood-induced failure is desired.

The risk of failure of a structure can be estimated with the following equation based on the binomial distribution:

$$J = 1 - (1 - p)^n$$

where J = risk, or probability, of failure
p = exceedance probability of one trial
n = number of trials

The risk of failure of a structure designed for an exceedance probability of .01 (100-year return interval) during the next 100-year period is determined with this equation as follows:

$$J = 1 - (1 - .01)^{100}$$

$$= .63$$

Thus, there would be a 63 percent chance of failure during any 100-year period—a rather high risk.

Using the equation to design for a 1 percent risk of failure during the next 100-year period yields the following:

$$p = 1 - (J - 1)^{.01}$$

$$= .0001005$$

This probability of .0001005 corresponds to a return interval of 9950 years and underscores the fact that to achieve a low level of risk requires designing for a very rare and extreme event. Extrapolation of a frequency curve, generally based on 50 to 70 years of observed data, to determine the magnitude of an event with this return interval is impractical. There is substantial uncertainty in determining the 100-year event from 70 years of record; determining a 10,000-year event in this way would be very unreliable. Thus, the PMS is estimated to represent the physical upper limit of storm severity. In the simulation of the PMS with a rainfall-runoff computer program such as HEC-1, the unit hydrograph, loss rate, and base flow parameters representing the severest conditions are used.

Probable maximum precipitation

Probable maximum precipitation (PMP), which is the basis for deriving a PMS, is theoretically the greatest depth of precipitation physically possible at a location for a given set of conditions. The conditions include a given duration, area, and time of year. The basis for developing PMP is an analysis of maximum storms of record in a particular region.

In the eastern United States, where orographic influences are relatively minor, three processes are used in determining PMP: moisture maximization, transposition, and envelopment. "Moisture maximization" involves increasing storm precipitation amounts recorded for major events to reflect the maximum amount of moisture that could exist in the atmosphere at the time and location of the storm. This adjustment is based on the assumption that if there had been more moisture in the atmosphere when the storm occurred, there would have been proportionately greater precipitation. "Transposition" refers to the movement of a storm of record from the location where it occurred to another location within a homogeneous region to maximize potential precipitation at a location of interest. "Envelopment" involves the construction of smooth curves on precipitation maps to envelope maximum precipitation amounts for various durations and areas to compensate for data gaps. Geographic smoothing is performed to ensure regional consistency.

In the western United States, precipitation conditions are more complex, and a variety of regional criteria, accounting primarily for orographic effects, are applied. See the references cited in the next paragraph.

Hydrometeorological reports (HMRs) of the National Oceanic and Atmospheric Administration (NOAA) provide generalized PMP criteria for different basins, states, and regions of the United States. An example of these criteria, taken from HMR No. 51[9] for states east of the 105th meridian, is shown in Fig. 8.15. PMP rainfall depths in inches for a 6-h-duration, 10-mi^2 storm are presented. PMP depths applicable to all seasons for storms ranging in duration from 6 to 72 h and in size from 10 to 20,000 mi^2 are provided in HMR No. 51. Criteria for regions of the western United States are found in other HMRs such as Nos. 36, 38, 43, and 49. NOAA Technical Report NWS 25[10] provides comparative information on observed rainfall and PMP for selected conditions. This information provides a useful frame of reference for PMP use in developing a PMF.

Probable maximum storm determination with computer program HMR52

Program description.   National Weather Service criteria for developing a PMS for locations in the eastern United States from PMP estimates in HMR No. 51[9] are specified in HMR No. 52.[11] The criteria require the determination of four conditions that will produce maximum peak discharge or runoff volume at a given location: (1) the location of the storm center, (2) the size of the storm area, (3) the storm orientation, and (4) the temporal arrangement of precipitation amounts. These are determined in a trial-and-error procedure that has been incorporated in an HEC computer program named HMR52.[12]

The program HMR52 computes the PMS for a basin and a set of subbasin hyetographs. Input for the program includes the following: (1) x and y coordinates defining the boundaries of the basin and each of the subbasins, normally determined with a digitizer, (2) PMP depth-area-duration data from HMR No. 51, (3) preferred storm orientation from HMR No. 52, (4) x and y coordinates of the storm center, (5) storm area size, (6) storm orientation, (7) temporal arrangement of 6-h rainfall depths, and (8) time interval for the hyetographs.[13]

Maximization of the average depth of precipitation over a basin generally produces maximum peak discharge and volume of runoff. The program HMR52 has an option for optimizing storm area size and orientation to maximize average depth. The program centers the storm at the centroid of the basin area if no other location is specified. The major axis of the storm is oriented so that the moment of inertia of the basin area about this axis is a minimum.

Precipitation maximization first determines basin average precipitation depths from an array of standard storm area sizes, holding the

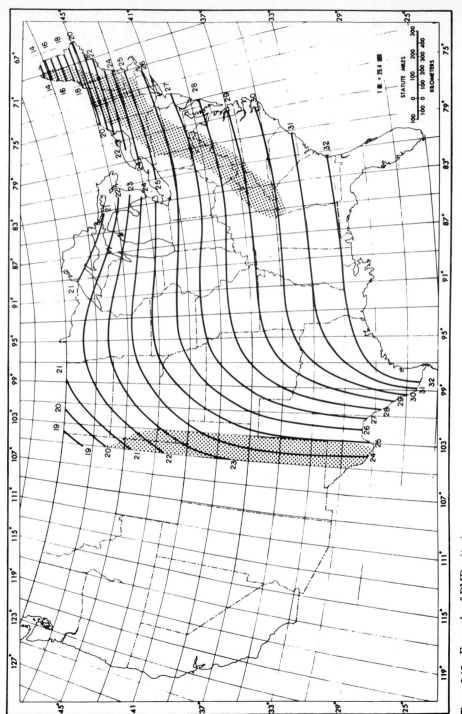

Figure 8.15   Example of PMP criteria.

orientation constant. The area size which produces the maximum depth is identified as the critical storm area size. Next, an array of orientations of this critical storm in 10° increments is analyzed to determine the maximum average precipitation depth. After the orientation that produces the maximum depth is determined, additional computations are made at orientations 5° less and 5° greater than this one. Of these three, the one that produces the maximum average depth is selected as the critical orientation.

The time interval Δt used for defining the PMS is specified by the user in the range of 5 min to 6 h. Incremental precipitation is computed for the total number of Δt intervals from the basin average depth vs. duration data. Precipitation is computed with uniform intensity within each 6-h period outside of the 24-h period of maximum precipitation.

The user may specify the location of the maximum 6-h period in any position after the initial 24 h of the storm, or the program will place it in the seventh position (hours 37 to 42) by default. The largest Δt increment of precipitation is located in the middle of the largest 6-h period, and the remaining Δt increments in that period are arranged alternately before and after the largest increment. The remaining 6-h periods, with decreasing precipitation magnitude, are arranged alternately before and after the largest 6-h period, except that the second, third, and fourth largest increments cannot be placed in the first 24 h of the storm. These are located after the largest increment if they would normally fall in the first 24-h period. A program-generated hyetograph for a PMS with a Δt of 1 h is shown in Fig. 8.16.

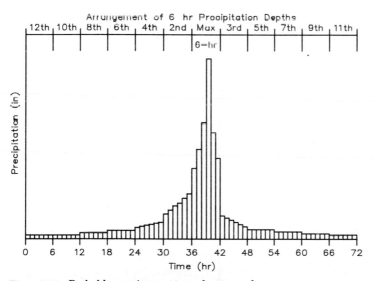

Figure 8.16 Probable maximum storm hyetograph.

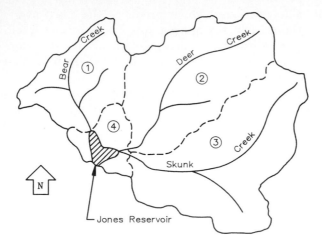

Figure 8.17 Watershed above Jones Reservoir.

Example problem: probable maximum storm for the watershed above Jones Reservoir.   An inflow hydrograph representing a PMF is required for the Jones Reservoir, shown in Fig. 8.17, and this hydrograph is to be routed through the reservoir. The program HMR52 is used to develop PMS hyetographs for the four subbasins above the reservoir, and HEC-1 is used to perform runoff and routing computations. Excerpts from the output of the two programs are included in this example. Complete input and output files are presented in the HMR52 user's manual.[12] Since no values were specified for storm center, storm area size, orientation, and temporal arrangement, default values in the program are used. The storm is centered at the centroid of the basin, and storm area size and orientation are obtained from optimization procedures which maximize the depth of precipitation. A 2-h time interval is used.

Output from HMR52 summarizing precipitation depths for various storm sizes is shown in Fig. 8.18, revealing that a storm area size of 300 mi^2 produces the greatest depth. Output summarizing depths obtained by varying the orientation of the 300 mi^2 storm incrementally is shown in Fig. 8.19. Since the best orientation occurred at 285°, the program changed this value by −5° and +5° to obtain storms of 280° and 290° for the final two computations. These are shown in the last two lines of Fig. 8.19, and, by coincidence, they happen to be repeats of storms already calculated.

With the PMS 6-h precipitation values defined for the basin, the program computes 2-h hyetographs for each of the four subbasins. For example, the output for subbasin 2 is shown in Fig. 8.20. The four

VARYING STORM AREA SIZE AND FIXED ORIENTATION

STORM AREA	ORIEN- TATION	BASIN-AVERAGED INCREMENTAL DEPTHS FOR 6-HR PERIODS												SUM OF DEPTHS FOR 3 PEAK 6-HR PERIODS
10.	285.	9.62	1.35	0.71	0.48	0.37	0.30	0.25	0.21	0.19	0.17	0.15	0.14	11.69
25.	285.	12.06	1.96	1.05	0.71	0.54	0.44	0.37	0.32	0.28	0.25	0.22	0.20	15.07
50.	285.	13.82	2.40	1.28	0.88	0.67	0.54	0.45	0.39	0.34	0.30	0.27	0.25	17.50
100.	285.	14.81	2.76	1.48	1.01	0.77	0.62	0.52	0.45	0.39	0.35	0.32	0.29	19.05
175.	285.	15.23	2.99	1.59	1.09	0.83	0.67	0.56	0.48	0.42	0.38	0.34	0.31	19.82
300.	285.	15.38	3.21	1.67	1.13	0.86	0.69	0.58	0.50	0.44	0.39	0.35	0.32	20.25
450.	285.	15.07	3.32	1.68	1.13	0.85	0.69	0.57	0.49	0.43	0.39	0.35	0.32	20.07
700.	285.	14.40	3.40	1.66	1.11	0.83	0.67	0.56	0.48	0.42	0.37	0.34	0.31	19.46
1000.	285.	13.76	3.42	1.63	1.08	0.81	0.65	0.54	0.47	0.41	0.36	0.33	0.30	18.82
1500.	285.	13.04	3.33	1.58	1.04	0.78	0.63	0.52	0.45	0.39	0.35	0.32	0.29	17.95
2150.	285.	12.23	3.22	1.51	0.99	0.75	0.60	0.50	0.43	0.38	0.33	0.30	0.27	16.95
3000.	285.	11.31	3.05	1.42	0.93	0.70	0.56	0.47	0.40	0.35	0.32	0.28	0.26	15.78
4500.	285.	10.84	3.05	1.41	0.92	0.70	0.56	0.47	0.40	0.35	0.31	0.28	0.26	15.30
6500.	285.	10.45	3.00	1.37	0.89	0.67	0.54	0.45	0.39	0.34	0.30	0.27	0.25	14.82
10000.	285.	9.76	2.97	1.31	0.85	0.63	0.51	0.42	0.36	0.32	0.28	0.26	0.23	14.05
15000.	285.	9.04	2.86	1.30	0.85	0.64	0.52	0.43	0.37	0.33	0.29	0.26	0.24	13.21
20000.	285.	8.34	2.78	1.30	0.85	0.64	0.52	0.44	0.38	0.33	0.29	0.26	0.24	12.41

Figure 8.18 HMR52 program output: precipitation depths for various storm sizes.

FIXED STORM AREA SIZE AND VARYING ORIENTATION

STORM AREA	ORIEN-TATION	BASIN-AVERAGED INCREMENTAL DEPTHS FOR 6-HR PERIODS										SUM OF DEPTHS FOR 3 PEAK 6-HR PERIODS
300.	140.	14.85	3.14	1.64	1.11	0.84	0.68	0.57	0.49	0.43	0.38	0.34	0.31	19.62
300.	150.	14.60	3.10	1.62	1.10	0.83	0.67	0.56	0.48	0.42	0.38	0.34	0.31	19.31
300.	160.	14.37	3.07	1.60	1.09	0.83	0.66	0.56	0.48	0.42	0.37	0.34	0.31	19.04
300.	170.	14.18	3.04	1.59	1.08	0.82	0.66	0.55	0.48	0.42	0.37	0.33	0.30	18.81
300.	180.	14.03	3.02	1.58	1.07	0.81	0.66	0.55	0.47	0.41	0.37	0.33	0.30	18.63
300.	190.	13.96	3.01	1.58	1.07	0.81	0.65	0.55	0.47	0.41	0.37	0.33	0.30	18.54
300.	200.	13.96	3.01	1.57	1.07	0.81	0.65	0.55	0.47	0.41	0.37	0.33	0.30	18.54
300.	210.	14.04	3.02	1.58	1.07	0.81	0.65	0.55	0.47	0.41	0.37	0.33	0.30	18.54
300.	220.	14.19	3.03	1.59	1.08	0.82	0.66	0.55	0.47	0.42	0.37	0.33	0.30	18.64
300.	230.	14.40	3.07	1.60	1.09	0.82	0.66	0.56	0.48	0.42	0.37	0.34	0.30	18.81
300.	240.	14.63	3.10	1.62	1.10	0.83	0.67	0.56	0.48	0.42	0.38	0.34	0.31	19.06
300.	250.	14.88	3.14	1.63	1.11	0.84	0.68	0.57	0.49	0.43	0.38	0.34	0.31	19.34
300.	260.	15.11	3.17	1.65	1.12	0.85	0.68	0.57	0.49	0.43	0.38	0.34	0.31	19.65
300.	270.	15.28	3.19	1.66	1.13	0.85	0.69	0.58	0.50	0.43	0.39	0.35	0.32	19.93
300.	280.	15.37	3.20	1.66	1.13	0.86	0.69	0.58	0.50	0.44	0.39	0.35	0.32	20.13
300.	290.	15.37	3.21	1.67	1.13	0.86	0.69	0.58	0.50	0.44	0.39	0.35	0.32	20.24
300.	300.	15.28	3.19	1.66	1.13	0.86	0.69	0.58	0.50	0.44	0.39	0.35	0.32	20.25
300.	310.	15.10	3.17	1.65	1.12	0.85	0.68	0.57	0.49	0.43	0.39	0.35	0.32	20.14
300.	280.	15.37	3.20	1.66	1.13	0.86	0.69	0.58	0.50	0.44	0.39	0.35	0.32	19.92
300.	290.	15.37	3.21	1.67	1.13	0.86	0.69	0.58	0.50	0.44	0.39	0.35	0.32	20.24

Figure 8.19 HMR52 program output: precipitation depths for various storm orientations.

Probable Maximum Storm for Subbasin 2

DAY 1

TIME	INCR	TOTAL	TIME	INCR	TOTAL	TIME	INCR	TOTAL	TIME	INCR	TOTAL
0200	0.11	0.11	0800	0.13	0.44	1400	0.16	0.87	2000	0.23	1.42
0400	0.11	0.21	1000	0.13	0.57	1600	0.16	1.03	2200	0.23	1.65
0600	0.11	0.32	1200	0.13	0.70	1800	0.16	1.19	2400	0.23	1.88
6-HR TOTAL	0.32		6-HR TOTAL	0.39			0.49			0.68	

DAY 2

TIME	INCR	TOTAL	TIME	INCR	TOTAL	TIME	INCR	TOTAL	TIME	INCR	TOTAL
0200	0.33	2.21	0800	0.81	3.81	1400	4.13	10.36	2000	0.66	22.64
0400	0.37	2.58	1000	1.04	4.85	1600	8.94	19.29	2200	0.54	23.18
0600	0.42	3.00	1200	1.38	6.23	1800	2.69	21.98	2400	0.47	23.64
6-HR TOTAL	1.12		6-HR TOTAL	3.23			15.75			1.66	

DAY 3

TIME	INCR	TOTAL	TIME	INCR	TOTAL	TIME	INCR	TOTAL	TIME	INCR	TOTAL
0200	0.28	23.92	0800	0.19	24.68	1400	0.14	25.21	2000	0.12	25.61
0400	0.28	24.21	1000	0.19	24.87	1600	0.14	25.35	2200	0.12	25.73
0600	0.28	24.49	1200	0.19	25.06	1800	0.14	25.50	2400	0.12	25.84
6-HR TOTAL	0.85		6-HR TOTAL	0.57			0.43			0.35	

Figure 8.20 HMR52 program output: PMS precipitation for subbasin 2.

subbasin hyetographs, along with runoff and routing parameters, are used as input for HEC-1 so that the program can compute discharge hydrographs at locations of interest.

Because of the hydrologic characteristics of a basin, the storm that produces the greatest depth of precipitation may not result in the largest flood. When HMR52 and HEC-1 are used to estimate the PMF for basins with highly unusual shapes or runoff characteristics that are quite spatially heterogeneous, a number of trial computations may be warranted. These trials should consist of a sensitivity analysis of several factors that affect the size of the PMF. The analysis should include the position of the peak 6-h interval, storm area size, storm orientation, and storm center.

Five of the trials used in the example computation for the basin above the Jones Reservoir are as follows:

Trial 1.   Storm center, area size, orientation, and temporal distribution are selected by the program.

Trial 2.   Same as trial 1, except that the peak 6-h interval is shifted to the tenth position (hours 54 to 60). This change increased the peak flow slightly, so it is used in subsequent trials.

Trial 3.   Same as trial 2, except that the isohyetal pattern is manually centered on the watershed.

Trial 4.   Same as trial 3, except that a storm area size of 175 mi^2 is specified.

Trial 5.   The storm center is positioned on the basis of only subbasins 1, 2, and 3, which produce most of the runoff.

The storm pattern for trials 1 and 2 is shown in Fig 8.21. The summary of PMF calculations for the five trials is shown in Table 8.4. Obviously, there is very little difference in the results. Recent studies indicate that default values in HMR52 will generally suffice in developing the PMF.[6]

Workshop Problem 8.1: Developing Design Storms and Using the HEC-1 Depth-Area Option

Purpose

The purpose of this workshop problem is to provide an understanding of the procedures for developing a design storm for the Rahway River basin using precipitation data and criteria from National Weather

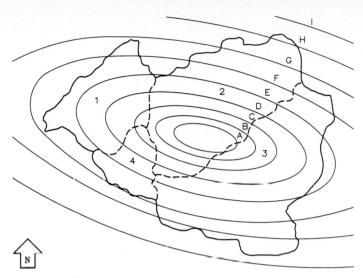

Figure 8.21 Storm pattern for trials 1 and 2 of Jones Reservoir calculations.

Service publications. Experience will be gained in applying the depth-area option of HEC-1.

Problem statement

A consistent set of discharge hydrographs from a 4 percent chance, 24-h design storm is to be developed for various locations in the Rahway River Basin. A time step of 1 h is to be used. Point precipitation amounts in inches from NOAA publications TP-40 and HYDRO-35 are as follows:

15 min	30 min	1 h	2 h	3 h	6 h	12 h	24 h
1.30	1.84	2.42	3.15	3.40	4.30	5.10	5.90

Tasks

1. Use the depth-area curves shown in Fig 8.2 to develop a plot of accumulated precipitation depth vs. duration. Make a plot on log-log graph paper similar to Fig. 8.5.
2. Manually compute the rainfall depths for the design storm, using

TABLE 8.4 Summary of PMF Calculations

Trial	Position of peak 6-h interval, mi²	Storm area, mi²	Orientation degrees, mi	Storm center x, mi	Storm center y, in	Total rainfall, ft³/s	Peak inflow, ft³/s	Peak outflow
1	7	300	285	32.2	83.8	25.50	127,500	94,650
2	10	300	285	32.2	83.8	25.50	128,000	95,300
3	10	300	285	31.0	83.6	25.44	127,650	95,250
4	10	175	290	31.0	83.6	24.86	125,200	91,650
5	10	300	296	32.7	84.0	25.41	127,200	93,900

TABLE P8.1   Storm Calculations

Time, h	Accu- mulated point rain- fall, in	Point rain- fall factor	Areal ad- justed accu- mulated rain- fall, in	Partial- series adjust- ment factor	Partial to an- nual adjust- ment	Ad- justed rain- fall, in	Ar- ranged rain- fall, in
1							
2							
3							
.							
.							
.							
24							

headings in Table P8.1 as a pattern for setting up a tabulation of computations.

3. Modify the HEC-1 input file developed in previous workshop problems so that it will compute runoff for the 4 percent chance design storm using the precipitation data given above. Use a time step of 15 min, and compute 150 hydrograph ordinates. Specify an initial loss of 1.0 in and a constant loss rate of 0.25 in/h for all subbasins. Base flow and unit hydrograph parameters used in the analysis of the August 1971 storm are assumed to be applicable to this analysis.

4. Make two HEC-1 runs as follows:
   a. For the first run, use a PH line of input to develop a 4 percent chance storm for a storm area size equal to the drainage area at Springfield, which is 25.49 mi^2. Apply the storm uniformly to all subbasins.
   b. For the second run, use the depth-area option with JD and PH lines of input to develop a consistent set of discharge hydrographs for all subbasins.

5. On the basis of the results of the two HEC-1 runs, answer the following questions:
   a. Compare subbasin peak discharges for the two runs. Why are peak discharges for one run consistently higher than peak discharges for the other? For which subbasin is there the greatest percent difference?
   b. What percent difference in peak discharge is there at Springfield between the two runs? Why?
   c. How do precipitation amounts for run 1 compare with the results of your hand computations?

d. Should the peak discharges obtained with run 2 be regarded as having an exceedance frequency of 4 percent? Why?

References

1. Department of the Army, *Standard Project Flood Determination,* EM 1110-2-1411 (formerly EB 52-8), Office of the Chief of Engineers, Washington, D.C., March 1952 (rev. March 1965).
2. National Weather Service, *Rainfall Frequency Atlas of the United States, 30-Minute to 24-Hour Durations, 1- to 100-Year Return Periods,* Technical Paper No. 40, U.S. Department of Commerce, 1961.
3. National Weather Service, *Two- to Ten-Day Precipitation for Return Periods of 2 to 100 Years in the Contiguous United States,* Technical Paper 49, U.S. Department of Commerce, 1964.
4. National Weather Service, *Five- to 60-Minute Precipitation Frequency for the Eastern and Central United States,* NOAA Technical Memorandum NWS HYDRO-35, Silver Spring, Md., 1977.
5. *Precipitation Frequency Atlas of the Western United States, Atlas 2,* vol. 1–13, National Oceanic and Atmospheric Administration, Washington, D.C., 1973.
6. Hydrologic Engineering Center, *HEC-1 Flood Hydrograph Package, Users Manual,* U.S. Army Corps of Engineers, Davis, Calif. 1981 (rev. 1987).
7. Hydrologic Engineering Center, *Hydrologic Analysis of Ungaged Watersheds Using HEC-1,* Training Document 15, U.S. Army Corps of Engineers, Davis, Calif. 1982.
8. Philip B. Bedient and Wayne C. Huber, *Hydrology and Floodplain Analysis,* Addison-Wesley, New York, 1988.
9. *Probable Maximum Precipitation Estimates, United States East of the 105th Meridian,* HMR No. 51, National Oceanic and Atmospheric Administration, Washington, D.C., 1978.
10. *Comparison of Generalized Estimates of Probable Maximum Precipitation with Greatest Observed Rainfall,* Technical Report NWS 25, National Oceanic and Atmospheric Administration, Washington, D.C., 1987.
11. *Application of Probable Maximum Precipitation Estimates, United States East of the 105th Meridian,* HMR No. 52, National Oceanic and Atmospheric Administration, Washington, D.C., 1982.
12. Hydrologic Engineering Center, *HMR52 Probable Maximum Storm (Eastern United States), Users Manual,* U.S. Army Corps of Engineers, Davis, Calif., 1984.
13. Paul B. Ely and John C. Peters, "Probable Maximum Flood Estimation—Eastern United States," *Water Resources Bulletin,* AWRA, vol. 20, no. 3, 1984.

Analysis of Urbanizing Basins

Effects of Urbanization on Runoff

Introduction

Generally, the effects of urbanization on runoff from a basin include higher volume, higher peak discharge, and shorter time of concentration. These changes are associated with the increased imperviousness and more efficient drainage that are characteristic of constructed drainage systems. Areas of the natural land surface are covered with buildings and pavement, resulting in a greater proportion of impervious area and shorter and smoother flow paths for runoff. The effects on the discharge hydrograph are depicted in Fig. 9.1.

Accounting for the effects of urbanization in rainfall-runoff analysis

General. The effects of urbanization are generally characterized by reduced precipitation loss rates due to increased imperviousness and changes in runoff response reflected in the shape of the unit hydrograph. A unit hydrograph (UH) represents an empirical relationship accounting for a multitude of complex physical processes that occur in the transformation of rainfall to runoff. Since some UH parameters are not directly associated with physical characteristics of the basin, it is difficult to adjust these parameters to reflect the effects of urbanization. The most common approach is to develop and apply regional regression relationships between the parameters and measurable basin characteristics. As basin characteristics change, they can be measured, and UH parameters can be adjusted through the regression relationships.

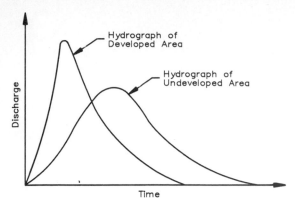

Figure 9.1 Effects of urbanization on a discharge hydrograph.

Loss rates. A basin area can be broken down into different types of surfaces representing different degrees of imperviousness. Imperviousness ranges from 12 percent in low-density residential areas to 85 percent in commercial areas according to SCS data in Table 2.4. The percentages to be used in computing runoff should be estimated from land-use surveys and aerial photographs taken at locations of interest. In some cases, actual areas associated with different degrees of imperviousness have been measured on a map.

Impervious areas do not always drain directly into the conveyance system. Some areas drain into adjacent pervious areas and do not contribute significantly to runoff. This condition should be considered in determining the effects of imperviousness and in estimating percent-imperviousness factors used in runoff computations. The SCS rainfall loss method described in Chap. 3 provides for computing composite curve numbers that account for impervious areas not directly connected to the drainage system.

Disturbance of a soil profile by urbanization may significantly change the soil's infiltration characteristics. Native soil profiles may be mixed or removed and fill material from other areas introduced. To compensate for this, the SCS curve number method includes a disturbed-soils classification based on soil texture.[1]

Loss rates often are higher for pervious areas after urban development has taken place, but the higher loss rates are counterbalanced in some locations by high soil moisture levels due to lawn and garden watering. Because of this, it may be reasonable to apply the same loss rates to pervious areas for the analysis of conditions both before and after development.

Unit hydrograph parameters. Regression equations have been developed for specific basins to relate UH parameters to basin characteristics, including degree of imperviousness in some cases. A study of several basins in the vicinity of Philadelphia, Pennsylvania, developed a relationship between Clark UH parameters T_c and R and basin characteristics:[2]

$$T_c + R = 19.46 I^{-0.40}(DA/S)^{0.24}$$

$$T_c = 12.98 I^{-0.42}(DA/S)^{0.27}$$

where I = percentage of impervious area
DA = drainage basin area
S = slope

The equations can be adjusted for urbanization by changing the variable I. A study of urban watersheds in Texas and Oklahoma found a relationship between the Clark UH parameters and basin area:[3]

$$\frac{R}{T_c + R} = 0.4$$

$$T_c + R = 0.75 DA$$

In this study, no substantial correlation was found between the Clark parameters and imperviousness. A study of basins in Colorado related Snyder UH parameters to percentage of impervious area:[4]

$$C_t = \frac{7.81}{I^{0.78}}$$

$$C_p = 0.89(C_t)^{0.46}$$

Equations obtained from multiple regression analysis of data for 13 watersheds in southeastern Michigan[5] are shown in Table 9.1. Population density and drainage area were found to be the most significant variables affecting UH peak discharge and period of rise. A plot of observed peaks vs. peaks computed with equation 3 is shown in Fig. 9.2.

A recent study of a watershed in Louisville, Kentucky, examined the use of a trend relationship of Clark UH parameters in predicting future flood conditions in an urbanizing area. Flood peak data from 1940 to 1973 was used to develop regression relationships such as the following:

TABLE 9.1   Results of Multiple Regression Analysis of 13 Watersheds in Southeastern Michigan

Equation	Standard error of estimate	Correlation coefficient R	Coefficient of determination R^2
1. $Q_P = 24.6A^{0.64}P_D^{0.24}$	.163	.944	.891
2. $Q_P = 77.7A^{0.77}$	.209	.905	.819
3. $Q_P = $ antilog $[1.85 + 0.694$ log $A + 0.000099P_D]$	.065	.991	.983
4. $T_R = 6.13A^{0.35} \times P_D^{-0.077}$	.197	.650	.423
5. $T_R = 4.25A^{0.31}$	.193	.667	.445
6. $T_R = $ antilog $[0.66 + 0.359$ log $A - 0.000067P_D]$	.144	.831	.690

Q_P = unit hydrograph peak discharge
T_R = period of rise of unit hydrograph
A = drainage area, mi^2
P_D = population density, persons/mi^2

$$T_c = 220.20748 - 0.11070968 * \text{YEAR}$$

$$\text{SUM} = 666.23229 - 0.33469876 * \text{YEAR}$$

where T_c = time of concentration, h
    SUM = $(T_c + K)$, h
    K = storage attenuation constant
    YEAR= actual projection year starting from 1940

A comparison of observed flood peaks from 1973 to 1983 with flood peaks simulated through the use of projected Clark parameters showed a close correspondence.[6]

Caution should be exercised in applying regression equations developed for one area to locations outside of that area. Reports describing the development of the regression relationships should be reviewed. Conditions under which the equations were derived should be carefully examined and compared with conditions at the location under investigation to ensure that they are similar.

Techniques for developing regression relationships are discussed in Chap. 6. Some additional watershed characteristics particularly associated with urban environments that might be included in the analysis are as follows:

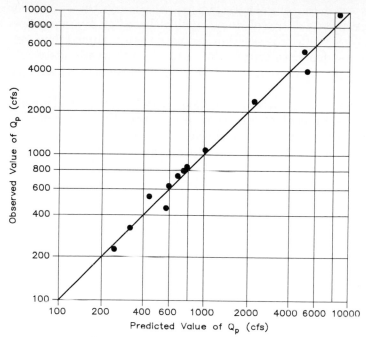

Figure 9.2 Plot of observed peaks vs. peaks computed from regression equation.

L_t Total length of drainage channel in basin, including all storm sewers 36 in or larger and all drainage channels large enough to appear on topographic maps (with L_t/A representing "drainage density")

P Percentage of drainage area that has sewers

n Index of roughness of conveyance system

Effects of urbanization on routing

The building of bridges, the development of buildings in the flood-plain, and the construction of channel improvements and levees are some of the changes accompanying urbanization that are likely to have a significant effect on flood routing. The hydraulic characteristics of flood waves moving through the reach and storage-outflow relationships may be affected by such changes.

Muskingum method.   Generally, channels with greater hydraulic efficiency are used in urban development, resulting in decreased travel time. In the Muskingum routing method, the parameter K, related to travel time, sometimes can be redefined for modified conditions on the basis of new estimates of velocity and travel time in the reach. For example, an estimate of travel time might be based on average velocity computed from new cross-section geometry and assumed flow conditions of normal depth. Of course, the accuracy of this estimate would depend to a large extent on how well these assumptions fit actual conditions.

Channelization frequently associated with urban development tends to reduce flood peak attenuation, as does the elimination of storage in the floodway caused by building construction. Increasing the value of the parameter X in the Muskingum method produces less attenuation of the flood peak, but there are no direct relationships with channel characteristics with which to determine the magnitude of the adjustment.

Modified Puls method. The storage-outflow relationship in the modified Puls method is a key factor in making adjustments for urbanization when this method is used. When the storage-outflow curve is redefined on the basis of modified flow conditions, the urbanized system can be modeled. The computation of gradually varied flow profiles with a program such as HEC-2 is frequently done for this purpose. Bridges, encroachments in the floodplain, and channel cross section modifications all can be modeled. Clearing and lining the channel can be accounted for by changing Manning's n values. Using HEC-2 to compute storage-outflow data is discussed in Chap. 15.

Kinematic-Wave Rainfall-Runoff Modeling

Introduction

Concept. The kinematic-wave method is an alternative to the unit hydrograph approach to rainfall-runoff modeling. The parameters of this model are developed from physical characteristics of the basin, and equations of motion are used to simulate the movement of water through the system. Parameters such as catchment length and area, roughness, slope, and channel geometry are used to define the flow of water conceptually over basin surfaces, into stream channels, and through the channel network. This method is particularly useful in urban studies because the effects of urbanization can be accounted for by changing the measurable physical parameters of slope, catchment length, surface roughness, and so forth.

Surface features of the basin are represented with two basic types of elements: overland flow and channel flow. One or two overland-flow elements are combined with one or two channel-flow elements to represent a subbasin. An entire basin is modeled by linking the various subbasins together in a network.

A typical urban drainage pattern is shown in Fig. 9.3. Overland flow occurs on two general types of surfaces: pervious and impervious. Pervious surfaces include lawns, gardens, and open areas; impervious surfaces include roofs, driveways, and streets. Water from rainfall initially travels as sheet flow for distances typically ranging from 100 ft to several hundred feet over pervious surfaces and from 30 to 100 ft over impervious surfaces.[7] The water collects in street gutters and usually travels a few hundred feet at most before entering catch basins connected to local storm sewers. These local sewers, typically 1.5 to 2 ft in diameter and sometimes connected in turn to larger and larger collector drains, carry the water to the main storm drain. It is assumed that the water flows through this collector system and out the main storm drain as open channel flow. Although the collector drains are normally pipes, and the main drain may be a large closed conduit rather than an open channel, they are usually designed to be only partially full.

Limitations. The kinematic-wave method is generally more applicable to the analysis of urban hydrology because of its inherent limitations. As basin area increases, the assumptions required for application become more tenuous. Since kinematic-wave theory does not provide for attenuation of flood waves, there is less potential for overestimating peak flows in a small urban basin with well-defined, relatively steep smooth channels and short travel times for the flood waves.

The method does not have explicit provision for dealing with the surcharging of storm drains and local storage of water that frequently occurs during major storm events. The ponding and overbank storage associated with large storms may not occur as much in smaller storms available for calibrating the model. Thus, if a model has been calibrated for a relatively small storm, program output from an analysis of a larger storm should be examined to determine if significant storage exists.

Kinematic-wave equations

The kinematic-wave equations are the continuity equation for unsteady open channel flow with lateral inflow [Eq. (9.1)] and a simplified Manning's equation [Eq. (9.2)].

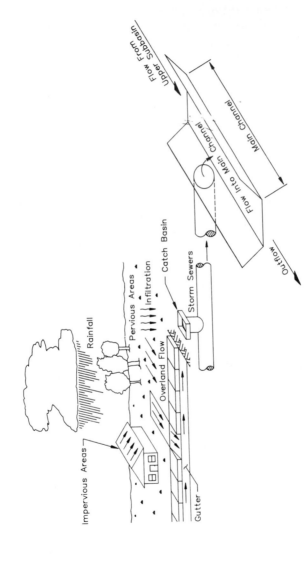

Figure 9.3 Typical urban drainage pattern.

$$\frac{\partial A}{\partial t} + \frac{\partial Q}{\partial x} = q_o \qquad (9.1)$$

where Q = channel flow, ft^3/s
A = cross-sectional area of flow, ft^2
q_o = lateral inflow, ft^3/(s)(ft)
x = distance along the flow path
t = time, s

Manning's equation is written in the form

$$Q = \alpha A^m \qquad (9.2)$$

The coefficient is found with the general equation:

$$\alpha = \frac{KS^{1/2}}{n} \qquad (9.3)$$

where K = constant that depends on geometry
S = slope
n = Manning's roughness coefficient

These same basic equations are used in computations of flow both for overland-flow elements and for channel-flow elements. A detailed discussion of the derivation of these equations and the finite difference methods of solution used in HEC-1 are presented in reference.[7]

Elements used in kinematic-wave computations in HEC-1

Three basic elements, shown in Fig. 9.4, are used to model the runoff process with HEC-1: (1) one or two typical overland-flow planes, (2) a typical collector channel, and (3) a typical main channel. The collector channel (element 2) may also be modeled with an additional subcollector channel. Each element represents average conditions in the basin. The relationships between flow elements are depicted in Fig. 9.5.

Since land-use patterns and development practices are usually similar within a hydrologic unit, the assignment of a single parameter, such as a typical length for an overland-flow plane, has been found to give good results.[7] The estimation of these typical or average values does require a good knowledge of basin conditions and sound engineering judgment. Basin maps, aerial photographs, and on-site visits are

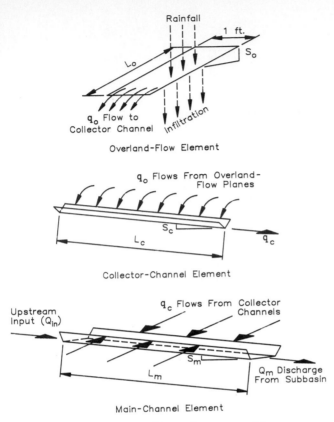

Figure 9.4   Basic elements in kinematic-wave model.

used to determine typical parameters. Experience gained in estimating kinematic parameters is likely to be helpful in this process.

Overland-flow planes.   An overland-flow element is a rectangular plane of unit width. Some of the rain falling on the plane is lost to infiltration; the remainder flows over the surface and runs off the lower edge into a gutter or collector channel. Infiltration losses may be held constant or caused to vary with time. The fraction of the element's surface area that is impervious is specified.

Either one or two overland-flow elements can be used, but generally two are used: one to represent impervious areas (such as roofs, driveways, and street surfaces) and the other to represent pervious areas (such as lawns, open fields, and wooded areas). The impervious element generally has a smoother surface and shorter flow length than the pervious element. The following input data for each overland-flow element is required by HEC-1:

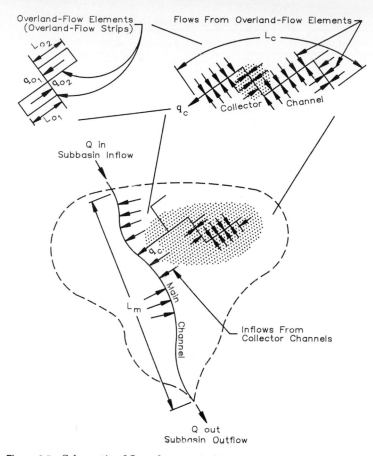

Figure 9.5 Schematic of flow elements in kinematic-wave model.

Typical overland-flow length L_o

Representative slope S_o

Roughness coefficient N

Percentage of subbasin area A_o this element represents

Infiltration and loss-rate parameters

The flow length L_o has the greatest influence on response of the overland-flow element. It can be considered the maximum length of the path taken by a representative water drop in traveling to the collector channel where it first becomes streamflow. An examination of several natural basins and urban catchments has indicated that small-scale drainage patterns are generally quite similar throughout an entire basin.[7] Although the value of L_o may not vary greatly over

TABLE 9.2   Catchment Roughness Factors for Overland Flow

Surface	Roughness coefficient N
Dense growth*	0.40–0.50
Pasture*	0.30–0.40
Lawns*	0.20–0.30
Bluegrass sod†	0.20–0.50
Short-grass prairie†	0.10–0.20
Sparse vegetation†	0.05–0.13
Bare clay—loam soil (eroded)†	0.01–0.03
Concrete/asphalt—very shallow depths* (depths less than ¼ in)	0.10–0.15
Concrete/asphalt—small depths* (depths on the order of ¼ in to several inches)	0.05–0.10

*From Crawford and Linsley.[8]
†From Woolhiser.[9]

the basin, the value that gives the correct runoff response should be verified by comparing model output with observed data.

The slope S_o is a representative slope of the path the water takes to the collector channel. Its value is greatly influenced by land use. In an urban setting, a single value for slope can be used for all areas of similar building practice, even when the mean ground slopes are found to vary considerably.

Roughness coefficients for sheet flow over rough surfaces at very shallow depths are much different from those for channel flow. Recommended values of N are presented in Table 9.2.

The area parameter for an overland-flow plane is specified as a percentage of the total area of the subbasin. If a single overland element is used, 100 percent is specified. Of course, if two elements are used, the sum of their percentages is 100.

Collector channel.   The collector channel models the flow from the uppermost point of channel flow—including the flow in street gutters—through the collector system to the main channel. Inflow is uniformly distributed along the entire length of the channel. This represents flow running off from overland-flow planes and approximates flows from individual catch basins and tributary pipes distributed along the channel. The following input data is required for the collector channel:

Surface area A_c drained by a single representative collector channel (e.g., gutter and storm drain)

Length L_c (gutter plus storm drain)

Channel slope S_c

Manning's roughness coefficient n

Channel shape (circular section or some variant of a trapezoid)

Pipe diameter or trapezoid bottom width and side slopes

The collector-channel parameters can be determined by examining a drainage map of the basin and selecting a typical collector system for each subbasin. Recommended guidelines for selecting the parameters are as follows:

1. The area associated with the collector system can be determined from the map. This is an area in square miles (square kilometers) rather than the percentage of the subbasin area. It does not have to be an integer multiple of the subbasin area.

2. The collector-channel length is taken as the longest flow path from the upstream end of the collector system to its outlet at the main channel. This length should include the distance the water will travel as gutter flow.

3. The channel shape and size will usually change along the length of the channel; however, a single shape must be chosen to represent the channel along its entire length. This is not as great a problem as might appear at first. As shown in Fig. 9.6, a triangle with a side slope of 1:1 matches reasonably well the area-discharge relationships for circular conduits for a given slope and roughness. The triangular shape is the one used by the computer if the shape is not specified in the input data. The selection of a channel shape is discussed in more detail in the section on the main channel, below.

4. If the representation is by a circular or trapezoidal shape, the channel dimensions chosen should represent the most commonly used size of channel in the system.

5. The channel slope can be estimated from a topographic map by taking the difference in elevation between the upstream and downstream ends and dividing by the length. If drop structures are used in the storm drains, the slopes should be adjusted accordingly.

6. A Manning's n which best represents the roughness of the major part of the channel should be used. Tables of n for various types of channels, such as concrete pipe and lined or unlined open channels, are available in hydraulic handbooks and other sources.[7]

Main channel. The main channel carries inflows from upstream subbasins as well as flows from collector channels within the subbasin. Inflow from the collector channels is considered uniformly distributed along the length of the main channel, approximating actual flows entering at a number of discrete points.

The upstream subbasin inflow is normally combined with the local subbasin inflows in the kinematic-wave computations. However, as an

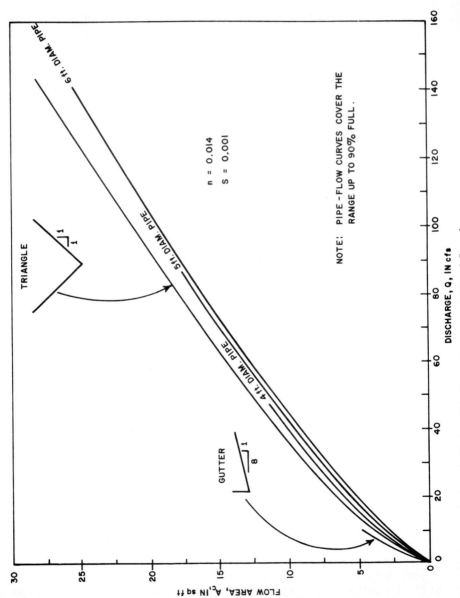

Figure 9.6 Area-discharge relationships for various cross-sectional flow shapes.

262

alternative, these can be computed separately and combined at the subbasin outlet. Any of the routing methods available in HEC-1 can be used for routing the upstream inflow to the subbasin outlet. The following input data is required for the main-channel element:

Channel or stream length L_m

Slope S_m

Manning's roughness coefficient n

Area A of subbasin

Channel shape (trapezoidal or circular)

Channel dimensions (diameter or width and side slopes)

The upstream hydrograph to be routed through the reach, if applicable

The main-channel parameters can be estimated from physically measurable characteristics of the channel and subbasin, with the exception of Manning's n. The following procedure can be used:

1. The channel length can be scaled from a drainage map of the basin.
2. The mean channel slope can be obtained from field measurements or estimated through the use of topographic maps.
3. Manning's n should be selected on the basis of the average channel conditions.
4. The subbasin area can be measured from topographic maps.
5. The selection of a channel cross section is discussed in the following section.
6. The channel dimensions follow from the preceding item.
7. An upstream hydrograph will not be routed through the channel reach unless the user specifically requests that the program do routing.[7]

Selection of a channel cross section.   Simulation of discharge with the kinematic-wave method is not very sensitive to the shape of the channel cross section, so only three shapes are used in HEC-1: trapezoidal, square, and circular. A triangular shape can be simulated by specifying a base width of zero for the trapezoid. As indicated in Fig. 9.6, a triangular shape provides a reasonably good approximation of small circular cross sections as well as street gutters.

In downstream reaches of the basin, the trapezoidal shape generally provides the best fit. For some natural cross sections, a trapezoid pro-

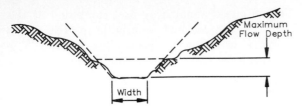

Representation by a Trapezoid

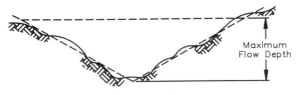

Representation by a Triangle

Figure 9.7 Example of varying channel shape with depth to improve approximation of natural conditions.

vides the best fit for low flows, while the triangle gives a better fit for high flows (Fig. 9.7).

The circular shape can be used for modeling storm sewers. The pipe flow is simulated for depths up to 90 percent of pipe capacity. For flows greater than 90 percent capacity, the program assumes that the capacity of the pipe increases as required without limit. In many cases this adequately represents the flow. When the flow exceeds the storm sewer's capacity, the water that does not enter the sewer at one catch basin flows over the surface until it finds another catch basin to enter.

Example problem: kinematic-wave application to Waller Creek

Introduction. An application of the kinematic-wave method to the Waller Creek basin in Austin, Texas,[10] is summarized here. This urban basin with an area of 4.13 mi^2 is primarily residential and commercial but also includes some large institutional areas, such as the University of Texas. The basin has a long, narrow shape with the main channel extending through its entire length (Fig. 9.8). One major tributary drains the western part of the basin. Ground slopes and the grade of the main channel are quite steep. Collector sewers typically are of constant diameter over the entire distance to the main channel. The availability of good precipitation and runoff data makes the basin well suited for an example application.

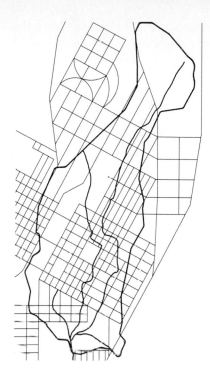

Figure 9.8   Waller Creek basin.

Determination of basin parameters. The basin was divided into two subbasins, one for the main channel and one for the major tributary. These subbasins, designated SUB1 and SUB2, have areas of 2.73 and 1.40 mi^2 respectively.

Overland-flow planes. A topographic map was used to determine the slope of pervious and impervious overland-flow planes, and an average slope of 0.02 was adopted for both. Overall basin imperviousness was determined to be 30 percent. Subbasin SUB2 was considered to have a higher percentage of imperviousness than SUB1 because of differences in land use. SUB2 is twice as large and contains a higher percentage of residential land than SUB1. Values for percent imperviousness of 25 and 40 and flow lengths of 75 ft and 150 ft were adopted for overland-flow planes in SUB1 and SUB2, respectively. Roughness values of 0.10 and 0.20 were used, and it was not necessary to adjust these during calibration.

Collector channel. A previous drainage study for the City of Austin[11] provided information for selecting parameters for the collector channel and the main channel. The typical collector channel was determined to be a pipe with a circular cross section. The pipe length, diameter, and contributing drainage area were determined from the study data. The pipe slope was assumed to be 0.02, the same as the ground slope, and the roughness coefficient n for the pipe was estimated to be 0.014.

Main channel. The slope and length chosen to represent the main channel were determined from a channel profile in the study report.[11] No information was available on the cross-sectional geometry of the main channel, so it was assumed to have a trapezoidal shape with a 6-ft bottom width and 2:1 side slopes. A Manning's n of 0.030 was assumed.

Loss rates. The SCS rainfall loss method was used in this study, and on the basis of land use, soil type, and percent imperviousness, the SCS procedure gave a curve number of 84. However, this number generated runoff volumes that were consistently too high, so individual curve numbers were estimated for each storm event (Table 9.3).

TABLE 9.3   SCS Curve Numbers (CNs) for Waller Creek

Storm no.	Impervious CN	Pervious CN	5-Day precipitation, in	10-Day precipitation, in
2	92	70	0.69	0.77
4	92	75	0.11	0.85
6	92	70	0.17	2.16
7	92	65	1.97	2.53
8	92	80	3.14	5.41
11	92	75	0.03	0.03
12	92	80	1.87	2.36

Data input file. An excerpt from the data input file for the Waller Creek kinematic-wave model is shown in Fig. 9.9. The data input for an HEC-1 model using unit hydrographs is exactly the same as for a kinematic-wave model except for the lines defining precipitation loss and UH parameters. These are replaced with lines specifying kinematic-wave parameters.

Since there are two overland-flow planes for each subbasin, there are two sets of rainfall loss parameters, one for each plane on separate LS lines. The first three entries on an LS line are the loss-function parameters for the overland-flow plane defined on the first UK line which follows. Entries 4 through 6 are the loss parameters for the plane defined on the succeeding UK line. The data on each UK line includes overland-flow length, representative slope, roughness coefficient, and percentage of subbasin area for the plane represented. The percentages specified on the two UK lines for a subbasin must sum to 100.

```
ID.....1.......2.......3.......4.......5.......6.......7.......8.......9......10

KK   SUB1
BA   2.73
PR   GAGE1   GAGE2   GAGE3
PW      0       1
PT   GAGE1   GAGE2   GAGE3
PW     .81     .19
LS      0      80       0       0      92       0
UK    150     .02     .20      75
UK     75     .02     .10      25
RK   1600     .02    .014     .07    CIRC       2
RK  20000   .0032     .03       0    TRAP       6       2

KK   SUB2
KO      4       2
BA   1.14
PR   GAGE1   GAGE2   GAGE3
PW      0       1
PT   GAGE1   GAGE2   GAGE3
PW     .13     .66     .21
LS      0      80       0       0      92       0
UK    150     .02     .20      60
UK     75     .02     .10      40
RK   2600    .020    .014     .04    CIRC     2.5
RK  13200    .011     .03       0    TRAP       0       2
```

Figure 9.9   Excerpt of K-group data from input file for kinematic-wave model.

The collector-channel parameters and the main-channel parameters for each subbasin are specified on two RK lines of input. The first is for the collector channel, and the next is for the main channel. The parameters specified in fields 1 to 7 include the length, slope, roughness, contributing area, shape, bottom width or diameter, and side slopes if applicable. The contributing area for the main channel is the total subbasin area indicated on the BA line. The program uses this total area irrespective of what is entered in field 4 of the RK line— zero is entered in this example. Also, the RK line for the main channel would have an entry of *YES* in field 8 if an upstream hydrograph were to be routed.

Simulation results.   Seven of the largest storm events for which observed data was available were selected for testing the model. Computed and observed values for peak discharge and total runoff volume are compared in Table 9.4. Hydrographs for one of the storms, no. 12, are shown in Fig. 9.10. Only the SCS loss-rate parameters were varied between storms; the kinematic-wave parameters were not changed.

TABLE 9.4   Waller Creek Simulation Results

Storm no.	Date	Peak discharge, ft³/s			Sum of flows		
		Ob-served	Simu-lated	% Dif-ference	Ob-served	Simu-lated	% Dif-ference
2	4–26–58	1700	1958	15.2	25,545	17,536	− 31.4
4	9–22–59	1910	2057	7.7	19,311	23,320	20.8
6	9–27–64	2280	2462	8.0	51,179	54,261	6.0
7	5–16–65	2320	2562	10.5	33,127	30,296	− 8.5
8	5–18–65	1980	1671	− 15.6	33,157	17,603	− 46.9
11	6–21–71	1560	1643	5.3	16,443	23,017	40.0
12	5–01–72	2160	2002	− 7.3	29,614	29,762	0.5

Although there is variation from storm to storm in the accuracy of the results, the overall response of the watershed is considered adequately reflected by the model.

Selection of computational increments Δx and Δt

The accuracy of the finite-difference solution to the kinematic-wave equations depends to a significant extent on the difference increments Δx and Δt used for each overland-flow or channel element. A wide range of modeling results is possible depending on the size of these increments.[12] Versions of HEC-1 prior to June 1988 computed Δx by

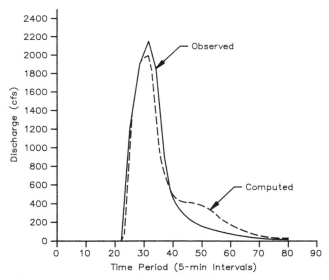

Figure 9.10   Comparison of hydrographs for storm 12.

approximately satisfying the stability condition:

$$\Delta x = c \, \Delta t$$

where c is wave celerity in feet per second.

Although HEC-1 automatically reduced Δt if Δx was greater than half the length of an element, it was necessary to check the accuracy of a kinematic-wave solution by comparing computed discharges for several progressively shorter computational time intervals. The effect of reducing the time interval on the accuracy of the solution is demonstrated in Fig. 9.11. Approximate solutions with three different sets of Δx and Δt increments are compared with an exact solution computed by the method of characteristics.[13] It can be observed that as Δx and Δt become smaller, the approximate solution approaches the true solution. Also, the difference between the computed hydrographs for $\Delta t = 1$ and $\Delta t = 3$ is smaller than that between hydrographs for $\Delta t = 3$ and $\Delta t = 6$. Thus, the amount of change between successive approximations can be used as a limit for testing solution accuracy. A change within a tolerable limit indicates a suitable Δt.

Limitations in versions of HEC-1 released prior to June 1988 on the minimum computational time interval (1 min) and the maximum number of ordinates (600) may preclude a kinematic-wave solution under some conditions. It may not be possible to reduce the time interval enough to obtain satisfactory accuracy within these limitations. Conditions under which this may occur include (1) subbasins with a very short time of concentration and (2) long-duration rainfall events. For a very small, highly responsive basin, a 1-min interval may be too

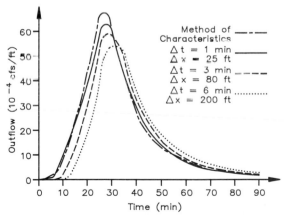

Figure 9.11 Comparative effect of different computational time and distance intervals on kinematic-wave solution.[13]

long to define the hydrograph. For a long-duration storm, such as a probable maximum storm, the number of ordinates is likely to exceed 600.

In the version of the program released in June 1988, Δx and Δt are automatically selected by the program on the basis of depth of flow in the catchment. A variable time step is used through each element to preserve computational accuracy. The hydrograph computed with the variable time step is interpolated to the computational time interval specified by the user in program input, and the amount of interpolation error is defined in output. If the interpolation error is considered to be too large, the user can reduce the computational time interval specified in input.

Development of Frequency Curves in Urbanizing Areas

Effects of urbanization on frequency curves

Generally, urbanization increases the mean and reduces the standard deviation of a frequency curve for a drainage basin. Studies have found that the effect decreases as the magnitude or recurrence interval of the flood increases.[14,15,16] The increased imperviousness and other factors associated with urbanization tend to have less impact on large floods. However, there are exceptions to this tendency. Sometimes constructed obstructions, such as bridges and severe encroachments, can restrict flow and cause a reduction in peak discharge. The pattern of development in a basin also can reduce the peak discharge at the outlet by altering the timing of flows from different areas within the basin. For example, development in one area of the basin may change the timing of the peak flow from that area so that it no longer coincides with peaks from other areas, thus flattening the hydrograph.

Methods of developing frequency curves in urbanizing areas

Analytic methods for developing frequency curves in urbanizing areas may be classified under two general categories: the statistical analysis of flow data and the analysis of design storms. Statistical methods include the analyses of (1) observed historical flows with theoretical frequency distributions such as the Log Pearson Type III, (2) reconstructed historical streamflows (adjusted for current conditions), (3) period-of-record rainfall-runoff simulation, (4) relationships developed from synthetic precipitation, and (5) empirical relationships developed from regional flood data. Design storm methods include (1) the conversion of rainfall intensities of a given frequency into peak runoff of the same frequency by means of empirical methods, such as the rational formula, and (2) the simulation of flood hydrographs from design storms. A large number of published and unpublished papers on urban flood flow frequency procedures have been reviewed and evaluated by the Agricultural Research Service.[17]

Statistical analysis of historical streamflows.   Time-series data such as historical streamflows are considered stationary if the properties of the data do not change with time. This condition is essential in statistical analysis to ensure that the data is from the same population and that inferences drawn from the data are not biased by changed conditions. In urbanizing areas, periods of stationary data are likely to be relatively short due to the influence of rapidly changing land use. Nevertheless, the analysis of such limited data provides a valid estimate of the frequency curve that can be compared with curves developed by other methods in the process of adopting a representative curve.

Reconstructed historical streamflows.   Nonstationary peak discharges can be converted into a stationary series representative of existing conditions with the following approach:

1. Develop and calibrate a rainfall-runoff model (e.g., HEC-1) for existing basin conditions and for basin conditions at several other points in time during the period of record.[18]

2. Develop a balanced design storm for the basin using generalized rainfall criteria, such as criteria contained in Weather Service Technical Paper 40. To fix the magnitude of the storm, a 25-year recurrence interval can be used. The recurrence interval is arbitrary, as it is not assumed in this approach that runoff frequency is equal to rainfall frequency. The purpose of adopting a specific magnitude is to establish a base storm to which ratios can be applied for subsequent steps in the analysis.

3. Apply a series of ratios (say six) to the design storm developed in the previous step such that the resulting calculated peak discharges at the gage will cover the range desired for frequency analysis. Input the balanced storms into the rainfall-runoff model for each of the basin conditions selected in step 1, and determine peak discharges at the gaged location.

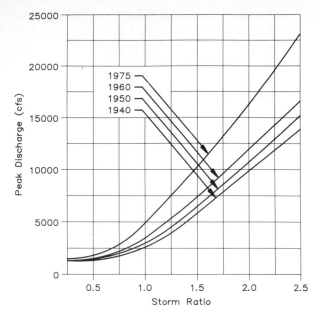

Figure 9.12 Peak discharge vs. storm ratio for four sets of basin conditions corresponding to the years 1940, 1950, 1960, and 1975.

4. From the results of step 3, plot curves representing peak discharge vs. storm ratio for each basin condition. This is illustrated in Fig. 9.12.

5. Use the curves developed in step 4 to adjust the observed annual peak discharges, as shown in Table 9.5. For example, the observed annual peak discharge of 6150 ft^3/s that occurred in 1950 is adjusted by entering the "1950" curve with that discharge and moving vertically to the "1975" curve to obtain the adjusted peak of 9200 ft^3/s. The adjusted peak thus obtained is assumed to be the peak discharge that would have occurred in 1950 if 1975 basin conditions had existed at that time.

The series of annual peak discharges determined with this procedure can be analyzed with statistical methods, such as the Log Pearson Type III distribution recommended in Bulletin 17B,[19] under the assumption that the adjusted data is stationary for the entire period of record from 1947 to 1976.

TABLE 9.5   Observed Annual Peak Discharges Adjusted to Existing Conditions

Year	Observed peak Q, ft^3/s	Peak Q adjusted to existing conditions, ft^3/s
1947	9000	13200
1948	8200	12800
1949	3880	6200
1950	6150	9200
1951	3400	5700
1952	3340	5590
1953	1460	2700
1954	3900	6400
1955	2290	3600
1956	5160	8000
1957	3870	6000
1958	2000	3700
1959	2610	4680
1960	2940	3990
1961	2160	2920
1962	3850	5320
1963	2250	3000
1964	1360	1790
1965	4420	6130
1966	2790	3800
1967	6420	8640
1968	7080	9500
1969	4600	5940
1970	2810	2810
1971	3470	3470
1972	2700	2700
1973	4390	4390
1974	3810	3810
1975	5560	5560
1976	5920	5920

Period-of-record rainfall-runoff simulation and synthetic precipitation analysis with a continuous simulation model. A continuous simulation model, such as the NWSRFS model,[20] can be used to develop flood-frequency relationships by simulating rainfall and runoff for the entire period of record, using historical or synthetic precipitation data. The continuous model is calibrated through the use of concurrent precipitation and streamflow data for a period representing existing conditions of urbanization. The calibrated model is then used to compute runoff, with period-of-record rainfall data used as input. Conventional frequency analysis is applied to the resulting annual peaks to obtain an "existing conditions" frequency curve.

Regression relationships developed from regional flood data. Regression relationships developed from a statistical analysis of historical flood data for specific areas range in form from very rough estimates based on a small amount of data to very detailed analyses based on extensive data. An example of a "rough estimate" approach to urban flood frequency is the set of flood-frequency equations developed for urban areas of Oklahoma.[21] Equations are provided for estimating peak discharges in basins 0.5 to 100 mi^2 in area for floods with return periods ranging from 2 to 100 years. Equation (9.4) is the general form of such equations.

$$Q_x(u) = \frac{7R_x Q_2 (R_L - 1)}{6} + \frac{Q_x(7 - R_L)}{6} \qquad (9.4)$$

where $Q_x(u)$ = urban peak discharge for recurrence interval x
$\quad\quad R_L$ = adjustment factor to account for the effect of urban development
$\quad\quad Q_x$ = natural peak discharge for recurrence interval x
$\quad\quad Q_2$ = mean annual flood discharge for natural conditions
$\quad\quad R_x$ = rainfall-intensity ratio for recurrence interval x

The relationships developed for estimating the magnitude and frequency of floods in New Jersey[22] exemplify a more detailed analytic approach. A total of 103 gaging stations in New Jersey and 2800 station years of data were analyzed. The analysis was limited to basins less than 1000 mi^2 in area which were not significantly affected by regulation or diversion. Equation (9.5), for determining the peak flow for a flood with a 100-year return interval, is an example of the relationships developed. Similar relationships were developed for other frequencies.

$$Q_{100} = 136A^{0.84}S^{0.26}St^{-0.51}I^{0.14} \qquad (9.5)$$

where Q_{100} = peak flow for flood with 100-year recurrence interval
$\quad\quad A$ = basin area
$\quad\quad S$ = main-channel slope
$\quad\quad St$ = surface storage index
$\quad\quad I$ = impervious-cover index: $I = 0.117D^{0.792 - 0.039 \log D}$
$\quad\quad D$ = population density, persons per square mile

The results of a nationwide study of flood frequency in metropolitan areas performed by the U.S. Geological Survey[23] provides equations for estimating the magnitude and frequency of floods at ungaged urban basins. Seven independent variables were used in the regression equations, but the less significant variables were dropped to form three-parameter equations exemplified by Eq. (9.6) for the flood with a 100-year return interval. Dropping the less significant variables increases the standard error of regression but also greatly reduces the amount of data and effort required for the application of the equations.

$$UQ100 = 7.70A^{0.15}(13 - BDF)^{-0.32}(RQ100^{0.82}) \qquad (9.6)$$

where $UQ100$ = urban peak discharge, in ft^3/s for flood with 100-year recurrence interval

$\quad\quad\;\; A$ = contributing drainage area

$\quad\;\; BDF$ = basin development factor

$\;\; RQ100$ = peak discharge, ft^3/s, for an equivalent rural drainage basin in the same hydrologic area as the urban basin

General equations such as Eq. (9.6) should be used with care. It is important to ensure that their application to a specific location is justified by the similarity of conditions there with the conditions under which the equations were developed.

Multiplan-Multiflood Analysis

Multiflood ratios

The multiplan and multiflood options in HEC-1 are useful in analyzing the effects of urbanization and other changes in a drainage basin on runoff. With the multiflood option, floods of several different sizes can be analyzed in one run. Each flood is computed as a ratio of a base flood event. This capability is particularly useful in comparing the effects of events of different sizes in floodplain studies. The ratios may be applied to either precipitation or runoff amounts. If they are applied to the ordinates of a base-event hyetograph, a complete set of runoff and routing computations is performed corresponding to each ratio. If the analysis does not involve changes in the rainfall-runoff response characteristics of the basin, the ratios are applied directly to the base-event discharge hydrograph.

HEC-1 input requirements for the multiflood option. One line, the JR line, is required in an input file to execute the multiflood option. It is located in the file immediately after the job initialization data. Up to nine ratios can be specified on this line, and these are applied to the precipitation or to the flows as indicated. The program applies the ratios to the base event and computes a hydrograph for the base event plus each of the events determined with the ratios.

Application of multiflood ratios to compute a modified frequency curve. A modified frequency curve reflecting urban development can be obtained by applying ratios to design storms. These steps are followed:

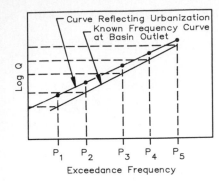

Figure 9.13 Modified frequency curve reflecting basin development.

1. Develop a peak discharge–frequency curve for natural watershed conditions from known gaged data or regional relationships (Fig. 9.13).

2. Develop design storms of various frequencies using the procedures outlined in Chap 8. In lieu of developing several design storms, use the multiflood option of HEC-1 to generate additional storms by applying ratios to one storm.

3. Develop a rainfall-runoff model (e.g., HEC-1) for natural watershed conditions. Calibrate the model with observed data wherever possible.

4. Use the model to compute the peak discharges for the storms developed in Step 2. Determine exceedance probabilities to associate with the design storms from the known frequency curve for natural conditions. Enter Fig. 9.13 with computed discharges, and obtain corresponding frequencies—$P_1, P_2, \ldots, P_5$.

5. Modify the parameters of the rainfall-runoff model to reflect the current state of urbanization; then compute new peak discharges for the design storms.

6. Plot the results of the simulation, assuming that the frequency of each balanced storm is the same for both natural and urbanized conditions in the watershed. In other words, obtain a new frequency curve for urbanized conditions by plotting new values for discharge at the storm frequencies $P_1, P_2, \ldots, P_5$.

If the data for natural watershed conditions is insufficient for performing the design storm analysis outlined above, sometimes it may be possible to use data from a previous period during which the degree of urbanization remained fairly stable. The procedure would be essentially the same. If insufficient data is available for either of these approaches, it may be necessary to assume that the runoff has the same frequency as the design storm that produced it.

Multiplan analysis

The multiplan option in HEC-1 adjusts model computations to reflect changes such as channel improvements, the addition of reservoirs, and new land uses, all of which affect the basin's rainfall-runoff response characteristics. The hydrologic impact of several alternative flood control plans or other basin modifications can be computed in a single run.

Definition of a plan. A plan represents a proposed river basin configuration that may include development schemes, water control measures, damage reduction features, and other changes. In the analysis of alternative plans, it is customary to consider the basin in its current condition as plan 1. Other alternatives, designated plan 2, plan 3, etc., are compared with plan 1. Alternative plans may include several different basin changes separately or in combination. For example, one plan may consist of the addition of an area of future urban development, another plan may add a reservoir, and a third may include both the urban development and the reservoir.

It is important to recognize that a plan in the context of multiplan analysis defines a single configuration of the entire drainage basin being modeled. In other words, each plan is a basinwide plan, and the model computes runoff at all locations of interest in the basin, reflecting the effect of changes imposed by the plan.

Analytical efficiency of multiplan analysis. The multiplan option in HEC-1 provides an efficient method of performing several simulations, each representing an alternative plan, in a single execution of the program. Simulations in the model which are not changed by a plan are not recomputed for the next plan in the sequence; instead, the results are passed directly to the next computation. Since separate executions of the program for each plan are eliminated, computer input and output functions are reduced.

The simulation results for alternative plans can be readily compared in the detailed output and the summary table available with this option. This output provides an efficient way of performing sensitivity analysis on model parameters. For example, the effects on runoff of incremental changes in a unit hydrograph parameter can be easily compared.

Multiplan applications. The multiplan capability facilitates the analysis of a variety of problems. Its primary applications include the analysis of project impacts on basin hydrology, the modification of frequency curves to reflect urbanization, the computation of flood loss mitigation benefits, and the sizing of flood control system components.

HEC-1 input requirements for multiplan analysis. A line of data with a JP identifier is included in the input file to implement multiplan simulation and indicate the number of plans to be analyzed. The JP line follows the job initialization data (IT line, IO line, etc.). A KP line is included with each K group of data that differs from the base conditions represented by plan 1, and it assigns a plan identification number to this model component. Several KP lines may be included within a K group to separate different plans at one computational point.

Examples of K groups for three different types of two-plan analyses are as follows:

1. Existing and future land use:
   KK
   KP 1
   (BA, BF, P__, L__, and U__ lines for existing conditions)
   KP 2
   (BF, L__, and U__ lines to describe changed conditions)

2. Proposed reservoir (an additional routing reach must be included to simulate the reservoir):
   KK
   KP 1
   R (indicates no routing for this plan because the reservoir does not exist in the base conditions)
   KP 2
   RS
   SV
   SQ

3. Proposed channel improvement or levee:
   KK
   KP 1
   (RS, SV, and SQ lines for existing channel if modified Puls method is being used)
   KP 2
   (RS, SV, and SQ lines for improved channel if modified Puls method is being used)

Multiflood-multiplan example. The multiflood option is frequently used in conjunction with the multiplan option to analyze simultaneously the impacts of different plans on flood events that differ in severity. The results of the computations provide hydrographs for each ratio of the base event for each plan.

A two-plan, three-flood analysis is depicted in Fig. 9.14. In this illustration, plan 1 represents natural conditions. Plan 2 represents the addition of urban development in subbasin A and a new reservoir in subbasin B. The more rapid runoff response associated with urban conditions is reflected in the higher, more rapidly rising peaks of the plan 2 hydrographs at gage A. The attenuation of these peaks by the proposed reservoir is reflected in the flatter, slower-rising peaks of the plan 2 hydrographs at gages B and C.

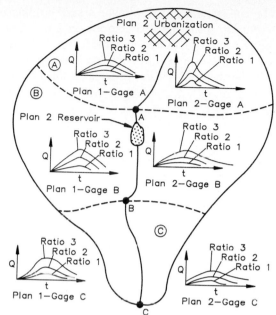

Plan 2 Urbanization

Ratio 3
Ratio 2
Ratio 1

Q

t

Plan 1—Gage A

Ⓐ

Ⓑ

Ratio 3
Ratio 2
Ratio 1

Q

t

Plan 2—Gage A

Plan 2 Reservoir

A

Ratio 3
Ratio 2
Ratio 1

Q

t

Plan 1—Gage B

Ratio 3
Ratio 2
Ratio 1

Q

t

Plan 2—Gage B

B

Ⓒ

Ratio 3
Ratio 2
Ratio 1

Q

t

Plan 1—Gage C

Ratio 3
Ratio 2
Ratio 1

Q

t

Plan 2—Gage C

C

Plan 1 — Natural Conditions in Subbasins A, B, and C
Plan 2 — Urbanization In A and Reservoir in B

Figure 9.14 Multiflood-multiplan example: three flood ratios, two plans.

Workshop Problem 9.1: Multiplan-Multiflood Analysis of the Rahway River Basin

Purpose

The purpose of this workshop problem is to provide an understanding of the multiplan and multiflood capabilities of HEC-1 and how to use them.

Problem statement

A land development company has recently won approval to develop much of subbasins 1 and 2 on the West Branch of the Rahway River. (Refer to Fig. 6.1.) The development was authorized only on the condition that a dam would be constructed to control the increased runoff to the extent that there would be no substantial increase in the 100-year flood at Millburn.

The development planned for the West Branch is anticipated to be about the same as that in the currently urbanized East Branch. The development is expected to be completed by the year 2000. The index of imperviousness for these basins is estimated to be 40 percent.

The proposed dam is to be constructed at the lower end of subbasin 2, and it is assumed that virtually all runoff from subbasin 2 will flow into the new reservoir. The storage of the new reservoir has been estimated by measuring water surface areas at various elevations on USGS topographic maps. The storage-elevation data is as follows:

Surface area, acres	0	4.59	73.44	143.24	249.76
Elevation, ft	200	214	220	240	260

The spillway for the dam will be a simple weir overflow with the following characteristics:

Spillway crest elevation: 220 ft

Effective length: 50 ft

Discharge coefficient: 3.3

The top of the dam will be level and have the following weir flow characteristics:

Top-of-dam elevation: 235 ft

Length: 400 ft

Discharge coefficient: 2.65

Before it receives final approval for the project from the state planning authority, the land development company must perform an engineering analysis to provide the information in Table P9.1

TABLE P9.1   Flood Peaks, ft³/s

	Millburn gage		Springfield gage	
	10-Year	100-Year	10-Year	100-Year
Existing conditions				
Year 2000 conditions without new dam				
Year 2000 conditions with new dam				

Tasks

Use the previously developed HEC-1 Rahway River basin model and flow-frequency curves to perform this analysis. The multiplan option in HEC-1 should be used to obtain the information required by the

```
EC
KKP.RESV
CN     1   DUMMY
FR         10   99.99    50    30    20    10    4    2    1
FR    .5   .2
QF              350   600   710   840   1060   1500   2000   2500
QF   3300  4450
QD          2   10  100000
DG        101
KKSPRFLD
CN     1   DUMMY
FR         10   99.99    50    30    20    10    4    2    1
FR    .5   .2
QF              400   1100   1450   1800   2420   3500   4800   6300
QF   8000  11000
QD          2   10  100000
DG        101
ZZ
```

Figure P9.1 ECON data set.

state planning authority. Three plans are to be analyzed: (1) existing conditions, (2) urbanized conditions in the year 2000 without the dam, and (3) urbanized conditions in the year 2000 with the dam.

1. Begin with the HEC-1 multibasin model as calibrated for the 25-year flood.

2. Set the output control to IPRINT = 5 and IPLOT = 1 on the IO line in the input file. Remove all other output control lines of input.

3. Use the multiflood option (JR line) to analyze the following eight ratios of the 25-year storm: 0.5, 0.75, 1.0, 1.1, 1.3, 1.4, 1.5, and 1.9.

4. Use the multiplan option (JP and KP lines) to analyze the three plans. Add the proposed urbanization data to subbasins 1 and 2, and add the proposed reservoir at Millburn.

5. The program analyzes modified frequency curves as part of flood damage analysis. The input frequency curve is used to determine the frequency of the peak discharge for each flood ratio. Add the data set labeled ECON in Fig. P9.1 to the HEC-1 input file just before the ZZ line. The economic data has been zeroed out. Add the adopted frequency curves on the FR and QF lines for the reservoir outflows at Millburn and at the Springfield gage. *Note:* The station name on the KK line in the ECON data must agree with the station name on the KK line in the model for the same location.

6. Execute HEC-1 using your input file, and complete Table P9.1 with the results.

References

1. Soil Conservation Service, *Urban Hydrology for Small Watersheds,* Technical Release 55, U.S. Department of Agriculture, June 1986.
2. D. L. Gundlach, *Direct Runoff Hydrograph Parameters versus Urbanization,* Technical Paper 48, U.S. Army Corps of Engineers, Hydrologic Engineering Center, Davis, Calif., 1976.
3. L. R. Beard and S. Chang, "Urbanization Impact on Streamflow," *Journal of the Hydraulics Division,* ASCE, vol. 105, no. HY6, June 1979.
4. Wright-McLaughlin Engineers, *Urban Storm Drainage Criteria Manual,* Denver Regional Council of Governments, Denver, Colo., 1969.
5. S. F. Daly and J. Peters, "Determining Peak Discharge Frequencies in an Urbanizing Watershed—A Case Study," *Proceedings of the International Symposium on Urban Runoff,* University of Kentucky, Lexington, July 1979.
6. Nageshwar Rao Bhaskar, "Projection of Urbanization Effects on Runoff Using Clark Instantaneous Unit Hydrograph Parameters," *Water Resources Bulletin,* AWRA, vol. 24, no. 1, February 1988, pp. 113–123.
7. Hydrologic Engineering Center, *Introduction and Application of Kinematic Wave Routing Techniques Using HEC-1,* Training Document 10, U.S. Army Corps of Engineers, Davis, Calif., May 1979.
8. N. H. Crawford and R. K. Linsley, *Digital Simulation in Hydrology—Stanford Watershed Model IV,* Technical Report 39, Stanford University, Department of Civil Engineering, Stanford, Calif., 1966.
9. D. A. Woolhiser, "Simulation of Unsteady Overland Flow," in K. Mahmood and V. Yevjevich (eds.), *Unsteady Flow in Open Channels,* Water Resources Publications, Fort Collins, Colo., 1975.
10. R. J. Cermak, *Application of HEC-1 Kinematic Wave,* U.S. Army Corps of Engineers, Hydrologic Engineering Center, Davis, Calif., 1981.
11. Espey, Houston and Associates, *Comprehensive Drainage Study, Waller Creek Drainage Basin,* City of Austin, Tex. 1976.
12. T. V. Hromadka II and J. J. DeVries, "Kinematic Wave Routing and Computational Error," *Journal of Hydraulic Engineering,* ASCE, vol. 114, no. 2, February 1988, pp. 207–217.
13. Gary W. Brunner, "Cascade vs. Single-Plane Overland Flow Modeling," master's thesis, Pennsylvania State University, University Park, May 1985.
14. G. E. Hollis, "The Effect of Urbanization on Floods of Different Recurrence Interval," *Water Resources Research,* vol. 11, no.3, 1975.
15. L. D. James, "Using a Digital Computer to Estimate the Effects of Urban Development on Flood Peaks," *Water Resources Research,* vol. 1, no. 2, 1965.
16. T. R. Hammer, *Procedures for Estimating the Hydrologic Impact of Urbanization,* report from the Regional Science Research Institute to the Office of Water Resources Research, U.S. Department of the Interior, 1971.
17. Science and Education Administration, *Review and Evaluation of Urban Flood Flow Frequency Procedures,* Bibliographies and Literature of Agriculture No. 9, U.S. Department of Agriculture, August 1980.
18. D. L. Gundlach, "Adjustment of Peak Discharge Rates for Urbanization," *Journal of the Irrigation and Drainage Division,* ASCE, vol. 104, no. IR3, September 1978.
19. U.S. Geological Survey, Interagency Advisory Committee on Water Data, *Guidelines for Determining Flood Flow Frequency,* Bulletin 17B, Office of Water Data Coordination, Reston, Va., 1982.
20. R. J. C. Burnash, R. L. Ferral, and R. A. McGuire, *A Generalized Streamflow Simulation System,* National Weather Service–State of California Joint River Forecast Center, March 1973.
21. V. B. Sauer, *An Approach to Estimating Flood Frequency for Urban Areas in Oklahoma,* Water Resources Investigations 23–74, U.S. Geological Survey, July 1974.
22. S. J. Stankowski, *Magnitude and Frequency of Floods in New Jersey with Effects of Urbanization,* Special Report 38, U.S. Geological Survey, 1974.
23. *Flood Characteristics of Urban Watersheds in the United States,* U.S. Geological Survey, Water Supply Paper 2207, 1983.

10

Water Surface Profile Analysis

Introduction

Purposes

Water surface profiles associated with peak discharges in a stream are needed for a variety of reasons. They are used to delineate areas of inundation and depths of flow associated with potential floods of various sizes. They are an important factor in determining the suitability of development proposals for parcels of land located in the floodplain. They reflect the effect of existing bridges on water levels and may affect the design of new bridges. And they are used for determining levee heights and flood insurance zones and for various other purposes.

Water surface profile analyses may be required both at the beginning and at the end of a floodplain H&H study. If the modified Puls method is being used for channel routing of flood flows, water surface profile computations may be performed at the beginning of the study with a range of flows to obtain the storage-outflow data needed for routing. Then, after the peak discharges are determined at locations of interest with the basin hydrology simulation model or frequency analysis, final water surface profiles may be computed on the basis of these peaks.

Computation of profiles with HEC-2

The HEC-2 computer simulation program uses a numerical method named the "standard step method" to compute changes in water surface elevation between adjacent cross sections on the basis of energy losses. The computations begin at one end of a study reach and proceed cross section by cross section to the other end of the reach. At bridge crossings where the flow hydraulics is more complicated, momentum and other equations may be used to compute the water surface elevation changes. The program also has an optional capability to compute profiles for reaches with encroachments, channel improvements, split flows, and ice-covered flows.

Theoretical Basis and Limiting Assumptions in HEC-2

General

The methodology incorporated in an HEC-2 model is based on several simplifying assumptions, but the model produces satisfactory results in many applications. The assumptions are as follows:

1. Steady flow

2. Gradually varied flow

3. One-dimensional flow with correction for horizontal velocity distribution

4. Small channel slope

5. Friction slope (averaged) constant between two adjacent cross sections

6. Rigid boundary conditions

As with other computer simulation programs, it would be a great mistake to treat HEC-2 as a black box and accept its output as fact merely because the numbers are printed out to two decimal places. Water surface elevations computed to the nearest hundredth of a foot, for example, may be 5 ft off. To use the program effectively, it is essential to understand the basis of its hydraulic computations and its limitations.

The hydraulics of open channel flow pertaining to the computation of water surface profiles is reviewed in this chapter to delineate the realm of applicability of the program. Classifications of open channel flow are discussed. Velocity distribution in a natural cross section and assumptions regarding pressure distribution are examined. Computation of energy losses between adjacent cross sections and use of the energy equation in calculating flow profiles are explained. An example problem is solved to illustrate how these concepts are applied.

Classifications of open channel flow

Steady flow vs. unsteady flow.   Time is the criterion for making the distinction between steady flow and unsteady flow. If depth, velocity, and

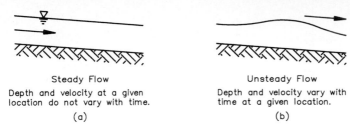

Figure 10.1 Steady flow and unsteady flow.

discharge remain constant with time at a particular location on a stream, the flow is steady (Fig. 10.1a); if any of these characteristics vary, it is unsteady (Fig. 10.1b). The passage of a flood wave through a reach of river is an example of unsteady flow because the depth, velocity, and discharge all are changing with time.

In a typical floodplain H&H study, the peak discharges used for calculating water surface profiles are obtained from hydrographs produced by a basin rainfall-runoff model or from frequency analysis. The peak discharge from a hydrograph is often used, even though it represents an unsteady-flow condition.

The reason that HEC-2, a steady-flow model, can be used for the unsteady flow represented by a flood hydrograph is that flood waves rise and fall gradually. An observer standing on the bank of a stream watching a flood wave pass would not see a distinct curvature of the wave; he or she would merely see evidence of the rise and fall of the water surface. Except in extreme cases, the change in flow occurs gradually, and adequate answers can generally be obtained with steady-flow analytic methods.

There are at least three cases in which a steady-flow model may not provide adequate results. One is when a wave is moving very rapidly, as from a dam breach, and the time-dependent term of the complete unsteady-flow [Eq. (4.1)] has a significant effect. Another is when backwater effects from downstream boundary conditions, such as tidal flows or water backing up a tributary, are significant. A third is when there is a pronounced "loop effect" in the relationship between discharge and elevation resulting from a flat channel slope. The loop is due primarily to the difference between water surface slopes associated with the rising and falling sides of the flood wave as it passes a point. The flatter the channel slope, the more pronounced this effect is. Since a steady-flow model cannot account for a looped relationship, it may not be suitable in extreme cases.

Uniform flow vs. varied flow.   When flow is uniform, the depth and velocity are constant along a length of channel (Fig. 10.2a). The force of gravity moving the water is exactly balanced by friction forces, and

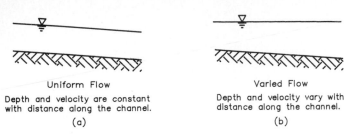

Uniform Flow

Depth and velocity are constant
with distance along the channel.

(a)

Varied Flow

Depth and velocity vary with
distance along the channel.

(b)

Figure 10.2 Uniform flow and varied flow.

the flow is neither accelerating nor decelerating. Is it reasonable to assume that flow is uniform in a natural channel? The cross-section geometry and area of flow are changing with distance, and the flow is accelerating or decelerating to fill the changing flow area. Strictly speaking, uniform flow would be possible only in a prismatic channel of constant cross section and constant slope. However, it is common to assume that flow is uniform if the water surface is approximately parallel to the bed of the stream. In this case, it is assumed that the energy grade line also is parallel to the streambed.

In varied flow (Fig. 10.2b), depth and velocity change with distance along the stream. The flow is considered "gradually varied" if the depth changes over a relatively long distance but "rapidly varied" if the depth changes abruptly. Gradually varied flow is a resistance-controlled phenomenon in which boundary friction has a paramount effect on the water surface profile. In rapidly varied flow, structural characteristics such as changes in the bottom slope and cross section are more important.

Zones of rapidly and gradually varied flow are illustrated in Fig. 10.3. Rapidly varied flow exists at the sluice gate and at the weir where the stream lines are curved and the depth is changing abruptly

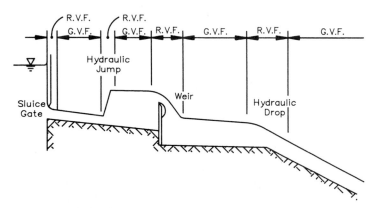

Figure 10.3 Rapidly varied flow and gradually varied flow.

for a short distance downstream. The hydraulic jump and the hydraulic drop also are good examples of rapidly varied flow. Gradually varied flow exists in the four reaches largely unaffected by structures or changing channel geometry. These distinctions are important because HEC-2 computes profiles for gradually varied flow. A hydrostatic pressure distribution, which does not hold for rapidly varied flow, is an assumption in steady-flow water surface profile computations.

Subcritical flow vs. supercritical flow. Flow in open channels may be classified as subcritical, supercritical, or critical. The determination of flow regime according to these classifications is important because of the differences in the way the flow behaves under each one and differences in the approach used to compute the water surface profile. Gravity waves created by disturbances or obstructions travel upstream in subcritical flow but do not in supercritical flow. This has an important influence on the starting point and sequence of water surface profile computations, as will be explained later.

The Froude number [Eq. (10.1)], used to distinguish flow regime, is a dimensionless ratio indicating the relative influence of gravitational forces vs. inertial forces. If the Froude number is less than 1, the flow is subcritical; if it exceeds 1, the flow is supercritical; and if it equals 1 the flow is critical.

$$\text{Fr} = \frac{V}{\sqrt{gD}} \tag{10.1}$$

where Fr = Froude number
V = mean velocity, ft/s = Q/A
g = gravitational acceleration, ft/s^2
D = hydraulic mean depth, ft = A/T
T = top width of channel

This definition of the Froude number assumes that there is a uniform distribution of velocity in the flow cross section. For a symmetrical channel—rectangular, triangular, trapezoidal, etc.—this assumption may be appropriate, but for a natural channel with an irregular cross section and overbank areas, such as the one shown in Fig. 10.4, it is not. Usually the depth of flow is smaller and the roughness greater in the overbank areas than in the channel; consequently, the velocities may be significantly smaller in the overbank areas. Thus, the velocity distribution is nonuniform, and the Froude number as generally defined cannot be used to determine flow regime. However, special Froude numbers that account for variation of discharge and velocity within a complex cross section have been developed.[1,2,3]

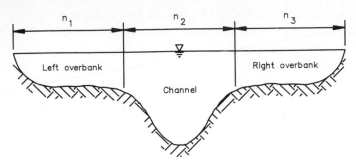

Figure 10.4 Cross section of a natural channel.

Critical depth and its significance

Critical depth is a very significant flow characteristic because it represents a criterion for determining flow regime. Flows at depths above critical depth are subcritical, and flows at depths below critical depth are supercritical. Flows at or near critical depth are termed critical flows; these are unstable because a minor change in specific energy, explained in the next paragraph, will cause a major change in depth.

Specific energy E at a cross section is the energy head above the low point in the channel. Thus, it is the sum of the depth y and the velocity head $V^2/2g$, expressed in Eq. (10.2). A specific energy curve is a plot of specific energy vs. depth for a given discharge (Fig. 10.5). The curve shows that for a given specific energy, there are two possible depths of flow, except at critical depth. Critical depth occurs at the point of minimum specific energy on the curve.

$$E = y + \frac{V^2}{2g} \tag{10.2}$$

Determination of critical depth is complicated by velocity distribution in irregular cross sections associated with floodplains. The velocity head in the specific energy equation can be multiplied by a Coriolis

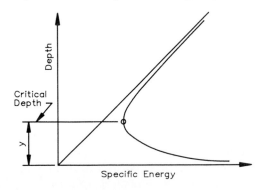

Figure 10.5 Specific energy curve.

or velocity distribution coefficient α to account for horizontal variations in velocity over the cross section and obtain a more accurate expression for specific energy [Eq. (10.3)]. Determination of this velocity coefficient will be discussed in a subsequent section.

$$E = y + \frac{\alpha V^2}{2g} \qquad (10.3)$$

where α is the velocity distribution coefficient.

Wide, flat floodplains can cause problems in computing critical depth. Disparity of flow in the channel and overbank areas causes multiple minimums and discontinuities in the specific energy curve and mixed flow regimes.[1,2,3] These problems, involving lateral and differential flows between the channel and the overbanks, may require special attention and, in some cases, solution by two-dimensional analysis.

Two minimums of specific energy may occur at cross sections with wide overbank areas (Fig. 10.6). The lower minimum occurs within the channel slightly below the top. As the depth of flow increases, filling the overbank areas, the velocity head decreases more rapidly than the elevation head increases, and a second minimum is reached above the top of the channel. If levees exist between the channel and the floodplain, the energy curve not only may have two minimums but

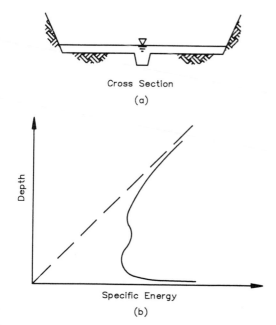

Cross Section

(a)

Depth

Specific Energy

(b)

Figure 10.6 Specific energy curve with two minimums.

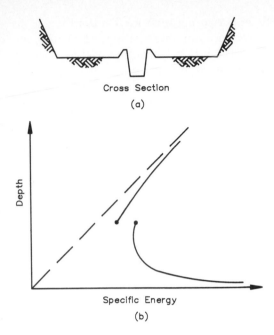

Cross Section

(a)

Depth

Specific Energy

(b)

Figure 10.7 Discontinuous specific energy curve.

also may be discontinuous (Fig. 10.7). As the levees are overtopped, the area of flow discontinuously increases and the specific energy decreases, resulting in a discontinuous specific energy curve. One minimum occurs at the discontinuity; another may occur on the continuous portion of the curve either above or below the top of the levees.

This type of floodplain may also have mixed-flow regimes, characterized by subcritical and supercritical flows occurring simultaneously in different parts of a cross section. Usually, when this occurs, flow in the channel is supercritical, and flow in the overbank areas is subcritical. A subdivision Froude number which can identify the flow regime in each of the two overbank areas and the channel has been developed and tested.[4] It can be used in identifying mixed-flow regimes and overbank flow too shallow to be modeled with the standard step method commonly used in water surface profile computations.

In HEC-2, the critical water surface elevation for a cross section is determined by calculating the elevation at which the total energy head is a minimum. This is done with an iteration procedure in which values for water surface elevation WS are assumed, and corresponding values of total energy head H are computed with Eq. (10.4) until a minimum for H is achieved.

$$H = \text{WS} + \frac{\alpha V^2}{2g} \tag{10.4}$$

To speed the iteration process, a parabolic interpolation procedure is employed which solves for values of H for three values of WS that are spaced at equal intervals.[5] The WS corresponding to the minimum value for H defined by a parabola passing through the three points is used as the basis for the next assumption of a value for WS.

HEC-2 will calculate subcritical or supercritical profiles. The user must specify the flow regime and set up the input file accordingly. Since it is possible to have both flow regimes in a study segment, it may be necessary to run the program for both regimes to determine the complete profile.

The profiles shown in Fig. 10.8 illustrate this problem. The upstream reach has a mild slope with a normal depth above critical, the intermediate reach has a steep slope with a normal depth below critical, and the downstream reach again has a mild slope with normal depth above critical. The critical-depth profile is shown with a dashed line.

A subcritical profile is computed starting at a downstream cross section and proceeding, cross section by cross section, upstream. A supercritical profile is computed starting at an upstream cross section and proceeding downstream. To illustrate, computation of a subcritical profile is described first, starting at the downstream end. The water surface profile is computed above critical depth on the lower reach with a mild slope. On the steep reach in the middle, the flow is actually supercritical, but in the subcritical mode, HEC-2 will not compute water surface elevations below critical depth. At the upper end of this reach, the slope changes to mild again, and the profile passes through critical. From this control section on upstream a subcritical-flow profile is computed.

A supercritical profile calculation is started at the upstream end

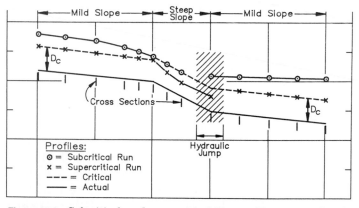

Figure 10.8 Subcritical- and supercritical-flow profiles computed with HEC-2.

and proceeds downstream. In the mild reach, the program cannot compute a profile above critical depth, so it assumes critical depth. In the steep reach, the program computes the supercritical profile below critical depth. At the lower reach with mild slope, if the program is unable to compute a water surface profile below critical depth, it assumes critical depth.

In this example, two profiles have been computed: a subcritical profile for the upper and lower reaches and a supercritical profile in the middle. Where the steep reach joins the lower mild reach, there is a hydraulic jump, but the program does not compute where it occurs. Its location would have to be determined with an external calculation using momentum principles.[6,7]

Velocity distribution in a cross section

Factors affecting velocity distribution. A natural channel cross section with lines of equal velocity is shown in Fig. 10.9. The highest velocities occur above the deepest portion of the river and slightly below the surface. The velocity drops off very rapidly near the boundary. The velocity distribution is affected mainly by three factors: the shape of the

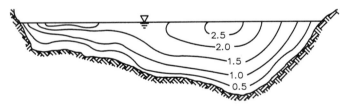

Figure 10.9   Velocity distribution in a natural channel.

cross section, the boundary roughness, and the longitudinal alignment of the channel (e.g., the presence of bends). Changes in roughness patterns and boundary geometry cause variations in the magnitude and direction of velocity. For example, flow around bends tends to produce a helicoid motion affecting velocity distribution.

In water surface profile calculations velocity is included in the kinetic energy term of the energy equation. The velocity head can be stated in energy units of foot pounds of energy per pound of fluid flowing. The energy equation is applied from cross section to cross section, and the total energy is evaluated for each cross section. Determination of a representative velocity for each cross section as a whole is critical to the precision of these calculations.

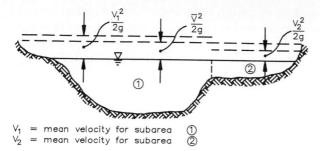

V_1 = mean velocity for subarea ①
V_2 = mean velocity for subarea ②

Figure 10.10   Determination of the mean velocity head for an irregular channel cross section.

Evaluation of velocity head for an irregular cross section.   The cross section in Fig. 10.10 is divided into two subareas to demonstrate how the mean velocity for a cross section is determined. The flow is divided into two parts because of the significant difference in velocity that usually exists between an overbank area and the channel. By comparison, the difference between velocities within each of the two subareas is small, so the velocity distribution within each subarea is assumed to be uniform. In more complex cross sections, further subdivision of an overbank area may be necessary for this assumption to be reasonable.

The kinetic energy head for a cross section can be represented by a discharge-weighted velocity head h_v as computed with Eq. (10.5). The kinetic energy h_v is set equal to the product of a velocity coefficient and a velocity head defined with a mean velocity (Q/A) in Eq. (10.6).

$$h_v = \frac{Q_1(V_1^2/2g) + Q_2(V_2^2/2g)}{Q_1 + Q_2} \qquad (10.5)$$

$$h_v = \frac{\alpha \overline{V}^2}{2g} \qquad \overline{V} = \frac{Q}{A} \qquad (10.6)$$

$$\alpha = \frac{2g[Q_1(V_1^2/2g) + Q_2(V_2^2/2g)]}{(Q_1 + Q_2)\overline{V}^2}$$

$$\alpha = \frac{Q_1 V_1^2 + Q_2 V_2^2}{(Q_1 + Q_2)\overline{V}^2}$$

In general,

$$\alpha = \frac{Q_1 V_1^2 + Q_2 V_2^2 + Q_3 V_3^2 + \cdots + Q_N V_N^2}{Q\overline{V}^2} \qquad (10.7)$$

The velocity coefficient α determined in the general case with Eq.

(10.7) is used in the energy equation to introduce the effects of velocity distribution in complex cross sections. The coefficient α makes it possible to use the mean velocity at each cross section in water surface profile computations. Typical values of α are given in Table 10.1.

TABLE 10.1   Typical Values of α

Channel type	Value of α		
	Minimum	Average	Maximum
Regular channels, flumes, spill-ways	1.10	1.15	1.20
Natural streams	1.15	1.30	1.50
Rivers under ice cover	1.20	1.50	2.00
River valleys, overflooded	1.50	1.75	2.00

The maximum value for α of 2.00 for "river valleys, overflooded," is low. The Corps of Engineers reportedly has encountered many locations where α exceeds 2.00 substantially. Thus, if velocity distribution is neglected, computations of velocity head may be more than 100 percent in error.

Determining flow distribution in a cross section. To find α for a cross section, it is necessary to know how the discharge is divided: how much of it is in the channel and how much in each of the overbank areas. Manning's equation [Eq. (10.8)] is used to make this determination. The values for conveyance K in Fig. 10.11 can be computed if values of n and the cross-section geometry are known. Assuming that the friction slope S_f is constant for a cross section, Q for each subarea can be found from the product of the conveyance and the square root of the friction slope.

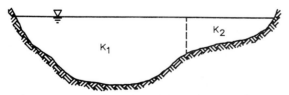

Figure 10.11 Division of an irregular channel to determine flow distribution.

$$V = \frac{1.5}{n} R^{2/3} S_f^{1/2} \tag{10.8}$$

$$Q = VA = \frac{1.5}{n} AR^{2/3} S_f^{1/2}$$

Let

$$K = \text{conveyance} = \frac{1.5}{n} AR^{2/3}$$

Then

$$Q = KS_f^{1/2} \tag{10.9}$$

and

$$Q_1 = K_1 S_f^{1/2}$$

$$Q_2 = K_2 S_f^{1/2}$$

or, in general,

$$Q_i = K_i S_f^{1/2}$$

where i is the subsection number.

In HEC-2, the hydraulic radius R of a subarea is computed by dividing the area by the wetted perimeter, which is defined as the actual wetted physical boundary of the subarea. The imaginary water boundary between two subareas is neglected. By substituting the value of Q from Eq. (10.9) in Eq. (10.7), an equation [Eq. (10.10)] for finding α with the subareas and their conveyances is derived.

$$\alpha = \frac{\left(\sum A_i\right)^2 \sum (K_i^3/A_i^2)}{\left(\sum K_i\right)^3} \tag{10.10}$$

Pressure distribution in a channel cross section

As indicated previously, a hydrostatic pressure distribution is a basic assumption in steady-flow profile computations. A hydrostatic pressure distribution may be assumed to exist in a moving fluid if the streamlines are essentially straight and parallel. Most natural channels with gradually varied flow meet these conditions, but they generally are not met in rapidly varied flow. Flow over a spillway is a good example. At point 1 in Fig. 10.12, the pressure distribution is less than hydrostatic due to the curvature of the streamlines and centrif-

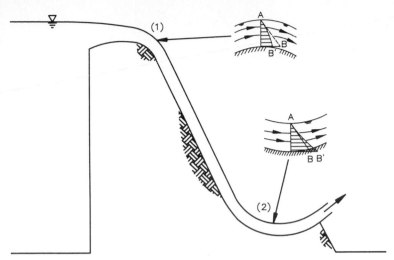

Figure 10.12   Spillway illustration of nonlinear pressure distribution.

ugal force. The centrifugal force at this point has a vertical component which tends to counteract the force of gravity and reduce the weight of the fluid. At point 2, in the bucket of the spillway, the reverse occurs. The vertical component of the centrifugal force acts in the same direction as gravity and accentuates the weight of the fluid. The pressure distribution in both cases is not hydrostatic. The nonlinear pressure distribution is represented by AB', as opposed to the straight-line distribution represented by AB.

Since errors may result if HEC-2 is applied under conditions of nonlinear pressure distribution, an evaluation of the effect of such conditions may be appropriate in questionable cases. This can be done by computing the centrifugal force at the bottom and comparing it with the static force.

Water Surface Profile Computations

Energy principles

Total energy head for a cross section.   The energy loss between two cross sections, taken perpendicular to the flow, is the basis for water surface profile computations. The total energy head H for the cross section of flow shown in Fig. 10.13 and expressed in Eq.(10.11) is

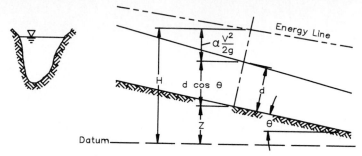

Figure 10.13 Total energy head for a flow cross section.

equal to the sum of the elevation head, the pressure head, and the velocity head. The term cos θ is usually

$$H = Z + d \cos \theta + \frac{\alpha V^2}{2g} \tag{10.11}$$

assumed to be equal to 1 in natural channels. For a bed slope of 1:10, which is a very steep slope for a natural channel, cos θ equals 0.99. If this slope were used in HEC-2, it would result in an error of only 1 percent. If HEC-2 is inappropriately applied to the analysis of spillways or channels with a greater slope, larger errors may result.

The energy equation. With the total energy head defined, the energy equation [Eq. (10.12)] written between two cross sections can be described with reference to Fig. 10.14.

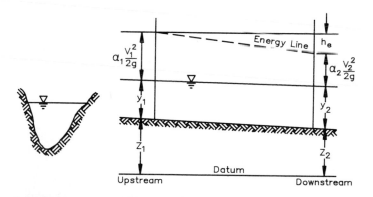

Figure 10.14 Energy loss between two cross sections.

$$Z_1 + y_1 + \alpha_1 \frac{V_1^2}{2g} = Z_2 + y_2 + \alpha_2 \frac{V_2^2}{2g} + h_e \qquad (10.12)$$

Let
$$H = Z + y + \frac{\alpha V^2}{2g}$$

$$H_1 = H_2 + h_e$$

The total energy head at the upstream cross section H_1 is equal to the total energy head at the downstream cross section H_2 plus the energy loss term h_e. This equation is used by HEC-2 repeatedly in steps, going from one cross section to the next, to compute the water surface profile.

Energy losses. Although all the basic assumptions of water surface profile computations listed at the beginning of the chapter could be met between two cross sections 50 mi apart, the energy losses computed over such a long distance would probably be grossly inaccurate. Additional computational points are required to compute the friction loss with the approximation equations that are used. These equations are discussed in subsequent sections. Cross sections also are required throughout a stream reach to account for friction and other energy losses at locations where changes occur in discharge, slope, shape, and roughness; at the beginning and end of reaches with levees; and at bridges, control structures, and other obstructions.

In HEC-2, total energy loss h_e in Eq. (10.13) is composed of two components: a friction component h_f due to boundary roughness and computed with Manning's equation, and a component h_o that represents other losses, mainly contraction and expansion losses. These losses are accounted for by multiplying either the contraction coefficient C_c or the expansion coefficient C_e by the absolute difference in velocity heads between two cross sections.

$$h_e = h_f + h_o \qquad (10.13)$$

where h_e = total energy loss
h_f = energy loss due to friction
h_o = other losses

$$h_o = C_c \left| \alpha_1 \frac{V_1^2}{2g} - \alpha_2 \frac{V_2^2}{2g} \right|$$

and
$$h_o = C_e \left| \alpha_1 \frac{V_1^2}{2g} - \alpha_2 \frac{V_2^2}{2g} \right|$$

$$\alpha_1 \frac{V_1^2}{2g} = \text{upstream velocity head}$$

$$\alpha_2 \frac{V_2^2}{2g} = \text{downstream velocity head}$$

The program determines whether the flow is contracting or expanding by subtracting the velocity head $(V^2/2g)$ at the downstream cross section from the velocity head at the upstream cross section. If the difference is negative, the flow is contracting; if it is positive, the flow is expanding. The program applies the appropriate coefficient on the basis of this determination. It is also possible to account for contraction and expansion losses and losses due to eddies, bends, and junctions by increasing the friction coefficients.

Determining friction loss in a channel reach

Given a reach of river as depicted in Fig. 10.15a, the slope of the energy line S_f at a cross section may be determined with Manning's equation given the water surface elevation, the discharge, the roughness coefficient n, and the cross-section geometry. If the friction slope were determined in this manner for several closely spaced cross sections and the values were plotted, the curve shown in Fig. 10.15b could be obtained. The energy loss due to friction between cross sections 1 and 2 is measured by the area under the curve, computed with Eq. (10.14).

$$\int_{X_1}^{X_2} S_f \, dx = \overline{S}_f L \tag{10.14}$$

where $\overline{S}_f$ is the representative friction slope and L is the reach length.

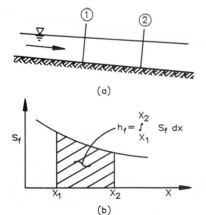

(a)

(b)

$$h_f = \int_{X_1}^{X_2} S_f \, dx$$

Figure 10.15 Friction loss in a channel reach.

Approximation of the friction loss h_f can be obtained by multiplying a representative friction slope $\overline{S}_f$ by L, the reach length between cross sections 1 and 2. There are four equations in HEC-2 for approximating friction loss between two cross sections. They are as follows:

$$\overline{S}_f = \left(\frac{Q_1 + Q_2}{K_1 + K_2}\right)^2 \qquad \text{average conveyance} \qquad (10.15)$$

$$\overline{S}_f = \frac{S_{f_1} + S_{f_2}}{2} \qquad \text{average friction slope} \qquad (10.16)$$

$$\overline{S}_f = \sqrt{S_{f_1} \cdot S_{f_2}} \qquad \text{geometric mean friction slope} \qquad (10.17)$$

$$\overline{S}_f = \frac{2S_{f_1} \cdot S_{f_2}}{S_{f_1} + S_{f_2}} \qquad \text{harmonic mean friction slope} \qquad (10.18)$$

If one of these equations is not specified in input, the program will default to Eq. (10.15). The program also has an option with which one of the four equations is selected automatically on a reach-by-reach basis depending on flow regime and profile type.[8] This option is discussed in the next chapter.

The choice of an appropriate method for friction slope averaging has been considered in several studies and discussed in the literature. One suggestion made recently[9] is that the true friction slope line for an irregular cross section can be approximated by a third-degree polynomial. After various approximation methods applied to irregular cross sections were compared, it was concluded that the average-friction-slope method produces the smallest maximum error, but not always the smallest error. Its general use was recommended, together with systematic location of cross sections.

A recent investigation based on the analysis of 98 natural-streambed data sets showed significant differences (more than 1 ft) in applying the different friction slope methods.[10] This study also found that 500-ft cross-sectional spacing eliminated the differences. It was recommended that interpolated cross sections be used, if necessary, to provide additional computational points to better integrate the friction loss–distance relationship.

In an earlier investigation of friction loss equations,[11] a geometric model was constructed by repeating the same flow cross section (Fig. 10.16) at closely spaced intervals for a total distance of 1½ mi. The channel was assumed to have a zero bottom slope. The starting water surface elevation was taken at critical depth as for a class H2 profile. The water surface profile computed with this model is essentially equivalent to the

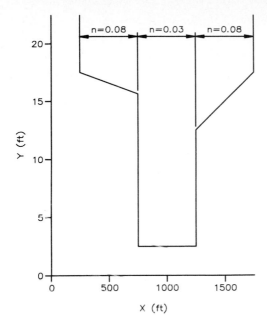

Figure 10.16 Cross section repeated for friction loss analysis.

integrating equation [Eq. 10.14] and is shown as the base curve in Fig. 10.17. This study found that for very short reach lengths, any of Eqs. (10.15) to (10.18) are satisfactory for approximating the area under the friction slope profile between two cross sections.

Sometimes, to minimize data requirements, it is desirable to use long reach lengths. As indicated in Fig. 10.17, from this same study,[11] when reach lengths of 0.5 mi are used, the equations produce quite different results. The water surface profile computed with the average-friction-slope equation [Eq. (10.16)] produces the greatest error because of its straight-line approximation of the friction slope curve between cross sections. This may be observed in Fig. 10.18 in the first 0.5-mi reach, where the shaded area under the straight-line approximation greatly exceeds the area under the true curve.

The best results from this analysis for an H2 profile obviously were obtained with Eq. (10.15). However, Eq. 10.15 does not produce the best results for all types of flow. In fact, the study found that Eq. (10.15) is less satisfactory for M1 profiles than the average-friction-slope [Eq. (10.16)].

Computing water surface profiles

Computational procedure. The steps in computing a subcritical-flow profile with the energy equation are described here to illustrate the

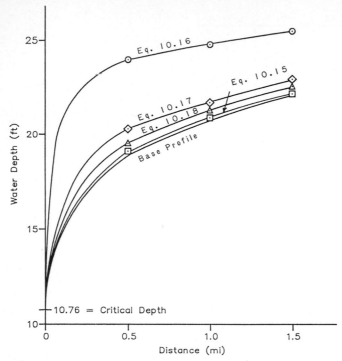

Figure 10.17 Type H2 profiles computed with different friction loss approximations.

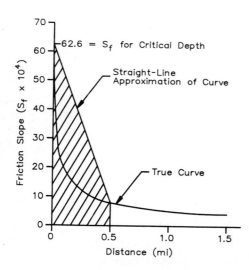

Figure 10.18 Curve depicting average-friction-slope computation.

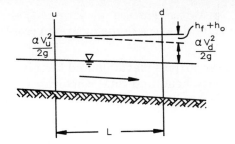

Figure 10.19 Computation of a subcritical water surface profile between two cross sections.

standard step method utilized by HEC-2. In Fig. 10.19, the water surface elevation at downstream cross section d is known, and the problem is to find the water surface elevation upstream at u.

On each side of the energy equation [Eq. (10.19)] written between these two cross sections, the sum of the pressure head y and the elevation head Z equals the water surface elevation WS.

$$\text{WS}_u + \frac{\alpha_u V_u^2}{2g} = \text{WS}_d + \frac{\alpha_d V_d^2}{2g} + h_e \qquad (10.19)$$

$$\text{WS}_u = Z_u + y_u$$

$$\text{WS}_d = Z_d + y_d$$

Assuming that the cross-section geometry is known at each location, the unknown terms in the equation are WS_u, V_u, and h_e. Since V_u can be found if WS_u is known, there are, in effect, only two unknowns. With two unknowns, another equation—the energy loss equation [Eq. (10.20)]—is needed for a solution.

$$h_e = h_f + h_o$$

$$h_e = L \left(\frac{2Q}{K_d + K_u} \right)^2 + C \left| \frac{\alpha_u V_u^2}{2g} - \frac{\alpha_d V_d^2}{2g} \right| \qquad (10.20)$$

where
$$\alpha = \frac{\sum [K(K/A)^2]}{K(K_t/A_t)^2} \qquad \begin{aligned} K_t &= \sum K \\ A_t &= \sum A \end{aligned}$$

Because of the irregularity of natural channels, a trial-and-error solution of Eqs. (10.19) and (10.20) is required. The computational steps are as follows:

1. A water surface elevation at the upstream cross section is assumed. A first trial can be made by assuming that the friction slope at the first cross section applies to the reach.[11] Thus, $\Delta\text{WS} = (Q/K)^2 L$,

where ΔWS is the change in water surface elevation, $Q =$ discharge, $K =$ conveyance, and $L =$ distance to the upstream cross section.

2. On the basis of the assumed water surface elevation and the given cross-section geometry, the velocity head and total conveyance at the upstream cross section are determined.

3. With the computed values of velocity head and conveyance from Step 2, the total energy loss h_e is computed with Eq. (10.20).

4. With the energy loss determined in Step 3, the upstream water surface elevation is computed with the energy equation [Eq. (10.19)].

5. The computed upstream water surface elevation from step 4 is compared with the assumed value in step 1. If the values do not agree within a selected tolerance, the steps are repeated.

Tabulation of computations.   A convenient form (Fig. 10.20) that can be used in calculating water surface profiles by the standard step method is used in the example problem which follows. With reference to this figure: columns (2) and (4) through (12) are for solving Manning's equation to obtain energy loss due to friction; columns (13) and (14) are for determining velocity distribution in the cross section; columns (15) through (17) are for determining kinetic energy; column (18) contains "other losses," which consist of contraction and expansion losses; and column (19) contains the computed water surface elevation.

Column (1), Cross Section No., is for the cross section identification number. Miles or kilometers upstream from the mouth is frequently used for identification.

Column (2), Assumed, is for the assumed water surface elevation, which must agree, within $+0.05$ ft or some other allowable tolerance, with the computed water surface elevation for trial calculations to be successful.

Column (3), Computed, is for the known or rating curve value for the first cross section, but thereafter it is for the value calculated by adding the incremental change in water surface elevation to the computed water surface elevation of the previous cross section.

Column (4), A, is for the cross-section area. If the cross section is complex and has been subdivided into several parts (e.g., left overbank, channel, and right overbank), one line of the form is used for each subsection. The values for the subsections are summed to obtain the total area for the cross section.

Column (5), R, is for the hydraulic radius. For a complex cross section, one line of the form is used for each subsection.

Water Surface Profile Calculations

Project: _____

Q = _____

C_e = _____ C_c = _____

Cross Section No.	Water Surface Elevation		Area A	Hydraulic Radius R	$R^{2/3}$	n	K	$\bar{K}$	$\bar{S}_f$ (10^{-3})	L	h_f	K^3/A^2 (10^9)	α	V	$\frac{\alpha V^2}{2g}$	$\Delta\left(\frac{\alpha V^2}{2g}\right)$	h_o	Δ(Water Surface Elevation)
	Assumed	Computed																
(1)	(2)	(3)	(4)	(5)	(6)	(7)	(8)	(9)	(10)	(11)	(12)	(13)	(14)	(15)	(16)	(17)	(18)	(19)

(8) $K = \dfrac{1.486\, AR^{2/3}}{n}$

(9) $\bar{K} = \dfrac{K_{upstream} + K_{downstream}}{2}$

(10) $\bar{S}_f = \left(\dfrac{Q}{\bar{K}}\right)^2$

(12) $h_f = L\,\bar{S}_f$

(14) $\alpha = \dfrac{(A_t)^2 \; \Sigma \; (K_i^3/A_i^2)}{(K_t)^3}$

Where: i = incremental value
       t = total value

(15) $V = Q/A_t$

(17) $\Delta\left(\dfrac{\alpha V^2}{2g}\right) = \left(\dfrac{\alpha V^2}{2g}\right)_{downstream} - \left(\dfrac{\alpha V^2}{2g}\right)_{upstream}$

(18a) $h_o = C_e\,\left|\Delta\left(\dfrac{\alpha V^2}{2g}\right)\right|$ for $\Delta\left(\dfrac{\alpha V^2}{2g}\right) < 0$

(18b) $h_o = C_c\,\left|\Delta\left(\dfrac{\alpha V^2}{2g}\right)\right|$ for $\Delta\left(\dfrac{\alpha V^2}{2g}\right) > 0$

(19) Δ (Water Surface Elevation) = $\Delta\left(\dfrac{\alpha V^2}{2g}\right) + h_f + h_o$

Figure 10.20 Form for hand computations of a water surface profile with the standard step method.

305

Column (8), K, is for conveyance. If the cross section is complex, K is computed for each subsection and entered on one line of the form. The values for the subsections are summed to obtain a total K for the cross section as a whole.

Column (9), $\overline{K}$, is for the average conveyance for the reach.

Column (10), $\overline{S}_f$, is for the representative friction slope through the reach.

Column (11), L, is for the distance between cross sections; different values may be used for the channel and two overbank portions.

Column (12), h_f, is for the energy loss due to friction through the reach.

Column (13), K^3/A^2, is for part of the expression relating distributed flow velocity to an average value. If the cross section is complex, one of these values is calculated for each subsection, and the subsection values are summed to obtain a total. If the cross section is not complex, i.e., not divided into subsections, nothing is entered in column (13), and the value in column (14) equals 1.

Column (14), α, is for the velocity distribution coefficient for the cross section as a whole.

Column (15), V, is for the average velocity for the cross section.

Column (16), $\alpha V^2/2g$, is for the velocity head corrected for flow distribution.

Column (17), $\Delta(\alpha V^2/2g)$, is for the difference found by subtracting the velocity head at the downstream cross section from the velocity head at the upstream cross section. A positive value indicates that velocity is decreasing and "other losses" should be computed with an expansion coefficient. A negative value indicates that a contraction coefficient should be used to compute "other losses."

Column (18), h_o, is for "other losses," computed with either C_c or C_e.

Column (19), ΔWS, is for the change in water surface elevation from the previous cross section. It is the algebraic sum of the values in columns (12), (17), and (18).

Example problem: standard step hand calculations. The data for the example is given in Fig. 10.21. The solution is presented in Fig. 10.22. Starting with a known water surface elevation at river mile 1.0, the problem is to find the water surface elevation at river mile 1.5.

To solve the energy equation, it is first necessary to determine the conveyance and the velocity head for the downstream cross section corresponding to the known water surface elevation. To take into account nonuniform velocity distribution, cross section 1.0 is subdivided into three subareas—left overbank, channel, and right overbank. The area, hydraulic radius, n value, and computed K for each of these sub-

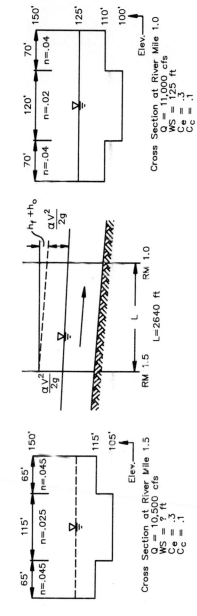

Figure 10.21 Cross-section and reach length data for example problem.

Cross Section at River Mile 1.5

70' n=.04   120' n=.02   70' n=.04   150'

125'

110'

100' ← Elev.

Q = 11,000 cfs
WS = 125 ft
C_e = .3
C_c = .1

$h_f + h_o$

$\dfrac{\alpha V^2}{2g}$

$\dfrac{\alpha V^2}{2g}$

RM 1.0

L

RM 1.5   L=2640 ft

Cross Section at River Mile 1.5

65' n=.045   115' n=.025   65' n=.045   150'

115'

105' ← Elev.

Q = 10,500 cfs
WS = ? ft
C_e = .3
C_c = .1

Water Surface Profile Calculations

Project: Sample Water Surface Profile Computation

$Q = 11{,}000$ cfs at RM 1.0; 10,500 cfs at RM 1.5

$C_e =$ 0.3 $C_c =$ 0.1

(1) Cross Section No.	(2) WSE Assumed	(3) WSE Computed	(4) Area A	(5) Hydraulic Radius R	(6) $R^{2/3}$	(7) n	(8) K	(9) $\bar{K}$	(10) $\bar{S}_f$ (10^{-3})	(11) L	(12) h_f	(13) K^3/A^2 (10^9)	(14) α	(15) V	(16) $\frac{\alpha V^2}{2g}$	(17) $\Delta\!\left(\frac{\alpha V^2}{2g}\right)$	(18) h_o	(19) Δ(Water Surface Elevation)
1.0		125.0	1050.0	12.35	5.35	0.040	208700					8.25						
			3000.0	21.40	7.70	0.020	1716300					561.80						
			1050.0	12.35	5.35	0.040	208700					8.25						
	Total		5100.0				2133700					578.30	1.55	2.16	0.11			
1.5	126.1		666.0	9.37	4.44	0.045	97650					2.10						
			2426.5	17.97	6.86	0.025	989400					164.50						
			666.0	9.37	4.44	0.045	97650					2.10						
	Total	125.06	3758.5				1184700	1659200	0.040	2640	0.106	168.70	1.43	2.79	0.17	−0.06	0.02	+0.07
	125.0		600.0	8.57	4.19	0.045	83000					1.59						
			2300.0	17.04	6.62	0.025	905000					140.12						
			600.0	8.57	4.19	0.045	83000					1.59						
	Total	125.05	3500.0				1071000	1602350	0.043	2640	0.113	143.30	1.43	3.00	0.20	−0.09	0.03	+0.05

(8) $K = \dfrac{1.486\,A\,R^{2/3}}{n}$

(9) $\bar{K} = \dfrac{K_{upstream} + K_{downstream}}{2}$

(10) $\bar{S}_f = \left(\dfrac{Q}{\bar{K}}\right)^2$

(12) $h_f = L\,\bar{S}_f$

(14) $\alpha = \dfrac{(A_t)^2\ \Sigma\ (K_i^3/A_i^2)}{(K_t)^3}$

Where: i = incremental value
       t = total value

(15) $V = Q/A_t$

(17) $\Delta\!\left(\dfrac{\alpha V^2}{2g}\right) = \left(\dfrac{\alpha V^2}{2g}\right)_{downstream} - \left(\dfrac{\alpha V^2}{2g}\right)_{upstream}$

(18a) $h_o = C_e\left|\Delta\!\left(\dfrac{\alpha V^2}{2g}\right)\right|$ for $\Delta\!\left(\dfrac{\alpha V^2}{2g}\right) < 0$

(18b) $h_o = C_c\left|\Delta\!\left(\dfrac{\alpha V^2}{2g}\right)\right|$ for $\Delta\!\left(\dfrac{\alpha V^2}{2g}\right) > 0$

(19) Δ (Water Surface Elevation) $= \Delta\!\left(\dfrac{\alpha V^2}{2g}\right) + h_f + h_o$

Figure 10.22 Standard step computations for example problem.

areas are entered in columns (4), (5), (7), and (8), respectively, on the computation form. The areas and the conveyances for the sub-areas and their sums are used to compute the velocity coefficient in column (14). The velocity head for cross section 1.0, computed with this velocity coefficient and the mean velocity, is entered in column (16).

The next step in the calculations is to assume a water surface elevation upstream at cross section 1.5 and compute the conveyance and velocity head corresponding to this assumed elevation. The elevation assumed in this example is 126.1 ft, and the computational procedure is exactly the same as that just described for cross section 1.0. The computed velocity head of 0.17 ft thus obtained is shown in column (16).

With the velocity heads computed at cross sections 1.0 and 1.5, the change in water surface elevation between the two cross sections based upon energy loss can now be computed. First, the difference in velocity heads of -0.06 ft is found by subtraction and entered in column (17). Since the velocity head is smaller at the downstream cross section, the flow is expanding, and the value for "other losses" entered in column (18) is found by multiplying the value in column (17) by the coefficient of expansion C_e. The friction loss between the two cross sections is obtained by multiplying the representative friction slope S_f in column (10) by the reach length L in column (11). The change in water surface elevation is found by adding the values in columns (12), (17), and (18). The total is entered in column (19).

The water surface elevation at cross section 1.5 computed in the first iteration is found by adding the change in water surface elevation in column (19) to the starting water surface elevation at cross section 1.0. The elevation of 125.06 ft. thus computed is entered in column (3). A comparison of this computed elevation of 125.06 ft. with the assumed elevation of 126.1 ft indicates that the difference is greater than 0.05 ft so another iteration is required.

For a second iteration, an assumed water surface elevation of 125.0 ft is used for cross section 1.5. Repeating the computations for this new elevation yields a computed water surface elevation of 125.05 ft, which meets the criterion of a 0.05-ft difference between the assumed and computed values.

A better estimate of the first-trial water surface elevation at cross section 1.5 would have led more quickly to a solution. If the first estimate had been based on the assumption that the friction slope at the first cross section also applied to the reach, the change in water surface elevation would have been calculated as

$$\Delta WS = (Q/K)^2 L$$

$$= (11{,}000/2{,}133{,}700)^2 (2640)$$

$$= 0.07 \text{ ft}$$

Adding 0.07 ft to the starting water surface elevation at cross section 1.0 would give an estimate for cross section 1.5 of 125.07 ft. This is much closer to the final value than 126.1, and it would have produced an answer within tolerance without the third iteration of computations.

In HEC-2, the criteria used to assume the water surface elevation at the next cross section varies from trial to trial. Generally, the first trial is based on projecting the previous cross section's water surface elevation by multiplying the reach length by the average of the friction slopes from the two previous cross sections. The second trial is based on an average of the computed and assumed elevations from the first trial. The third and subsequent trials are based on a "secant" method that projects the rate of change of the difference between computed and assumed elevations for the previous two trials to zero.[8]

Workshop Problem 10.1: Standard Step Hand Calculations for the Red Fox River*

Purpose

The purpose of this workshop problem is to provide an understanding of the standard step numerical method used in water surface profile computations.

Problem statement

A water surface profile computation is needed for a reach of the Red Fox River between cross sections 1 and 4 (Fig. P10.1). Cross-sectional plots are shown in Fig. P10.2. A discharge rating curve is shown in Fig. P10.3. Area-elevation curves and hydraulic radius–elevation curves are shown in Fig. P10.4. The distances between cross sections are as follows:

Cross sections	Distance, ft
1 to 2	500
2 to 3	400
3 to 4	400

---

*This workshop problem was originally adapted by the Corps of Engineers from material developed by the U.S. Bureau of Reclamation.[12]

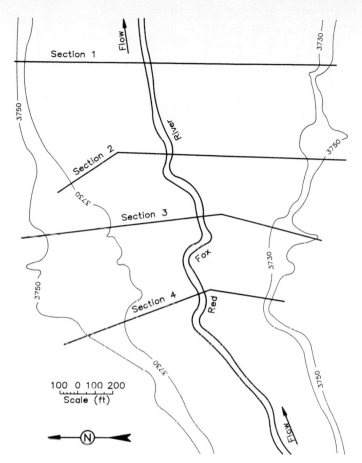

Figure P10.1 Map of the Red Fox River indicating cross sections for water surface profile analysis.

Tasks

1. Determine the water surface elevation at cross section 4 for a discharge of 6500 ft^3/s. Tabulate the calculations using a form similar to Fig. 10.20. Use expansion and contraction coefficients of 0.3 and 0.1, respectively. Calculated and assumed elevations should agree within 0.1 ft.

2. Plot profiles of the channel low point, the water surface, and the energy grade line for the reach of river from cross section 1 to cross section 4.

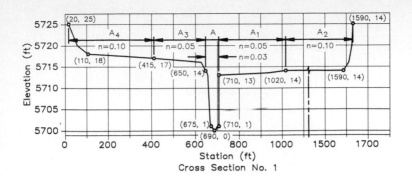

Cross Section No. 1

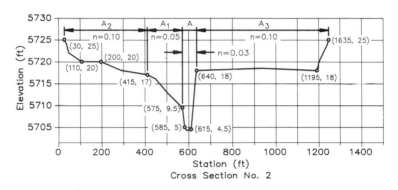

Cross Section No. 2

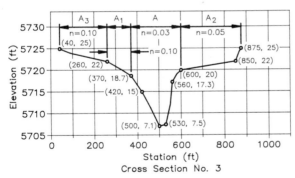

Cross Section No. 3

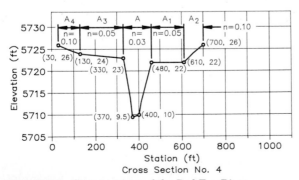

Cross Section No. 4

Figure P10.2 Cross sections of the Red Fox River.

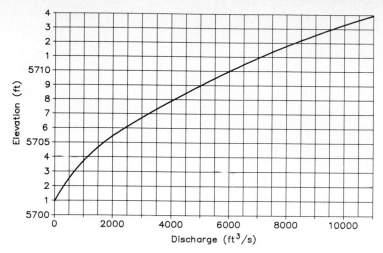

Figure P10.3 Discharge rating curve for cross section 1.

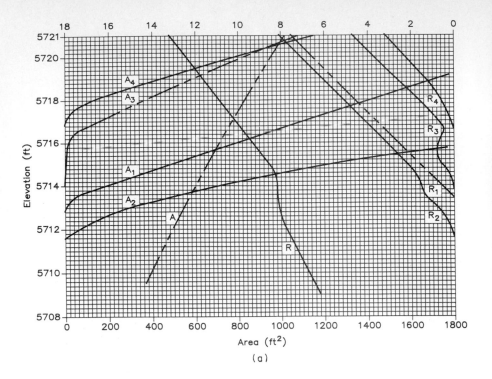

(a)

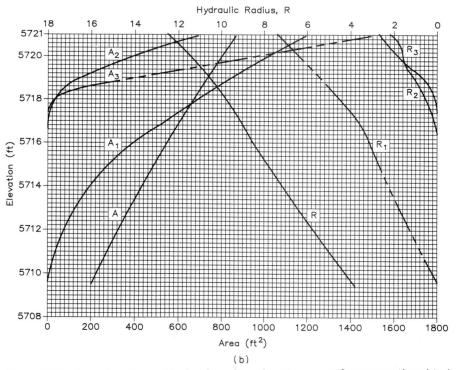

Hydraulic Radius, R

(b)

Figure P10.4 Area-elevation and hydraulic radius–elevation curves for cross sections 1 to 4. (a) Cross section 1; (b) cross section 2; (c) cross section 3; (d) cross section 4.

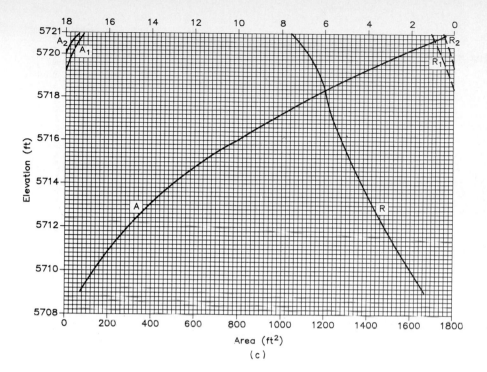

(c)

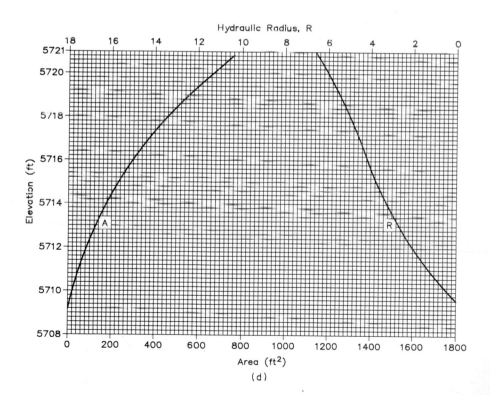

Hydraulic Radius, R

(d)

References

1. David H. Schoellhamer, "Calculation of Critical Depth and Subdivision Froude Number in HEC-2," master's thesis, University of California, Davis, 1982.
2. Merritt E. Blalock and Terry W. Sturn, "Minimum Specific Energy in Compound Open Channel," *Journal of the Hydraulics Division*, ASCE, vol. 107, no. HY6, June 1981.
3. S. Petryk and E. Grant, "Critical Flow in Rivers with Flood Plains," *Journal of the Hydraulics Division*, ASCE, vol. 104, no. HY5, May 1978.
4. David H. Schoellhamer, John C. Peters, and Bruce E. Larock, *Subdivision Froude Number*, Technical Paper 110, U.S. Army Corps of Engineers, Hydrologic Engineering Center, Davis, Calif., 1985.
5. Bill S. Eichert, *Critical Water Surface by Minimum Specific Energy Using the Parabolic Method*, Technical Paper 69, U.S. Army Corps of Engineers, Hydrologic Engineering Center, Davis, Calif., 1969.
6. Ven T. Chow, *Open Channel Hydraulics*, McGraw-Hill, New York, 1959.
7. F. M. Henderson, *Open Channel Flow*, MacMillan, New York, 1966.
8. Hydrologic Engineering Center, *HEC-2 Water Surface Profiles, Users Manual*, U.S. Army Corps of Engineers, Davis, Calif., 1982.
9. Eric M. Laurenson, "Friction Slope Averaging in Backwater Calculations," *Journal of Hydraulics Engineering*, ASCE, vol. 112, no. 12, December 1986.
10. Hydrologic Engineering Center, *Accuracy of Computed Water Surface Profiles*, U.S. Army Corps of Engineers, Davis, Calif., December 1986.
11. Hydrologic Engineering Center, *Water Surface Profiles*, vol. 6: *Hydrologic Engineering Methods for Water Resource Development*, U.S. Army Corps of Engineers, Davis, Calif., 1975.
12. *Guide for Computing Water Surface Profiles*, U.S. Bureau of Reclamation, November 1957.

Water Surface
Profiles Program HEC-2

Program Capabilities

General

The HEC-2 computer program has been developed for calculating water surface profiles for steady, gradually varied flow in natural or constructed channels. Both subcritical- and supercritical-flow regimes can be modeled, and the effects of obstructions to flow such as bridges, culverts, weirs, and buildings located in the floodplain may be included.

Generally, water surface profiles are calculated with the standard step method, which sequentially solves the one-dimensional energy equation between flow cross sections. At some bridges where more complex flow conditions exist, momentum and other hydraulic equations are used to determine changes in water surface elevation. The HEC-2 program has a wide variety of applications and numerous options for defining input and specifying output. Data requirements and basic input for the program are discussed in this chapter, and program output is described in Chap. 12.

A variety of optional analytic capabilities gives the program versatility in solving a wide range of problems. Some of the basic options are briefly described in the next section of this chapter; others are covered in subsequent chapters. The options and the chapters in which they are discussed are as follows:

Chapter 11

Multiple profile analysis

Critical-depth computation

Definition of effective-flow areas

Alternative friction loss computations

Interpolated cross sections

Chapter 13

Normal-bridge analysis

Special-bridge analysis

Chapter 14

Encroachment analysis

Channel improvement analysis

Chapter 15

Supercritical-flow analysis

Split-flow analysis

Ice-covered-flow analysis

Tributary stream profile computations

Determination of Manning's n

Computation of storage-outflow data

Basic options

Choice of metric or English units.   Either metric or English units of measure can be used in the program. English units are used unless otherwise specified in input. The selection of metric units is indicated on the job control line, J1, in the data input file, described later in this chapter. Distances used in the computations are measured in meters rather than feet when the metric system is selected, and this causes some minor differences that should be noted with caution. Tolerances normally used in the decision logic of the program, such as five-tenths of a foot, become five-tenths of a meter, approximately 3 times larger in order of magnitude. An example is the maximum allowable change in velocity head between cross sections, which may be specified in the seventh field of the J1 line. Adustments may need to be made in such tolerances, through program input, to obtain the desired accuracy of results.

Weir coefficients used in special-bridge and split-flow computations also are affected by the choice of units. The coefficient of discharge C used in the standard weir equation ranges from 2.5 to 3.1 with English units and from 1.39 to 1.72 with metric units.[1]

Multiple profiles.   HEC-2 can compute up to 14 profiles in a single run using the same cross-sectional data.

Critical-depth computation. Critical depth is calculated automatically for all cross sections in supercritical runs and for cross sections in subcritical runs whenever the calculated velocity head exceeds a specified tolerance. Critical depth can be obtained for all cross sections in a subcritical run with the proper input specification.

Definition of effective-flow areas. Several options are available to restrict flow to effective-flow areas of cross sections. Among these are options to simulate the hydraulic effect of sediment deposition, to confine flows to channels with levees, to block out road fills and bridge decks, and to model floodplain encroachments. These options are illustrated in Fig. 11.1. Effective-flow areas next to bridge openings also can be defined, as described in Chap. 13.

Sediment deposition. Sediment deposition is modeled by specifying the elevation of a horizontal line below which flow is blocked out in the cross section.

Confined flow within levees. If a levee is open at both ends, the flow can pass outside the levee without overtopping it. Normally the computations for cross sections with low overbank areas or levees are based on the assumption that the entire area below the water surface elevation is effective in conveying the discharge. However, if the water surface elevation at a particular cross section is less than the top-of-levee elevations, and if the water cannot enter the overbanks upstream or leave downstream, the flow areas in the overbanks should not be used in the computations. With the effective-area option specified, the program considers only flow confined by the levees unless the water surface elevation is above the top of one or both of the levees. Then the flow area or areas outside the levee(s) are included. The example problem at the end of this chapter describes this application.

If the effective-area option is employed and the water surface eleva-

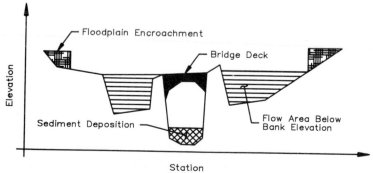

Figure 11.1   Restricting flow to effective-flow areas in HEC-2.

tion is close to the top of a levee, it may not be possible to balance the assumed and computed water surface elevations. This is due to the changing assumptions of flow area and other conditions that occur as the water surface changes from just below to just above the top of the levee. (Refer to the discussion in Chap. 10 of computing critical depth in a channel with levees.) When the program is unable to compute a water surface elevation under these conditions, it will print a note in the output that the assumed and computed water surface elevations cannot be balanced. A water surface elevation equal to the elevation that came closest to achieving a balance will be adopted. The user must evaluate the reasonableness of the adopted elevation and start the computations over again at this location, if necessary, to improve accuracy. Modifications of discharge, conveyance, etc., should be incorporated in the model if appropriate.

Alternative friction loss computations.   The friction loss between adjacent cross sections is computed as the product of the representative rate of friction loss (friction slope) and the weighted reach length. The user can select from the four friction loss equations discussed in Chap. 10: average conveyance [Eq. (10.15)], average friction slope [Eq. (10.16)], geometric mean friction slope [Eq. (10.17)], and harmonic mean friction slope [Eq. (10.18)]. Any of these friction loss equations will produce satisfactory estimates for short reach lengths, but the equations produce significantly different results for longer reach lengths. See Chap. 10 for a discussion of friction loss determination in channels.

HEC-2 has the capability to select an appropriate equation on a reach-by-reach basis depending on flow conditions. The criteria used to select the friction loss equation are shown in Table 11.1. It should be noted that these criteria cannot necessarily be used to select the best equation in reaches with significant lateral expansion, such as a reach below a contracted bridge opening.

TABLE 11.1   Criteria Used to Select the Friction Loss Equation

Profile type	Is friction slope at current cross section greater than friction slope at preceding cross section?	Equation used
Subcritical (M1, S1)	Yes	(10.16)
Subcritical (M2)	No	(10.18)
Supercritical (S2)	Yes	(10.16)
Supercritical (M3, S3)	No	(10.17)

Interpolated cross sections. Sometimes it is necessary to add cross sections to improve computational accuracy. For example, the change in velocity head may be too great between two cross sections for the program to accurately determine the energy gradient. Additional cross sections may be defined individually, or a program option can be used to automatically add cross sections by interpolation. The program can insert up to three interpolated cross sections between two adjacent cross sections. When this option is selected, a maximum allowable change in velocity head between cross sections must be specified by the user. The program then automatically generates additional cross sections when this criterion is exceeded.

The shape of the interpolated cross section is patterned after the shape of the upstream cross section. The minimum elevations of the interpolated cross sections are established by straight-line interpolation between the minimum elevations of the two cross sections between which they are added. The interpolation of horizontal stationing is based upon a ratio of the channel areas of these cross sections. Interpolated cross sections will not be added by the program if reach lengths between existing cross sections are less than 50 ft apart, if an encroachment option has been used in the run, or if the previous cross section is a special-bridge cross section.

Interpolated cross sections should be used with caution. Since their shape is determined by the upstream cross section, a substantial change in channel geometry may not be represented by interpolation. The number of interpolated cross sections added by the program may vary with discharge, so it is not advisable to exercise this option in multiple profile runs. Cross sections will be added automatically for some profiles and not for others because of differences in discharge, and it is usually desirable to do a multiple profile analysis using the same cross sections for all profiles.

Data Requirements for Water Surface Profile Computations

In this section an overview of the basic data required for computing water surface profiles is presented, with a description of what data is needed, where it is obtained, and the sensitivity of the analysis to the data. The following data is required:

Discharge

Flow regime

Starting water surface elevation

Roughness and other energy loss coefficients

The geometric model:

    Cross sections

    Reach lengths

Discharge

Flows for water surface profile computations usually are obtained from frequency analysis or design storms applied to a rainfall-runoff model, such as HEC-1. These may consist of 10-year, 50-year, 100-year, and 500-year peak discharges. Up to 19 different discharges can be specified in program input for each cross section. These can be used for starting discharges in a multiple profile run and to change the discharge at individual cross sections within a run.

Flow regime

A decision must be made at the outset whether to analyze the flow as subcritical or supercritical. The flow regime is subcritical in most natural channels; however, if this assumption is used and it is incorrect, the program output will provide a clue that a wrong choice may have been made. Profile calculations are not permitted to cross critical depth, so critical depth will be assumed and noted in the output at cross sections where the regime is different from that assumed. Since this result may be caused by other problems, such as too few cross sections or incorrect n values, it is important to determine why a solution for an assumed flow regime could not be obtained.

Profile computations begin at a cross section with known or assumed starting conditions and proceed upstream for subcritical flow (downstream for supercritical). In reaches where flow passes from one regime to another, as discussed in Chap. 10, a complete analysis requires computation of the profile twice, first assuming subcritical flow and then assuming supercritical flow.

Starting water surface elevation

The starting water surface elevation is frequently the most difficult starting condition to determine. There are at least three sources for this information: (1) a known water surface elevation, as from a rating curve, (2) normal depth based on slope-area computations, and (3) a condition of critical depth.

A known water surface elevation, if it is available, is the most preferred source. The elevation might be taken from a lake level, sea level, or a control section at a gage.

If a profile is to be started at a location where a known water surface elevation is not available, the slope-area method may be appropriate. Although natural channels do not meet uniform-flow conditions precisely because of uneven features, if the variability of channel slope and cross-sectional characteristics is not too great, the assumption of uniform flow may be reasonable. If so, the energy slope can be considered equal to the average bed slope, and the water surface elevation can be obtained from a normal-depth calculation. This would generally apply only to flows within the channel.

To employ the slope-area method, an estimate of the slope of the energy grade line and an initial estimate of the starting water surface elevation must be specified by the user. Assuming that there is uniform flow, the program computes a discharge for these initial conditions and compares this computed discharge with the given discharge. If there is a significant difference, the estimated elevation is adjusted and the discharge is computed again. This procedure is repeated until the computed discharge and the given discharge are within a 1 percent difference. The elevation thus computed is used as the starting water surface elevation.

If a control structure or some other condition exists which produces critical depth at the starting location, the critical water surface elevation is the obvious choice for the initial water surface elevation.

When there is considerable uncertainty as to what the starting water surface elevation should be at a particular location, it may be worthwhile to start the profile calculations downstream (in the case of subcritical flow, and upstream in the case of supercritical flow) and compute the water surface elevation over several cross sections leading to the starting point. An assumed water surface elevation will tend to converge with the true elevation unless the assumed elevation is on the wrong side of critical depth for the flow regime being analyzed.

A range of assumed values bracketing the correct starting elevation should be used to start the computation. Critical depth could be used as a value at one end of the range, normal depth as an intermediate value, and another depth for the other end of the range. For example, in Fig. 11.2, assuming that a starting water surface elevation is

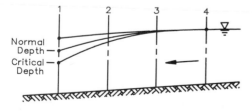

Figure 11.2 Determining a starting water surface elevation with preliminary profile computations for several additional downstream points.

needed at cross section 4, three water surface elevations might be selected to start the calculations farther downstream at cross section 1. As the standard step calculations progress from cross section 1 to cross section 4, the three profiles would tend to converge and to approach the true water surface elevation.

The distance downstream (or upstream) required for computed water surface elevations to converge at a particular location depends on several factors: discharge, roughness coefficients, channel slope, and cross-section geometry. Methods for determining study limits at bridges for subcritical flow can be used for estimating the appropriate distance downstream for starting computations. The nomographs shown in Figs. 11.3 and 11.4 for normal depth and critical depth, respectively, are based on regression equations relating this downstream distance to average reach hydraulic depth (1 percent chance flow) and average reach slope.

Roughness coefficient and other energy loss coefficients

Three types of loss coefficients are utilized by the program to evaluate head losses: (1) Manning's n value for friction loss, (2) contraction and expansion coefficients for transition (shock) losses, and (3) bridge loss coefficients for losses related to weir shape, pier configuration, and pressure flow conditions in special bridge analysis.

Manning's n.   Values for Manning's roughness coefficient n can be obtained from several different sources. Tables of n values are available in most hydraulics texts. Tables and photographs in Chow[2] have been used widely (see Table 11.2). Other methods for computing n include using formulas that utilize the results of field samples and laboratory analysis and using computer program HEC-2 with high-water marks.

HEC-2 can provide an estimate of n if high-water marks are available for a given reach. If the n-value option is used, the program computes an n value to obtain the known water surface elevation for a given discharge at each cross section. A problem with this approach is that all the errors associated with the data are reflected in the computed n values. For example, there may be considerable uncertainty in the accuracy of the assumed discharge and the elevation of the high-water marks. The discharge attributed to the known elevation may not be exactly that of the 100-year flood or some other event it is assumed to be associated with. The elevation of the water marks may have been affected by transitory factors such as snags or debris. When

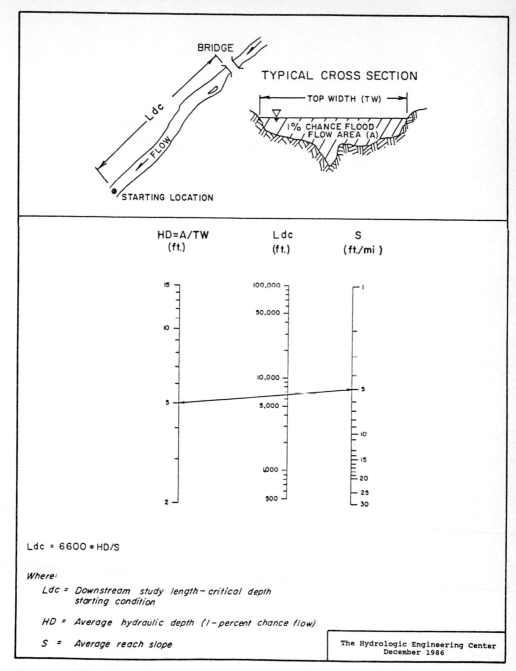

Figure 11.3 Estimation of downstream reach length with the normal-depth criterion.

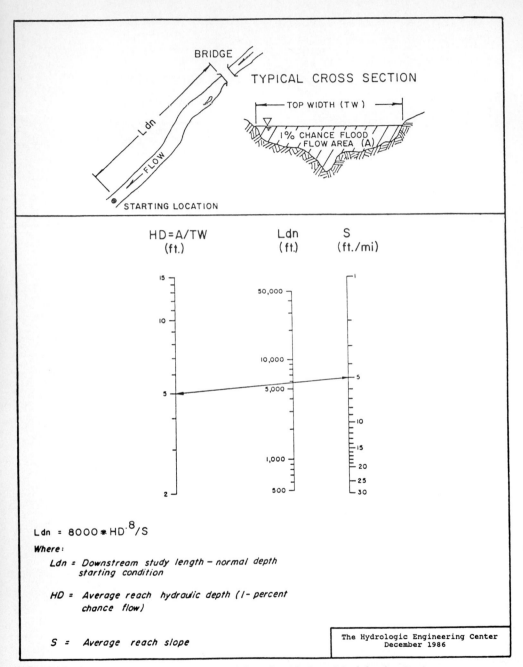

Figure 11.4 Estimation of downstream reach length with the critical-depth criterion.

TABLE 11.2   Values of Roughness Coefficient *n*

(Underscored Values Are Generally Recommended in Design)

Type of channel and description	Minimum	Normal	Maximum
A. Closed conduits flowing partly full			
A-1. Metal			
a. Brass, smooth	0.009	<u>0.010</u>	0.013
b. Steel			
1. Lockbar and welded	0.010	0.012	0.014
2. Riveted and spiral	0.013	0.016	0.017
c. Cast iron			
1. Coated	0.010	0.013	0.014
2. Uncoated	0.011	0.014	0.016
d. Wrought iron			
1. Black	0.012	0.014	0.015
2. Galvanized	0.013	0.016	0.017
e. Corrugated metal			
1. Subdrain	0.017	0.019	0.021
2. Storm drain	0.021	<u>0.024</u>	0.030
A-2. Nonmetal			
a. Lucite	0.008	0.009	0.010
b. Glass	0.009	<u>0.010</u>	0.013
c. Cement			
1. Neat, surface	0.010	0.011	0.013
2. Mortar	0.011	0.013	0.015
d. Concrete			
1. Culvert, straight and free of debris	0.010	0.011	0.013
2. Culvert with bends, connections, and some debris	0.011	<u>0.013</u>	0.014
3. Finished	0.011	0.012	0.014
4. Sewer with manholes, inlet, etc., straight	0.013	0.015	0.017
5. Unfinished, steel form	0.012	0.013	0.014
6. Unfinished, smooth wood form	0.012	<u>0.014</u>	0.016
7. Unfinished, rough wood form	0.015	0.017	0.020
e. Wood			
1. Stave	0.010	0.012	0.014
2. Laminated, treated	0.015	0.017	0.020
f. Clay			
1. Common drainage tile	0.011	<u>0.013</u>	0.017
2. Vitrified sewer	0.011	0.014	0.017
3. Vitrified sewer with manholes, inlet, etc.	0.013	0.015	0.017
4. Vitrified subdrain with open joint	0.014	0.016	0.018
g. Brickwork			
1. Glazed	0.011	0.013	0.015
2. Lined with cement mortar	0.012	0.015	0.017
h. Sanitary sewers coated with sewage slimes, with bends and connections	0.012	0.013	0.016
i. Paved invert, sewer, smooth bottom	0.016	0.019	0.020
j. Rubble masonry, cemented	0.018	0.025	0.030

TABLE 11.2   *(Continued)*

Type of channel and description	Minimum	Normal	Maximum
B. Lined or built-up channels			
B-1. Metal			
a. Smooth steel surface			
1. Unpainted	0.011	0.012	0.014
2. Painted	0.012	0.013	0.017
b. Corrugated	0.021	0.025	0.030
B-2. Nonmetal			
a. Cement			
1. Neat, surface	0.010	0.011	0.013
2. Mortar	0.011	0.013	0.015
b. Wood			
1. Planed, untreated	0.010	0.012	0.014
2. Planed, creosoted	0.011	0.012	0.015
3. Unplaned	0.011	0.013	0.015
4. Plank with battens	0.012	0.015	0.018
5. Lined with roofing paper	0.010	0.014	0.017
c. Concrete			
1. Trowel finish	0.011	0.013	0.015
2. Float finish	0.013	0.015	0.016
3. Finished, with gravel on bottom	0.015	0.017	0.020
4. Unfinished	0.014	0.017	0.020
5. Gunite, good section	0.016	0.019	0.023
6. Gunite, wavy section	0.018	0.022	0.025
7. On good excavated rock	0.017	0.020	
8. On irregular excavated rock	0.022	0.027	
d. Concrete bottom float finished with sides of			
1. Dressed stone in mortar	0.015	0.017	0.020
2. Random stone in mortar	0.017	0.020	0.024
3. Cement rubble masonry, plastered	0.016	0.020	0.024
4. Cement rubble masonry	0.020	0.025	0.030
5. Dry rubble or riprap	0.020	0.030	0.035
e. Gravel bottom with sides of			
1. Formed concrete	0.017	0.020	0.025
2. Random stone in mortar	0.020	0.023	0.026
3. Dry rubble or riprap	0.023	0.033	0.036
f. Brick			
1. Glazed	0.011	0.013	0.015
2. In cement mortar	0.012	0.015	0.018
g. Masonry			
1. Cemented rubble	0.017	0.025	0.030
2. Dry rubble	0.023	0.032	0.035
h. Dressed ashlar	0.013	0.015	0.017

TABLE 11.2   *(Continued)*

Type of channel and description	Minimum	Normal	Maximum
B. Lined or built-up channels *(cont.)*			
i. Asphalt			
1. Smooth	0.013	0.013	
2. Rough	0.016	0.016	
j. Vegetal lining	0.030		0.500
C. Excavated or dredged			
a. Earth, straight and uniform			
1. Clean, recently completed	0.016	0.018	0.020
2. Clean, after weathering	0.018	0.022	0.025
3. Gravel, uniform section, clean	0.022	0.025	0.030
4. With short grass, few weeds	0.022	0.027	0.033
b. Earth, winding and sluggish			
1. No vegetation	0.023	0.025	0.030
2. Grass, some weeds	0.025	0.030	0.033
3. Dense weeds or aquatic plants in deep channels	0.030	0.035	0.040
4. Earth bottom and rubble sides	0.028	0.030	0.035
5. Stony bottom and weedy banks	0.025	0.035	0.040
6. Cobble bottom and clean sides	0.030	0.040	0.050
c. Dragline-excavated or dredged			
1. No vegetation	0.025	0.028	0.033
2. Light brush on banks	0.035	0.050	0.060
d. Rock cuts			
1. Smooth and uniform	0.025	0.035	0.040
2. Jagged and irregular	0.035	0.040	0.050
e. Channels not maintained, weeds and brush uncut			
1. Dense weeds, high as flow depth	0.050	0.080	0.120
2. Clean bottom, brush on sides	0.040	0.050	0.080
3. Same, highest stage of flow	0.045	0.070	0.110
4. Dense brush, high stage	0.080	0.100	0.140
D. Natural streams			
D-1. Minor streams (top width at flood stage < 100 ft)			
a Streams on plain			
1. Clean, straight, full stage, no rifts or deep pools	0.025	0.030	0.033
2. Same as above, but more stones and weeds	0.030	0.035	0.040
3. Clean, winding, some pools and shoals	0.033	0.040	0.045
4. Same as above, but some weeds and stones	0.035	0.045	0.050
5. Same as above, lower stages, more ineffective slopes and sections	0.040	0.048	0.055

TABLE 11.2   *(Continued)*

Type of channel and description	Minimum	Normal	Maximum
D. Natural streams *(cont.)*			
6. Same as 4, but more stones	0.045	0.050	0.060
7. Sluggish reaches, weedy, deep pools	0.050	0.070	0.080
8. Very weedy reaches, deep pools, or floodways with heavy stand of timber and underbrush	0.075	0.100	0.150
b. Mountain streams, no vegetation in channel, banks usually steep, trees and brush along banks submerged at high stages			
1. Bottom: gravels, cobbles, and few boulders	0.030	0.040	0.050
2. Bottom: cobbles with large boulders	0.040	0.050	0.070
D-2. Floodplains			
a. Pasture, no brush			
1. Short grass	0.025	0.030	0.035
2. High grass	0.030	0.035	0.050
b. Cultivated areas			
1. No crop	0.020	0.030	0.040
2. Mature row crops	0.025	0.035	0.045
3. Mature field crops	0.030	0.040	0.050
c. Brush			
1. Scattered brush, heavy weeds	0.035	0.050	0.070
2. Light brush and trees, in winter	0.035	0.050	0.060
3. Light brush and trees, in summer	0.040	0.060	0.080
4. Medium to dense brush, in winter	0.045	0.070	0.110
5. Medium to dense brush, in summer	0.070	0.100	0.160
d. Trees			
1. Dense willows, summer, straight	0.110	0.150	0.200
2. Cleared land with tree stumps, no sprouts	0.030	0.040	0.050
3. Same as above, but with heavy growth of sprouts	0.050	0.060	0.080
4. Heavy stand of timber, a few down trees, little undergrowth, flood stage below branches	0.080	0.100	0.120
5. Same as above, but with flood stage reaching branches	0.100	0.120	0.160
D-3. Major streams (top width at flood stage > 100 ft). The *n* value is less than that for minor streams of similar description because banks offer less effective resistance.			
a. Regular section with no boulders or brush	0.025		0.060
b. Irregular and rough section	0.035		0.100

SOURCE: Chow.[2]

significant errors are introduced, the program produces erratic, impractical results, with computed n values fluctuating widely among the cross sections. A better approach may be to use a trial-and-error fitting of high-water marks with HEC-2.

When n values are estimated from known high-water marks of different flood events, the time of year when an event occurred is an important consideration. There is a big difference in roughness, which should be reflected in n values, between land that has just been planted with crops in the spring and the same land in the fall, when the crops are fully matured and ready for harvest. Even in an urban environment, bank growth at certain times of the year can have a significant impact on n values. Gage data also may be used to determine n values. Several flood events, of a small enough scale so that the flows remain within the channel, can be used to estimate channel n values—or NV values if bed forms, vegetation, etc., cause n to change with depth. NV is the variable name used in HEC-2 for n values that vary with depth. After a channel n value has been established, other available flood data can be used to estimate an average overbank n value.

Because the coefficient of roughness n depends on many factors, such as type and amount of vegetation, channel configuration, and stage, several options are available to vary n. If three n values are sufficient to describe the roughness of the channel and the two overbank areas, only three values are input for a cross section. If three are insufficient to describe the roughness variation, up to 20 n values that vary with horizontal distance across the cross section can be specified. As already indicated, variation of the channel n value with stage also can be specified.

Equivalent roughness k.   An equivalent roughness parameter k, commonly used in the hydraulic design of channels, was made an option in HEC-2 in 1988. Equivalent roughness, sometimes called "roughness height," is a measure of the linear dimension of the roughness elements but is not necessarily equal to the actual or even the average height. Two roughness elements with different linear dimensions may have the same k value if they produce identical roughness effects because of differences in shape and orientation.[2] Most channels (including concrete-lined channels) with appreciable velocity are "hydraulically rough," and for hydraulically rough channels, assumed for all computations in HEC-2, conversion of k to an equivalent Manning's n is performed with the following equation:[3]

$$n = \frac{ZR^{1/6}}{32.6[\log_{10} 12.2 + \log_{10} (R/k)]}$$ (11.1)

where n = Manning's roughness coefficient
k = equivalent roughness
R = hydraulic radius
Z = 1.486 (English units) or 1.00 (metric units)

Up to 20 values of k that vary with horizontal distance across the cross section can be specified in input. Tables and charts are available[3] for determining k values for concrete-lined channels. Values for riprap-lined channels may be taken as the theoretical spherical diameter of the stone of median size. Approximate k values for a variety of bed materials, including those for natural rivers, are shown in Table 11.3. The values of k for natural river channels (0.1 to 3.0 ft) are normally much larger than the actual diameters of the bed materials to account for boundary irregularities and sand waves.

TABLE 11.3   Approximate Values of Equivalent Roughness k

	k, ft
Brass, copper, lead, glass	0.0001–0.0030
Wrought iron, steel	0.0002–0.0080
Asphalted cast iron	0.0004–0.0070
Galvanized iron	0.0005–0.0150
Cast iron	0.0008–0.0180
Wood stave	0.0006–0.0030
Cement	0.0013–0.0040
Concrete	0.0015–0.0100
Drain tile	0.0020–0.0100
Riveted steel	0.0030–0.0300
Natural riverbed	0.1000–3.0000

SOURCE: Chow.[2]

Composite *n* for the channel. Subdivision of a channel between the right- and left-bank stations is sometimes desirable because of major changes in cross-section geometry or major changes in roughness; however, this may result in distortion of channel conveyance computations. A trapezoidal channel cross section having heavy brush and trees on the side slopes and a relatively smooth bottom may be subdivided for the purpose of obtaining a better definition of roughness variation. However, the computation of wetted perimeters and hydraulic radii for the subsections may produce distorted results, as the following example shows.[4]

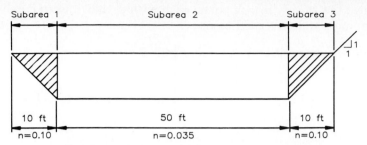

Figure 11.5 Subdivision of a channel for variation in roughness.

In Fig. 11.5, the total conveyance K_T would require a composite n_c value of 0.034, which is less than the n values of 0.035 and 0.10 for the subsections. Total conveyance is computed as the sum of conveyance values for the three subsections as follows:

$$A_1 = A_3 = 50 \qquad A_2 = 500$$

$$P_1 = P_3 = 14.14 \qquad P_2 = 50$$

$$R_1 = R_3 = 3.54 \qquad R_2 = 10$$

$$K_1 = K_3 = 1730 \qquad K_2 = 98,500$$

For the composite section, values for subsections 1, 2, and 3 are summed: $A_c = A_1 + A_2 + A_3$, $P_c = P_1 + P_2 + P_3$, and so on.

$$A_c = 600 \qquad R_c - 7.66$$

$$P_c = 78.3 \qquad K_c = 102,000$$

And

$$n_c = \frac{1.486 A_c R_c^{2/3}}{K_T} = 0.034$$

In this case, the cross section should not be subdivided but should be assigned a composite n somewhat higher than 0.035 to account for the greater roughness on the banks. HEC-2 was modified in 1988 to improve conveyance computations when different n or k values are specified for subdivisions of the channel cross section. The program determines if the cross section should be subdivided or if a composite n value should be applied. If a channel side slope is steeper than 5:1, the cross section will not be subdivided, and a composite roughness n_c will be computed with Eq. (11.2).[2] For the determination of n_c, the water area is divided fictitiously into N parts, each with a known wetted perimeter P_i and a roughness coefficient n_i.

$$n_c = \left[\frac{\sum_1^N (P_i n_i^{1.5})}{P} \right]^{2/3} \tag{11.2}$$

where n_c = composite or equivalent coefficient of roughness
$\quad P$ = wetted perimeter of cross section
$\quad P_i$ = wetted perimeter of imaginary subdivision i
$\quad n_i$ = coefficient of roughness for imaginary subdivision i

It is recommended that an n_c computed automatically by the program be checked for reasonableness with the given data and conditions.

The n values assigned to the first cross section in a data set apply to the remaining cross sections unless changed by the assignment of new values. The n values assigned to the first profile also can be modified (multiplied) by a factor to obtain the values for subsequent profiles.

Contraction and expansion coefficients. These coefficients are multiplied by the absolute difference in velocity heads between two adjacent cross sections to determine the energy loss caused by contraction and expansion of the flow. There is less guidance available for estimating these coefficients than there is for the selection of n values. Most of the published data is for channel design, not for natural channels. Suggested values for these coefficients in the HEC-2 user's manual[1] are as follows:

	Contraction	Expansion
Gradual transitions	0.1	0.3
Bridge sections	0.3	0.5
Abrupt transitions	0.6	0.8

The impact of expansion and contraction coefficients on the computation of energy loss varies. In a mild channel with small changes in velocity head, the impact on the water surface profile is small. In a steep mountain stream, the changes in velocity head are much greater, and the impact of these coefficients is critical to the solution. The solutions should be verified with sensitivity analysis applied to a range of values for the coefficients.

Development of the geometric model

Boundary geometry for the analysis of flow in natural streams is defined with cross sections and the distances (reach lengths) between them. Cross sections are located at intervals along a stream to characterize the flow capacity of the channel and its adjacent overbank areas. If the engineer spends time carefully visualizing how the flow will behave and locating the cross sections to properly simulate the existing physical conditions, the model's accuracy and reliability will be enhanced. Generally, reducing the distance between cross sections increases the accuracy of energy loss computations and the resulting water surface profile. But the closer together the cross sections, the greater the cost, particularly if field surveys are involved.

Guidelines for locating cross sections.   Cross sections should be located as follows:

1. They are needed where there is an appreciable change in cross-sectional area, roughness, or slope in a channel.
2. They should be located so that they are normal to the flow lines.
3. They should be located in detail at bridges—upstream, downstream, and within the structure.
4. They are needed at the head and tail of reaches with levees.
5. They are needed at all control sections.
6. They should be located immediately below a confluence on a main stem and above the confluence on a tributary.
7. In general, more cross sections are needed to define energy losses in urban areas, in channels with steep slopes, and in small streams.
8. According to Beasley,[5] reach lengths should be limited to a maximum of ½ mi for wide floodplains and for slopes equal to or less than 2 ft/mi, 1800 ft for slopes equal to or less than 3 ft/mi, and 1200 ft for slopes greater than 3 ft/mi.

Modeling examples

Cross-sections layout.   To illustrate techniques for locating cross sections, a map delineating a stretch of stream with floodplain limits, as might be defined by a contour on a topographic map, is shown in Fig. 11.6. A subcritical-flow regime is assumed, and the first cross section

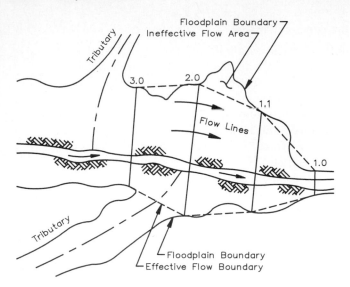

Figure 11.6   Example cross-section layout for water surface profile computations.

(1) is located at the downstream end of the reach. In deciding where the next cross section upstream should be located, it is important to understand how cross sections function in the computational process. Two adjacent cross sections define a step in the reach and also a step in the computations. The conditions at the upstream cross section are averaged with conditions at the downstream cross section to arrive at average conditions for the reach. In effect, there is a linear transition from one cross section to the next.

If the second cross section (2) is located just below the tributary as shown, does it provide for a linear transition between cross sections? The lower boundary is fairly straight in this segment and can be approximated with a straight line. However, for the upper boundary, because of its unevenness, straight-line approximation is questionable.

In deciding if another cross section needs to be added between cross sections 1 and 2, it is important to consider reach length. If cross sections 1 and 2 are close together, say only a few hundred feet or less apart, adding a cross section may be providing more detail than is necessary. For a longer reach length, it probably is appropriate to introduce an extra cross section, such as cross section 1.1, located midway between 1 and 2.

In order to model sharp breaks in a boundary, generally a cross section would be located at each end of the break. If the breaks are minor in size, too many cross sections would be required. The line on the upper boundary between cross sections 1.1 and 1, for example, tends to balance out the minor breaks instead of following the boundary precisely.

Modeling the flow boundary. Defining ineffective-flow areas is essential in modeling flow in natural channels, and the wide indentation in the upper boundary between cross sections 2 and 1.1 is an ineffective-flow area. Slack areas or eddies are not effective in carrying flow downstream, and a water-to-water boundary exists between such areas and the main flow in the channel. Drawing in effective-flow lines, which are perpendicular to cross sections, can be useful in determining which parts of the channel are effective and which are not. Determination of ineffective-flow areas in wide, flat floodplains is especially difficult.

There may be an inclination to extend cross section 3, which has been located between the two tributaries in Fig. 11.6, across the first tributary. However, in defining the flow in the main channel, one should not consider the tributary as an effective-flow area. If it were considered as such an area, the program would include the cross-sectional area of the tributary as part of the area of the main channel, thus introducing a much larger cross section than actually exists to carry the flow. Cross section 3 should be terminated at the bank of the main channel. An effective-flow boundary should then be extended across the mouth of the tributary, as shown.

Modeling meandering streams. A meandering stream, such as the one shown in Fig. 11.7, poses some unique problems in locating cross sections. Again, with the first cross section located at the downstream end of the reach for a subcritical profile calculation, where should the second one be located? There is a significant advantage in locating it at the first bend upstream. At this location, cross section 2 is approximately normal to the flow in the channel and to the flows in both of the overbank areas. On the other hand, the channel cross section at

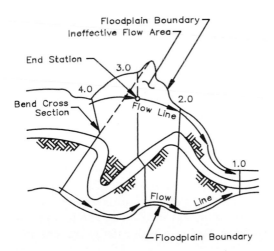

Figure 11.7 Example cross-section layout for a meandering stream.

this location may not be representative of the channel in the downstream reach. The channel is wider and usually deeper here and probably has some sandbars that do not exist downstream. The channel configuration could be redefined arbitrarily at the bend to be more representative, but this would introduce additional costs for manipulating and reentering the data.

Compromises frequently must be made between conflicting elements of model design. In deciding on the best solution, it is necessary to keep study objectives in mind. For example, if the model is to define the low-flow water surface profile for the stream, the greatest emphasis should be given to correctly modeling the channel. If the model's main purpose is to define the 100-year-flood profile, conditions in the floodplain should be given the greatest emphasis. In large flood flows, the flow lines in the channel may not parallel those in the overbank areas. Ideally, a model would be applicable to a range of discharges, perhaps from the 10-year event to the 500-year event. If this is not practical, the model should be made most applicable to the event that is most significant in the study—for example, the 100-year event in a flood insurance study.

Bending cross sections to fit complex flow patterns. Cross section 3 in Fig. 11.7 is located in a straight section of the stream where the flow lines in the channel would not parallel flow lines in the overbank areas in flood stage. To adjust for this, the channel portion of cross section 3 is made perpendicular to the channel flow lines, and the parts of the cross section on the right and left sides of the channel are bent to be approximately perpendicular to the flows in the respective overbank areas. Since cross section 3 extends into an ineffective-flow area, it could be arbitrarily terminated at the boundary of the effective-flow area. A new end station for this cross section is shown in the figure. To accomplish this, fictitious ground points could be defined in input.

Cross sections located where river bends are sharp and close together may converge and even tend to cross. Of course, intersecting cross sections are not valid because they would indicate areas of negative flow, which are physically impossible. A solution to this problem is to bend one or both of the cross sections so that they do not intersect. As an example, cross section 4 would intersect with cross section 3 if it were not bent as shown. To reiterate, drawing in the flow lines can be helpful in this process.

Defining a cross section in input. A cross section in an HEC-2 input file is defined by a set of points, each having a station number corresponding to the horizontal distance from an origin located on the left and an elevation. The left- and right-bank stations separating the channel from the overbank areas are specified. Left and right orientation is normally established by facing downstream. Up to 100 data points can be used to describe a cross section, but 15 to 20 are usually sufficient for all but very complex cross sections. When encroachment options are utilized, no more than 95 points should be used. Extra data points are generated automatically by the program to define encroachment limits. The channel improvement option also generates additional data points, four or more depending on the geometry.

Numerous program options are available for the modification of cross-section data. Cross sections can be repeated by the program at adjacent locations without change or with horizontal and/or vertical dimensions modified. The program can be instructed to introduce improved channel sections, encroachments, and effective-flow areas. End points of a cross section that are too low to contain the computed water surface elevation during water surface profile computations are extended vertically by the program. In some cases the end stations may be extended even though the water surface elevation, as finally computed, is below the end points. This would indicate that the water surface was above the end points at some stage in the iteration process but eventually dropped below them.

Reach Lengths.   Three reach lengths are defined in HEC-2, one for the channel and one for each of the overbanks. With these, the program determines a discharge-weighted average reach length between two cross sections and multiplies this distance by the average conveyance in energy loss computations.

The reach length of a meandering channel is likely to be much longer than the overbank reach lengths, depending on the location and spacing of cross sections. The channel reach length is determined by tracing the thalweg, which is the lowest part of the channel. Usually, it is assumed that the channel portion of the flow will continue to follow the channel even after the channel is inundated and water is flowing in the overbank areas, which are not parallel to the channel.

In reality, as the depth of a flow increases, it may reach a point where the water in the channel will tend to stop following the channel and will begin flowing across the bends. As the depth continues to increase, the channel's effect on the direction of flow and its contribution to total conveyance may diminish. After the channel is submerged and the water is flowing in the overbanks, a substantial part of the flow may travel a shorter, more direct route, as may be envisioned between cross sections 3 and 4 in Fig. 11.7. The portion of the flow that actually leaves the channel in a situation like this depends on a number of factors, including the size of the channel relative to the overbank areas and the n values in the channel relative to the overbank n values.

The reach lengths in the overbank areas will vary depending on the extent of flow in the overbanks. Reach lengths for a 10-year event may be different from those for a 100-year or 500-year event. Assuming that the engineer does not want to develop separate models with different reach lengths for the events, it becomes necessary to use the objectives of the study as a guide in selecting a representative length. If the 100-year event is the most significant event, overbank reach lengths based on that event should be used. These may be slightly short for the 500-year event and slightly long for the 10-year event, but they probably would be a reasonable compromise. It is possible, of course, to use a separate model for each event.

Checking of input data prior to the preparation of an input file

To ensure the validity of the results of a program application, it is important to check carefully the input data as well as the output. Profile accuracy is strongly dependent upon accurate representation of the flow geometry as defined by cross sections and reach lengths and the accurate selection of loss coefficients. It is possible for undetected input errors to produce output that appears to be valid but is, in fact, erroneous.

It is good practice to carefully check field survey data and cross-section layouts prior to data entry. Several points to check are as follows:

1. Has the stream been measured accurately in miles or kilometers from the mouth? Measuring the stream on a map with dividers and marking it off in increments of 0.1 mi, for example, is good practice (see Fig. 11.8). Distances to stream crossings, tributaries, and other features can be determined and tabulated (Table 11.4) for use in preparing comment lines for the input file.

2. Have the cross sections been surveyed at the locations and alignments requested? Making a layout of the cross sections on a USGS quadrangle map—which could be the same map as in step 1—is useful for checking the survey notes. Interpolated cross sections also can be laid out on this map to determine stations and elevations of right and left banks and invert and other key points for data entry.

3. Do cross-section elevations based on survey data appear reasonable when compared with quadrangle map elevations? If several temporary bench marks (TBMs) exist in a study area, the survey crew

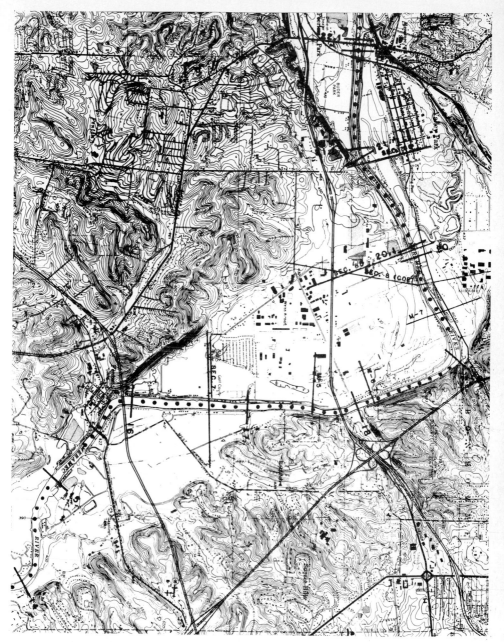

Figure 11.8 Topographic map with river miles marked in tenths of a mile. (*Corps of Engineers, St. Louis District.*)

TABLE 11.4   River Landmarks and Miles

Landmark	Mile
Creeks and Rivers	
Pomme	1.98
Mattese	4.70
Saline-Sugar-Romaine	10.31
Fenton	15.05
Grand Glaize	20.18
Fishpot	22.17
Williams	23.07
Keifer	24.18
Bridges	
Missouri-Pacific R.R.	0.50
Telegraph Road	2.03
St. Louis–San Francisco R.R.	4.78
Lemay Ferry Road	6.02
Interstate 55	6.90
Missouri Highway 21	10.15
Gravois Road	15.55
Missouri Highway 30	15.92
Interstate 44	17.85
St. Louis–San Francisco R.R.	21.92
Missouri Highway 141	22.05

may inadvertently use the wrong one, and the result will be significant errors in the survey data. The use of a wrong TBM may result in an entire cross section being recorded several feet too high. Although the error may not be obvious from an inspection of the invert profile, it is likely to be detected if cross section elevations are compared with quad map contours.

4. Are estimated n values consistent with land use and ground conditions? Have ineffective-flow areas been eliminated? And are cross-section data points consistent? Cross-section plots such as the one shown in Fig. 11.9 facilitate checking the consistency of the data and provide a valuable record for future reference. Land use and corresponding n values are recorded across the top, coordinates of significant ground points are recorded (only two points are recorded for illustration; other points would be recorded as well), and ineffective-flow areas (shaded) are eliminated. Essentially, all the data necessary for including this cross section in the input file is shown.

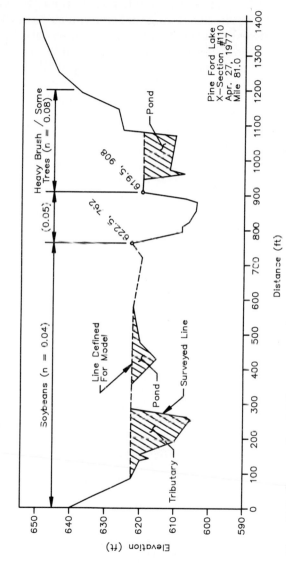

Figure 11.9 Cross-section plot from survey notes.

The HEC-2 Data Input File

Data organization and format

A single run of HEC-2 may range in scope from a one-profile, one-cross-section job (for example, to calculate normal or critical depth) to a job consisting of 14 profiles with 800 cross sections. It may be relatively complex or simple, depending on the complexity of the physical system being modeled and the number of program options being employed. Twenty-seven different types of input data, each associated with a specific line and line identifier in an input file, are needed to cover the numerous options and data requirements of the model. A line of input data is equivalent to the data provided on a single punched card used in earlier input-output operations. It is also called a "record" in some documents. As modifications are made to the HEC-2 program in the future, the number of different types of lines of input data is likely to change. Changes in input format and detail will be reflected in future editions of the HEC-2 user's manual.[1] A current edition should be a basic reference for model applications.

Consistent with the format for other HEC programs, data lines for HEC-2 are laid out in 10 fields of eight columns each (Fig. 5.2). One input variable is entered in each field except the first field, in which the first two columns are used for two line identification characters, such as T1, J1, and X1; the remaining six columns are each used for a variable. If decimal points are not indicated in the data, all numbers must be right-justified within the field. All blank fields are read as zeros. The program has an optional capability to accept free-format input and convert it into the 10-field format.

When a free-format data file is used, the program rewrites the file in fixed format and stores the reformatted file in a scratch file, Tape 10. This file can be saved by renaming it or copying it to a file with another name. The fixed-format version, under the new name, can then be used for subsequent runs, and conversion of the free-format file for each run is avoided.

The various data lines can be classified into six categories—documentation, job control, change, cross section, bridge, and split flow—as follows:

Documentation lines	Change lines	Bridge lines
T1	NC	SB
T2	NH	BT
T3	NV	
AC	QT	
C	ET	
	CI	
	IC	

Job control lines	Cross section lines	Split-flow lines
J1	X1	SF
JR	RC	JC
JS	X2–X5	JP
J2–J6	GR	TW
EJ		WS
ER		WC
		TN
		NS
		NG
		TC
		CR
		EE

Documentation lines permit the user to identify the stream name, the study location, the discharge frequency, the data sources, and other pertinent information. Job control lines control the processing of data, specify the level of output, select computational options, and terminate execution of the program. Change lines provide options to initialize and change values for Manning's n, discharge, encroachments, channel improvements, and ice analysis. Cross-section lines describe the geometric properties of the stream. Bridge lines and split-flow lines provide input data for bridge analysis and split-flow analysis, respectively.

Detailed information for lines of input is available in help files of the full-screen editor COED,[6] available with the HEC-2 program, or in the user's manual.[1] Reference should be made to one of these sources for the discussion of input file content and format which follows.

The required lines of data and their sequence for a basic one-profile, two-cross section model are T1, T2, T3, J1, NC, X1, GR, X1, GR, EJ, and ER. For versions of the model prior to 1988, three blank lines were required before the ER line. Multiple profile data sets are constructed by introducing successive sets of T1, T2, T3, J1, and J2 lines for the second and succeeding profiles in a position immediately following the EJ line for the first profile.

Example problem: development of a data input file for HEC-2

Two water surface profiles for a subcritical-flow regime are to be computed for the stream reach shown in Fig 11.10. A discharge of 200 ft^3/s and a starting water surface elevation of 13 ft are to be used for the

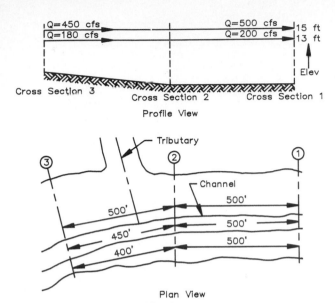

Figure 11.10 Plan and profile of stream reach with three cross sections.

first run, and a discharge of 500 ft³/s and a starting water surface elevation of 15 ft are to be used for the second run. It might be noted that for a subcritical run, the computations are started at the downstream end of the reach at cross section 1. In both cases, the discharge between cross sections 3 and 2 is increased due to tributary inflow, and the same expansion and contraction coefficients of 0.3 and 0.1, respectively, are used throughout. The three cross sections in the model are defined in Fig. 11.11.

A complete input file for this two-profile run is shown in Fig. 11.12. The first three lines constitute a set of title lines, the next two lines contain job control data, and the following lines down to the EJ line contain cross-section data. This completes the data for the first profile. The second profile is specified by the title and job control lines which follow. Three blank lines and an ER line indicate the end of the run. The three blank lines are required only for program versions dated prior to mid-1988. Each group of lines will now be considered in detail.

The first three lines—T1, T2, and T3—are for job identification. These contain alphanumeric title information, which is entered on the lines without regard to the 10-field format commonly used for HEC program input.

The next two lines—job control lines J1 and QT—are shown in Fig. 11.13. The "2" in the second field of the J1 line indicates the field on the QT line specifying the discharge for the first profile computation

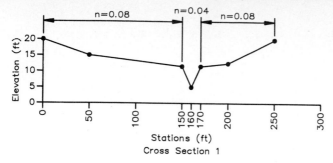

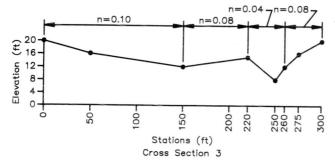

NOTE: Cross section 2 was not surveyed. Assume that it is 10% wider and 0.4 ft. higher than cross section 1, based on information from U.S.G.S. topographic map.

Figure 11.11   Three stream cross sections for the example problem.

The QT line is the discharge table for the run, and in this example, it contains a "2" in the first field, indicating that the table has two discharges. The first discharge value, 200, appears in the second field, corresponding to the "2" on the J1 line. The "0" in field 4 of the J1 line indicates that subcritical flow is assumed, and the "0" in field 5 indicates that a known water surface elevation is to be used as a starting elevation. This known starting elevation, 13, for the first profile is entered in field 9 of the J1 line.

Data for the first cross section is entered on the next four lines, as shown in Fig. 11.14. On the NC line, Manning's n values for the left overbank, the right overbank, and the channel are located in fields 1, 2, and 3 respectively. Contraction and expansion coefficients are located in fields 4 and 5 of this line. On the X1 line, the cross section number is shown in the first field. The "7" in the second field indicates the number of data points entered on the GR lines, which follow, defining the cross-section configuration. The left- and right-bank stations 150 and 170 for the channel are entered in fields 3 and 4. Fields

```
ID.....1.......2.......3.......4.......5.......6.......7.......8.......9......10

T1 SAMPLE PROBLEM HEC-2
T2 FIRST PROFILE
T3 SAMPLE CREEK
J1            2            0    0                              13
QT    2      200    500
NC    .08    .08    .04    .1    .3
X1    1      7      150    170    0      0      0
GR    20     0      15     50     12     150    5      160    12     170
GR    15     200    20     250
X1    2      0      0      0      500    500    500    1.1    .4
NH    4      .10    150    .08    220    .04    260    .08    300
QT    2      180    450
X1    3      8      220    260    500    400    450
X3    10
GR    20     0      16     50     12     150    16     220    8      250
GR    12     260    16     275    20     300
EJ
T1 SAMPLE PROBLEM HEC-2
T2 SECOND PROFILE
T3 SAMPLE CREEK
J1            3                                              15
J2    15



ER
```

Figure 11.12 Input file for two-profile run.

```
ID.....1.......2.......3.......4.......5.......6.......7.......8.......9......10

J1            2            0    0                              13
QT    2      200    500
```

Figure 11.13 Job control lines for profile 1.

```
ID.....1.......2.......3.......4.......5.......6.......7.......8.......9......10

NC    .08    .08    .04    .1    .3
X1    1      7      150    170    0      0      0
GR    20     0      15     50     12     150    5      160    12     170
GR    15     200    20     250
```

Figure 11.14 Roughness coefficients and cross-section data for the first cross section.

5, 6, and 7 are for the left overbank, right overbank, and channel reach lengths to the next cross section downstream. Since this is the first cross section in the run, the reach lengths on the first X1 line are equal to zero.

The coordinate points of the cross section are recorded on the GR lines, which follow the X1 line. Elevation and station coordinates in that order are entered for each point used to define the cross section. For the first cross section, seven pairs of values are entered. The stationing normally starts with the leftmost point of the cross section, as viewed from a position upstream, facing downstream, and proceeds from left to right. If the number of points is too large to be accommodated by one GR line, additional GR lines are used.

The input data for cross section 2 is next (see Fig. 11.15). Since this cross section is a repeat of cross section 1, the values in fields 2, 3, and 4 of the X1 line are zeros. Zeros in these fields indicate that there is no change from the previous cross section; i.e., no new ground points (GR data) will be entered, and the left- and right-bank stations are the same. Equal reach lengths of 500 ft obtained from the plan for the left overbank, the right overbank, and the channel are entered in fields 5, 6, and 7.

```
ID.....1.......2.......3.......4.......5.......6.......7.......8.......9......10

X1    2     0     0     0    500    500    500    1.1     .4
```
Figure 11.15 Cross-section data for the second cross section.

According to the cross section descriptions in Fig. 11.11, cross section 2 is to be widened 10 percent and raised in elevation 0.4 ft with respect to cross section 1. This change is specified in fields 8 and 9 of the X1 line. The "1.1" in field 8 is a factor by which the program will multiply x coordinates (stations) in cross section 1 to arrive at the 10 percent expanded cross section at 2. The program will also add the 0.4 ft it reads from field 9 to the y coordinates (elevations) in cross section 1 to obtain the elevation of coordinate points at cross section 2. It should be noted that these changes reflected in cross section 2 might be considered permanent in the sense that if cross section 2 is repeated, these new values will be repeated.

The final cross section to be considered is cross section 3 in which a number of changes are indicated. The NH and QT lines, preceding the cross-section data, indicate changes in roughness coefficients and discharge at this location. Two n values are specified for the left-overbank part of the cross section in Fig. 11.11, so an NH line must be used (Fig. 11.16). On the NH line, n values are assigned by stationing, beginning on the extreme left. The stations indicate where the rough-

```
ID.....1.......2.......3.......4.......5.......6.......7.......8.......9......10

NH     4     .10     150     .08     220     .04     260     .08     300
QT     2     180     450
X1     3      8      220     260     500     400     450
X3    10
GR    20      0       16      50      12     150      16     220      8     250
GR    12     260      16     275      20     300
EJ
```

Figure 11.16 Data for cross section 3 with new roughness and discharge values specified on the NH and QT lines.

ness changes. The "4" in field 1 in this case indicates that four n values are to be used. The first n value of .10 in field 2 applies to the leftmost part of the cross section, between station 0 and station 150, as indicated by the "150" in field 3. The value of .08 in field 4 applies from station 150 to station 220, and so on.

Due to the tributary entering the stream between cross sections 2 and 3, the flow in the main stream is 10 percent smaller at cross section 3 than it is at cross section 2. This 10 percent decrease in discharge is indicated on the QT line; the two values of discharge entered in fields 2 and 3 are the values from the preceding QT line reduced by 10 percent.

By comparing cross sections in Fig. 11.11, it can readily be seen that the cross-section geometry changes significantly at cross section 3. The changes must be reflected on the X1 and GR lines for this cross section. On the X1 line, eight GR data points are indicated in field 2; new left- and right-bank stations 220 and 260 are entered in fields 3 and 4; and new reach lengths of 500, 400,and 450 for the two overbanks and the channel, respectively, are indicated in fields 5, 6, and 7.

An X3 line is included to implement the effective-area option. Cross section 3 has an area in the left overbank that is lower than the left bank of the channel. The program will consider flow to exist in this low overbank area when the water surface elevation rises above the overbank low point unless the effective-area option is used. For divided flow to occur as depicted in Fig. 11.17, the flow must be able to

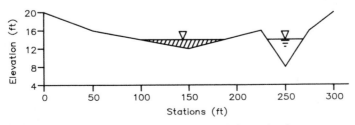

Figure 11.17 Cross section with low area in left overbank.

enter the left overbank upstream and return to the channel somewhere downstream. If this is not the case, then the flow should be contained within the channel until the water surface elevation exceeds the lowest bank elevation.

The effective-area option, implemented by placing a "10" in field 1 of the X3 line, confines the flow to the channel until it overtops one or both of the banks. The program tests the computed water surface elevation with the right- and left-bank elevations independently to determine when overtopping occurs. After the water surface rises above a bank elevation, the program considers the flow to no longer be confined to the channel, but to exist in all areas of that overbank that are lower than the water surface elevation.

Following the X3 line are the GR lines, required to redefine the cross-section geometry at cross section 3. Eight sets of elevation-station coordinates are specified.

Since cross section 3 is the last cross section in the model, the next line is the EJ line, signaling to the program the end of data for the first profile computation. If this were a single profile run, the job could be terminated by adding an ER line (preceded by three blank lines in earlier versions of the program). However, in this example an additional profile is to be computed, so three more title lines are added (Fig. 11.18). Then a new J1 line is

```
ID.....1.......2.......3.......4.......5.......6.......7.......8.......9......10

T1 SAMPLE PROBLEM HEC-2
T2 SECOND PROFILE
T3 SAMPLE CREEK
J1              3                                                    15
J2    15


ER
```

Figure 11.18 Title and job control lines for second profile and end-of-run line.

added to indicate (in field 2) the field on the QT line where the new discharge for the second profile is read and to specify a new starting water surface elevation (in field 9). A J2 line with "15" in field 1 signals to the program that this is the last profile of a multiple profile run and instructs the program to print out a summary table in the output. The run is terminated with an ER line.

Checking an input file for errors

After an input file has been completed, it should be checked with an error checking program, such as EDIT2, so that any errors can be identified. This is a very important step—one that inexperienced users may tend to overlook. The EDIT2 program processes an input file and provides a listing of data errors found, not only "fatal" errors that would cause a run to abort but also data inconsistencies and other problems that could affect computational accuracy. Three levels of data checks and error messages provided by the program are (1) errors, (2) cautions, and (3) notes. "Errors" usually are serious enough to stop an HEC-2 run or produce erroneous results. "Cautions" indicate inconsistencies in the data for usual applications of the program; they alert the user to possible problems that may need attention. "Notes" are basically informational, identifying unusual data entries or inconsistencies of relatively minor significance.

Workshop Problem 11.1: Water Surface Profile Computations with HEC-2

Purpose

The purpose of this workshop problem is to provide an understanding of how an input file for a basic water surface profile model is constructed and executed with HEC-2.

Problem statement

Three water surface profiles are to be computed for the same reach of the Red Fox River described in Workshop Problem 10.1. The same rating curves and cross-section data are to be used. An incomplete data input file is shown in Fig. P11.1 for a multiple profile run with discharges of 1000, 6500, and 10,000 ft^3/s.

Tasks

Add the data necessary to complete this input file. Assume that the contraction and expansion coefficients are 0.3 and 0.1, respectively, for all reaches. Assume that the overbank reach lengths are equal to the channel reach length. No interpolated cross sections are to be used, and cross section-plots are not required. Standard output should be specified, including an input listing and a summary output table. Use the program COED or some other suitable editor to make the entries in the input file.

After the input file is completed, use the program EDIT2 or a similar error-checking program to check it for errors, and correct any errors that are identified. Run HEC-2 using the completed input file,

T1	RED FOX RIVER								
T2	WATER SURFACE PROFILE COMPUTATIONS WITH HEC-2								
T3	Q = 1000 CFS								
J1	0	2	0	0	0	0	0	0	3.8	0
QT	3	1000	6500	10000	0	0	0	0	0	0
NC	0	0	0	.1	.3	0	0	0	0	0
NH	5	.1	415	.050	650	.030	710	.050	1020	.1
NH	1635	0	0	0	0	0	0	0	0	0
X1	1	11	650	710	0	0	0	0	0	0
GR	25	20	18	110	17	415	14	650	1	675
GR	0	690	1	710	13	710	14	1020	14	1590
GR	25	1635	0	0	0	0	0	0	0	0
NH	4	.100	415	.050	575	.030	640	.100	1250	0
X1	2	10	575	640	500	500	500	0	0	0
GR	25	30	20	110	20	200	17	415	10	575
GR	4	580	4	615	18	640	18	1195	25	1250
T1	RED FOX RIVER								
T2	WATER SURFACE PROFILE COMPUTATIONS WITH HEC-2								
T3	Q = 10000 CFS								
J1	0	4	0	0	0	0	0	0	13.4	0
J2	15.0								

ER

Figure P11.1 Incomplete HEC-2 input file for Red Fox River water surface profile computation.

and compare the results for the 6500 ft^3/s profile with the results from Workshop Problem 10.1.

References

1. Hydrologic Engineering Center, *HEC-2 Water Surface Profiles, Users Manual,* U.S. Army Corps of Engineers, Davis, Calif., 1982.
2. Ven T. Chow, *Open Channel Hydraulics,* McGraw-Hill, New York, 1959.
3. Department of the Army, Corps of Engineers, *Hydraulic Design of Flood Control Channels,* Engineer Manual EM 1110-2-1601, Office of the Chief of Engineers, 1970.
4. Jacob Davidian, "Computation of Water-Surface Profiles in Open Channels," Chap. A15 in *Techniques of Water-Resources Investigations of the United States Geological Survey,* Book 3: *Applications of Hydraulics,* U.S. Department of the Interior, 1984.
5. James G. Beasley, "An Investigation of the Data Requirements of Ohio for the HEC-2 Water Surface Profiles Model," master's thesis, Ohio State University, 1973.
6. Hydrologic Engineering Center, "COED Corps of Engineers, Editor," U.S. Army Corps of Engineers, Davis, Calif., February 1987. (Draft.)

Chapter 4

Review of Computed Water Surface Profiles

Introduction

Analysis of program output is an essential part of both modeling and the subsequent review of computed profiles. Modeling is a process of iteration in which input parameters are adjusted and readjusted on the basis of computed output until reasonably accurate results are assured. Project review is a challenging task which amounts to someone else verifying the application of the program. The purpose of both of these functions is to ensure reliable and accurate results. Important decisions are made on the basis of the calculated profiles, and it is not uncommon for the results to be challenged in court.

Description of HEC-2 Program Output

A wide variety of output is available, as reflected by the output control options shown in Table 12.1.

Standard output

Standard output consists of a program banner, a listing of input data, detailed output by cross section, summary tables, and line-printer profile plots. Any of these except the banner can be suppressed at the user's option. Ordinarily, the output is directed to the console (monitor) or to a line printer, but it is also convenient to direct it to a file, which can be saved and used for review later.

Program banner.   The banner at the beginning of the output file gives the program's release date, the date of last update, and a list of error

TABLE 12.1   Output Control Options

Output	Control (lines)
Commentary	C
Input Data Listing*	J1.1
Detailed Output by Cross Section*	J5
Flow Distribution	J2.10, X2.10
Traces	J2.10, X2.10
Summary Tables*	J2.1, J3, J5
Profile Plots*	J2.3
Cross-Section Plots	J2.2, X1.10
Archival Tape	AC
Program Storage Tapes	J6
Storage-Discharge Data	J4

*This data is normal program output but may be suppressed.

corrections and program modifications. Error corrections correct program deficiencies, and modifications incorporate major enhancements to the program. This information is useful in reconciling differences in output produced by different versions of the program.

Output labels.   Unique labels are printed at the beginning of each set of profile data and cross-section data (for example, *PROF 1 or *SECNO 1.000) to facilitate the location of particular data in the file, using the search features of word processors or text editors.

Cross-section data.   Data for 40 variables is tabulated for each cross section. A heading displaying abbreviations for the 40 variable names, arranged in the same spatial order as the data, is provided at the beginning of every fourth or fifth set of cross-section data to serve as a frame of reference for identifying output values. Definitions for these variables are given in Table 12.2. Frequently, special notes and error messages produced by the program are included with the cross-section output to explain various assumptions and options used in the computations.

The names of the variables are generally abbreviations of the terms they represent. For example, in the first line, SECNO represents "cross-*section no.*," CWSEL represents " computed *water surface elevation*," and CRIWS stands for "*critical water surface elevation*." In the second line, QLOB is flow, or Q, in the *left overbank*, QCH is Q in the *channel*, and so on. Cross-section output for the first profile of the example problem in Chap. 10 is shown in Fig. 12.1. The note with cross section 1.000, "CCHV = .100   CEHV = .300," indicates the values for contraction and expansion coefficients entered on the NC

TABLE 12.2   Cross-Section Output Variables

Variable	Description
ACH	Cross-section area of the channel.
ALOB	Cross-section area of the left overbank.
AROB	Cross-section area of the right overbank.
BANK ELEV LEFT/RIGHT	Left- and right-bank elevations.
CORAR	Area of the bridge deck subtracted from the total cross-section area in the normal-bridge method.
CRIWS	Critical water surface elevation.
CWSEL	Computed water surface elevation.
DEPTH	Depth of flow.
EG	Energy gradient elevation for a cross section which is equal to the computed water surface elevation CWSEL plus the velocity head HV.
ELMIN	Minimum elevation in the cross section.
ENDST	Ending station where the water surface intersects the ground on the right side.
HL	Energy loss due to friction.
HV	Discharge-weighted velocity head for a cross section.
ICONT	Number of trials to determine the water surface elevation by the slope-area method, or the number of trials to balance the energy gradient by the special-bridge method, or the number of trials required to calculate encroachment stations by encroachment methods 5 and 6.
IDC	Number of trials required to determine critical depth.
ITRIAL	Number of trials required to balance the assumed and computed water surface elevations.
OLOSS	Energy loss due to minor losses such as transition losses.
Q	Total flow in the cross section.
QCH	Amount of flow in the channel.
QLOB	Amount of flow in the left overbank.
QROB	Amount of flow in the right overbank.
SECNO	Identifying cross-section number. Equal to the number in the first field of the X1 line.
SLOPE	Slope of the energy grade line for the current section.
SSTA	Starting station where the water surface intersects the ground.

TABLE 12.2   (Continued)

Variable	Description
TIME	Travel time in hours from the first cross section to the current cross section.
TOPWID	Width at the calculated water surface elevation.
TWA	Cumulative surface area (acres or thousands of square meters) of the stream from the first cross section.
VCH	Mean velocity in the channel.
VLOB	Mean velocity in the left overbank.
VOL	Cumulative volume (acre-feet or thousands of cubic meters) of water in the stream from the first cross section.
VROB	Mean velocity in the right overbank.
WSELK	Known water surface elevation; for example, a high-water mark.
WTN	Length-weighted value of Manning's n for the channel. Used when computing Manning's n from high-water marks.
XLCH	Distance in the channel between the previous cross section and the current cross section.
XLOBL	Distance in the left overbank between the previous cross section and the current cross section.
XLOBR	Distance in the right overbank between the previous cross section and the current cross section.
XNCH	Manning's n for the channel area.
XNL	Manning's n for the left overbank area.
XNR	Manning's n for the right overbank area.

line. Notes "1490" and "3495" with cross section 3.000 data indicate that an NH line and the effective-area option were used.

Special notes.   Special notes are printed at various locations in the detailed cross-section output to provide information on computational assumptions and options that have been used. The notes have numbers that refer to a Special Note Listing in the user's manual,[1] which provides additional comments and explanation. Many of the notes identify problems in input that need to be corrected.

SECNO	DEPTH	CWSEL	CRIWS	WSELK	EG	HV	HL	OLOSS	BANK ELEV
Q	QLOB	QCH	QROB	ALOB	ACH	AROB	VOL	TWA	LEFT/RIGHT
TIME	VLOB	VCH	VROB	XNL	XNCH	XNR	WTN	ELMIN	SSTA
SLOPE	XLOBL	XLCH	XLOBR	ITRIAL	IDC	ICONT	CORAR	TOPWID	ENDST

*PROF 1

CCHV= .100 CEHV= .300
*SECNO 1.000

1.00	8.00	13.00	.00	13.00	13.07	.07	.00	.00	12.00
200.	5.	194.	1.	17.	90.	5.	0.	0.	12.00
.00	.28	2.15	.28	.080	.040	.080	.000	5.00	116.67
.000590	0.	0.	0.	0	0	0	.00	63.33	180.00

*SECNO 2.000

2.00	7.88	13.28	.00	.00	13.35	.06	.28	.00	12.40
200.	4.	195.	1.	15.	97.	4.	1.	1.	12.40
.07	.25	2.02	.25	.080	.040	.080	.000	5.40	132.38
.000517	500.	500.	500.	1	0	0	.00	64.40	196.79

1490 NH CARD USED
*SECNO 3.000

3495 OVERBANK AREA ASSUMED NON-EFFECTIVE,ELLEA=   16.00 ELREA=   12.00

3.00	5.55	13.55	.00	.00	13.61	.06	.26	.00	16.00
180.	0.	178.	2.	0.	93.	4.	2.	1.	12.00
.14	.00	1.92	.39	.080	.040	.080	.000	8.00	229.22
.000649	500.	450.	400.	1	0	0	.00	36.56	265.78

Figure 12.1   Typical cross-section output.

Profile plots. Line-printer profile plots are printed following the detailed cross-section data for jobs having five or more cross sections. Elevation profiles are plotted for water surface, critical depth, energy grade line, channel invert, left and right banks, and lowest end station. Different horizontal and vertical scales can be specified. Since separate plotting capability is available in PLOT2 and other graphics programs, it is probably more efficient to redirect the output to one of these programs. Producing the line-printer plots can be very time-consuming.

Summary table. The Standard Summary Table 150, consisting of two parts, is provided with multiple profile runs unless otherwise specified by the user. For a single profile run it is necessary to put the value −1 in the field 1 of the J2 line to obtain this table. An example of this summary table is shown with the example analysis of output included in the "Output Analysis" section of this chapter.

Optional output

Several options are available to obtain additional output, including comments, flow distribution, computational traces, and a variety of tables and plots.

Comments.   Comments may be included in both input files and output files to document data sources and study assumptions and to label specific cross sections. They appear immediately ahead of the cross section to which they refer and are especially useful in long computer runs for locating bridges, gages, cross sections adjusted for skew, tributaries, ends of routing reaches, cross sections at which a high-water mark is being used, and other features of special interest. The liberal use of comments not only helps individuals running the program sort out the results and make necessary adjustments to input but also facilitates later use (possibly several years later) of the results by individuals unfamiliar with the model and provides a valuable record. Comments and the cross sections to which they apply are specified on C lines, located at the beginning of the input file.

Additional capability to include messages in the input file listing was added to the program by a revision in 1988. Messages, notes, explanation of data, etc., can be inserted anywhere in the input data set by placing an asterisk, "*," in the first column of the line containing the information. This information will be printed in the input listing. Blank lines may also be included in the input file, and these will be shown in the listing of input but disregarded by the program in execution.

Flow distribution.   When flow distribution is specified, the program prints out the lateral distribution of area, velocity, and percentage of total flow for up to 13 subdivisions of the cross section. Program versions distributed prior to 1988 did not subdivide the channel, even when conveyance calculations did. In a new version released in 1988, the flow distribution in the channel is shown if it is subdivided, and the average depth for each subdivision is included. Considerable computer time may be required to produce the extra detail provided by the flow distribution option, so judicious use of this option is recommended. An example of flow distribution is provided in the example analysis of output in the "Output Analysis" section of this chapter.

TABLE 12.3   Predefined Summary Tables

Code	Table
100	Cross-section output at bridges (special bridge only)
105	Four-cross-section output at bridges (special bridge only)
110	Encroachment data
120	Channel improvement data
150	Standard summary (two tables produced)
200	Floodway data
201	Flood insurance zone data

Traces. With this option, the program prints out the values for key variables in the order in which they are computed to aid the user in checking, debugging, or gaining a better understanding of the program. Two levels of program trace are available. A minor trace includes computations for interpolated cross sections, Manning's n, computed water surface elevation, critical water surface elevation, and weir flow. A major trace includes, in addition to the computations of the minor trace, computations for the hydraulic properties of each subdivision of a cross section.

Tables and plots. The predefined tables listed in Table 12.3 and user-defined tables composed of up to 13 variables selected from the 90 computed by the program can be requested with J3 lines. In 1988, 12 flow-under-ice output variables, numbered 68 to 79 in the user's manual,[1] were eliminated from HEC-2. Fifteen of the remaining variables were renumbered as follows:

Old numbers	New numbers
89–90	64–65
64–67	70–73
80–88	74–82

The following four hydraulics variables were added:

Number	Name	Definition
66	CUMDS	Cumulative channel distance from the first cross section
67	SHEAR	Boundary shear stress within the channel
68	POWER	Channel stream power
69	FRCH	Channel Froude number for uniform-flow conditions

Plots of all cross sections or of individual cross sections can be requested with J2 and X1 lines, respectively. However, as mentioned earlier, plots from other graphics packages may be more useful.

A scratch file (Tape 95) in binary code is created for each run. This file contains values for the 90 variables mentioned above, which can be used in predefined and user-defined tables generated by the program SUMPO and in cross-section and profile plots generated by the program PLOT2. The Tape 95 file can be saved by renaming or copying if subsequent use of the data is contemplated. In the 1988 version of the program, the same output variables that are written to the Tape 95 file are also written to a formatted file (Tape 96).

Output Analysis

Output analysis is broader than the term "output" implies. Except for the possible effect of errors in the program code, the quality of output is determined entirely by the quality of input. Thus, examination of the input listing and identification of any needed adjustments to input are integral parts of output analysis, whether the purpose is for model development or project review.

Changes in key variables that may reflect errors in input

An examination of the summary output table and detailed cross-section output is useful in identifying errors. Key variables such as computed water surface elevation, top width, velocity, energy slope, distribution of discharge, and conveyance ratio should be compared from cross section to cross section. Extreme variations frequently indicate erroneous input.

Computed water surface elevation.   An unreasonably large change in the computed water surface elevation is one of the most obvious indicators of an error. Since it is the end product of the computations, it could result from any of a number of deficiencies. Examination of other output variables and input data is needed to isolate the source of the problem.

Top width.   Extreme changes in the top width of the flow may indicate that cross-section geometry as defined by GR points is in error. GR points can be verified by examining survey notes, cross section plots, and topographic maps. The computed flow boundaries should be fairly consistent with the topography at a given elevation, allowing for noneffective-flow areas.

Incorrect constraining elevations specified in the effective-area option at a bridge location may also cause abrupt changes in top width. If the effective-area option is constraining the flow to the channel at one face of the bridge but not at the other, elevation adjustments (data adjustments on the X3 line) are probably needed.

Velocity.   Significant changes in velocity between cross sections may indicate that the cross-section geometry is not adequately defined or that friction losses and minor energy losses are insufficient.

Energy slope.   A large change in the slope of the energy grade line between cross sections is another indication that energy losses have not been modeled correctly. A large change in energy slope indicates a large change in conveyance. If this occurs in a long reach, it may mean that additional cross sections are needed to more accurately compute the energy losses. As a rule of thumb, a decrease in energy slope of more than 30 percent or an increase of more than 40 percent indicates that the reach length is too long for accurate loss calculations.

Discharge.   Evidence of significant redistribution of flow, say from a major portion in the right overbank to a major portion in the left overbank, in a short distance may merit investigation of the channel shape and alignment to make certain that the redistribution is physically possible. Consistency can be checked by examining the location of the cross sections on a topographic map and visualizing where the flow would go. Expected flow patterns are compared with those indicated in output.

Conveyance.   Values for conveyance are not included in normal program output; however, a ratio of conveyance between cross sections K_R can be obtained in user-specified output. This ratio, like energy slope, indicates the adequacy of cross-section spacing. If K_R is beyond certain limits—for example, if it is outside of the range of $0.7 < K_R < 1.4$ for a long reach[2]—additional cross sections may be needed.

Example problem: analysis of output

In the summary output table, values for key variables such as top width and channel velocity can be compared. In Fig. 12.2 the top widths for the first profile are 472 ft, 468 ft, 330 ft, and 382 ft. Is the change in top width from 468 to 330 reasonable? Since the reach length is 425, the change is not abrupt, and one probably might conclude that a 30 percent contraction in that distance is reasonable. The

SUMMARY PRINTOUT TABLE 150

SECNO	XLCH	ELTRD	ELLC	ELMIN	Q	CWSEL	CRIWS	EG	10K**S	VCH	AREA	.01K
5150.000	.00	.00	.00	707.20	5600.00	722.30	.00	722.50	9.95	4.75	2524.75	1774.90
5150.000	.00	.00	.00	707.20	16000.00	729.88	720.57	730.09	7.95	5.71	7034.49	5675.10
5170.000	1200.00	.00	.00	707.40	5600.00	723.54	.00	723.76	10.95	5.18	2346.76	1692.50
5170.000	1200.00	.00	.00	707.40	16000.00	730.92	721.81	731.18	10.15	6.56	6123.60	5021.89
5192.000	425.00	.00	.00	706.60	5600.00	724.06	.00	724.18	11.24	4.40	2930.37	1670.18
5192.000	425.00	.00	.00	706.60	16000.00	731.40	721.87	731.52	7.80	5.13	7762.85	5730.55
5195.000	1825.00	.00	.00	707.65	5600.00	725.98	.00	726.08	9.65	4.26	3053.57	1802.63
5195.000	1825.00	.00	.00	707.65	16000.00	732.88	723.07	733.01	8.46	5.40	7360.14	5499.3€

SUMMARY PRINTOUT TABLE 150

SECNO	Q	CWSEL	DIFWSP	DIFWSX	DIFKWS	TOPWID	XLCH
5150.000	5600.00	722.30	.00	.00	.00	381.65	.00
5150.000	16000.00	729.88	7.58	.00	-10.12	729.65	.00
5170.000	5600.00	723.54	.00	1.24	.00	329.71	1200.00
5170.000	16000.00	730.92	7.38	1.04	.00	598.16	1200.00
5192.000	5600.00	724.06	.00	.53	-7.34	467.98	425.00
5192.000	16000.00	731.40	7.34	.48	.00	766.75	425.00
5195.000	5600.00	725.98	.00	1.91	.00	472.13	1825.00
5195.000	16000.00	732.88	6.90	1.47	.00	704.84	1825.00

Figure 12.2 Summary output table.

```
CCHV=    .200 CEHV=    .400
*SECNO 5150.000

3265 DIVIDED FLOW

   5150.00   15.10   722.30     .00   722.30   722.50      .20      .00      .00   713.20
    5600.    2138.   2903.     559.   1514.    610.       401.      0.       0.   709.20
      .00     1.41    4.75    1.40    .120     .055       .120     .000   707.20   191.15
   .000995     0.       0.      0.       0        0          0      .00   381.65   593.75


*SECNO 5170.000

3265 DIVIDED FLOW

   5170.00   16.14   723.54     .00     .00   723.76      .22     1.25      .01   713.40
    5600.    2277.   2725.     599.   1426.    526.       395.     67.      10.   709.40
      .10     1.60    5.18    1.51    .120     .055       .120     .000   707.40   146.15
   .001095   1200.   1200.    1200.      2        0          0      .00   329.71   486.18

FLOW DISTRIBUTION FOR SECNO=    5170.00          CWSEL=     723.54

STA=    146.    192.     252.     268.     372.     408.     432.     486.
 PER Q=    4.1     3.1      3.1     30.4     48.7      8.0      2.6
 AREA=   189.8   171.6    114.2    950.2    525.7    243.3    152.0
 VEL=     1.2     1.0      1.5      1.8      5.2      1.9      1.0


*SECNO 5192.000

                 USGS GAGE BISON CREEK AT LA PORTE

   5192.00   17.46   724.06     .00   731.40   724.18      .12      .40      .02   709.70
    5600.    1248.   1672.    2680.    676.    380.      1875.     88.      13.   717.70
      .14     1.85    4.40    1.43    .120     .055       .120     .000   706.60   154.30
   .001124    175.    425.     475.      2        0          0      .00   467.98   622.28


*SECNO 5195.000
   5195.00   18.33   725.98     .00     .00   726.08      .11     1.90      .00   710.75
    5600.    1213.   1587.    2800.    715.    372.      1966.    213.      33.   718.75
      .36     1.70    4.26    1.42    .120     .055       .120     .000   707.65   109.83
   .000965   1825.   1825.    1825.      2        0          0      .00   472.13   581.96
```

Figure 12.3 Cross-section output for four cross sections.

channel velocities of 4.26, 4.40, 5.18, and 4.75 ft/s do not exhibit major differences either, so it appears that the model is producing reasonable results.

An examination of the cross-section output in Fig. 12.3 indicates that the flow distribution might be in error. Most of the flow is in the channel and the right overbank at the two upstream cross sections but shifts to the channel and left overbank at the two downstream cross sections. The magnitude of the changes does not indicate a dramatic shift in flow, but an examination of the topography might be considered to assure that the shift is physically possible.

The flow distribution option was used for cross section 5170. This option can be specified in input for any cross section where a more detailed breakdown of flow and velocity distribution within the cross section is desired. The percentage of discharge, the area, and the velocity are distributed between the ground stations specified on the GR lines. This additional information is particularly useful in floodway analysis.

The note "3265 DIVIDED FLOW" at cross section 5170 indicates that an island or some other protrusion is dividing the flow in the cross section. A comparison of the distance between the starting station SSTA and the end station ENDST with the value for TOPWID indicates that divided flow does indeed exist. The value for the top width is smaller than the difference between the starting and ending stations for the water surface.

A plot of cross section 5150 is shown in Fig. 12.4. There are advantages to viewing a plot for determining errors; the eye frequently can more easily scan data and recognize inconsistencies when the data is in plot form rather than in a tabulation. The plot can quickly be examined for dips and rises in the ground surface that do not make sense. In this particular cross section, the ground points rise above the water surface in the left overbank, indicating an island. Since the rise is defined by more than one point, the plot tends to verify that there is an island at this location and that the special note revealing divided flow is correct. In the examination of cross-section plots, the logical location of bank stations also should be verified.

The printer plot of the profile, which is produced as standard output by HEC-2, is shown in Fig. 12.5. Several profiles are included in the plot: elevations of energy grade, water surface, invert, critical water surface, left and right banks, and lower end station. Several features of the profile plot should be examined. One is the invert profile, which should be relatively uniform and sloping in the right direction. The water surface profile should be compared with the lower end station profile, which is the profile of the lowest elevation at either of the end stations specified on the GR lines. If the water surface profile rises above the lower end station profile, as it does in this example, it indicates that the cross sections in this portion of the river are not modeled adequately. The computed water surface is above the defined cross-section limits.

Project Review

The points to be considered in an analysis of output for the purpose of project review should be basically the same as for model development.

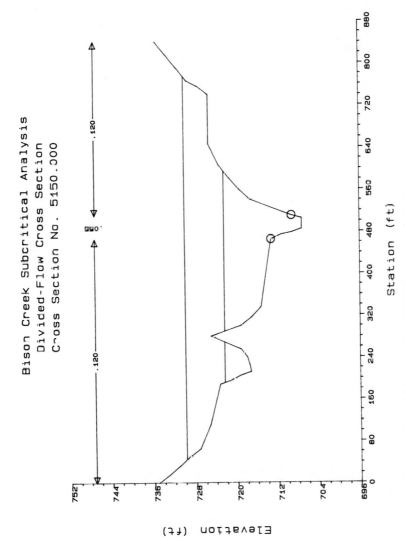

Figure 12.4 Cross-section plot (made with a PLOT2 program plotter).

367

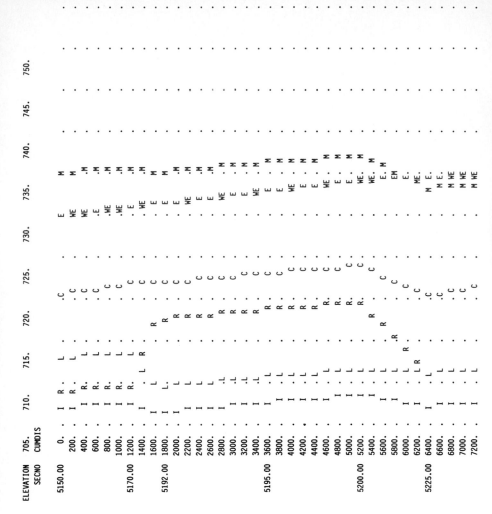

Figure 12.5   Profile plot (made with a line printer).

The perspective of the reviewer may be broader, with greater emphasis given in the analysis to an examination of input for data errors that may not be reflected in output. The modeler may be confident that the input has been developed correctly and so may be less concerned about such errors. In any case, the analysis of output in project review should include a careful examination of the input along with the computed output.

Although the method of analyzing the data is largely a matter of personal choice, the following steps and set of questions might be used as guidelines.

Steps

1. *Examination of the summary output table and summary of errors:* Significant changes in key variables and error messages are noted, and the cross sections where these occur are marked both in the input data listing and in the detailed output for each profile. A user-defined output table showing KRATIO and other selected variables also may be helpful in identifying problem areas.

2. *Review of the input data listing:* A detailed review of the entire input list, cross section by cross section, is conducted. Reference is made to topographic maps showing the location and alignment of all surveyed and interpolated cross sections, the individual cross-section plots, photographs, and other materials used in preparing the input. The data entries for valley and bridge cross sections, the selection of appropriate bridge routines, the inclusion of discharge changes (QT lines) at stream junctions, and numerous other features are checked in this review.

3. *Review of the detailed output for one or two selected profiles:* A cross section–by–cross section review of the detailed output for at least one complete profile is done to identify problems not noticed in the other steps.

Questions

Manning's *n* values. How were *n* values selected and verified? Are they consistent with land use in the reach? Are similar *n* values used for different reaches where conditions are about the same? The *n* values should be checked with land-use descriptions contained in the survey notes, aerial photographs, and field inspections, supplemented by publications on the subject, such as Chow.[3] The selection of an *n* value for a reach of channel or overbank is somewhat arbitrary. Estimates by different individuals can be expected to vary by as much as 25 percent. The review should concentrate on identifying gross errors in estimates.

Expansion and contraction coefficients. Have these coefficients been selected on the basis of HEC-2 user's manual guidelines or other valid criteria? Have the values been adjusted at bridges? Usually expansion and contraction losses are not significant except at bridges and other locations where drastic changes in cross-section geometry occur. Higher coefficients should be used at such locations. For example, values of 0.6 and 0.8 or higher may be appropriate for computing losses at box and circular culverts, with values of 0.1 and 0.3 upstream and downstream from the culvert.

Valley cross sections.   Several questions should be addressed in checking detailed cross-section data:

1. Should the cross section be adjusted for skew? An adjustment should be made if the skew angle is 18° or greater.

2. Are there a sufficient number of interpolated cross sections between surveyed cross sections? Are the interpolated cross sections realistic, considering the surveyed cross sections and the contour map?

3. Is a valley cross section typical of the reach it represents?

4. Is there is a large change in velocity or water surface elevation in the output, and if so, can it be explained by changes in the minimum and bank elevations of the cross sections? If there is a large change in the distribution of flow in the channel and overbanks between cross sections, is the change consistent with the topography? A large change in the distribution of the flow between the channel and the overbanks may indicate that additional cross sections are needed for the transition.

5. Is the cross section number SECNO on the X1 line for the cross section consistent with previously developed river mileages or distances?

6. Are overbank reach lengths the same as the channel reach length at locations where longer or shorter paths of flow in the overbanks are appropriate? There is a tendency to incorrectly specify reach lengths that are the same under these conditions.

7. Are the elevations for the first and last GR points in each cross section higher than the computed water surface elevation so that end-point extensions are not added by the program? In floodway computations, such extensions are not allowed.

8. For repeated cross sections, has the variable PXSECE on the X1 line for adjusting cross-section elevations been determined from an invert profile plot to ensure that the minimum elevation ELMIN is correct? Is the repeated cross section valid for representing conditions at the location specified?

9. Does the output indicate that critical depth has been assumed at two or more consecutive cross sections in the computations? If so, an additional run of the program with the input data file reversed for analyzing supercritical flow may be necessary to define the profile.

10. Are there a significant number of buildings in the floodplain that will affect flow and storage? If so, how have they been modeled? Is the modeling approach appropriate for the results desired? See

Chap. 14 for information on the modeling of flood flows past buildings in the floodplain.

11. Have complicated cross sections been plotted? Do the plots reveal any problems with the data?

Bridges. Many errors in water surface profile computations are related to bridge modeling. Sometimes the modeling of flow patterns through a bridge is very difficult; therefore, data for a bridge should be carefully examined to ensure that the profiles through the bridge are reasonable for the range of flows being used. The correct choice of an analytic method, the normal-bridge or special-bridge method, is very important. See Chap. 13 for a discussion on selecting a bridge method.

Normal-bridge method. Questions to consider in checking the input data for the normal-bridge method are as follows:

1. Have the n values been reduced for the two cross sections within the bridge? The n value for roadway surface should probably be 0.02 or less; the roughness through the bridge opening is frequently less than the channel roughness upstream and downstream from the bridge. Have the n values been changed back to reflect channel roughness at the first cross section upstream from the bridge?

2. Have contraction and expansion coefficients been increased over those in the valley cross sections to account for additional losses through the bridge? Have these coefficients been changed back upstream from the bridge?

3. Is the entry of the bridge geometry correct? The approach roadway can be defined on GR lines, if desired. Identical sets of GR data normally should be used for the two cross sections through the bridge and the cross sections immediately upstream and downstream from the bridge.

4. Is there a GR station for every BT station? This is required in the normal-bridge method.

5. If piers are present, have they been defined on the GR lines? Piers should not be overlooked.

6. Are a sufficient number of cross sections included for the bridge analysis? The recommended minimum number is six, as discussed in Chap. 13. There should be a valley cross section located 1 to 10 ft outside each face of the bridge, one located just inside each face of the bridge, one located a short distance from the bridge upstream, and one located a short distance from the bridge downstream. Additional cross sections may be required in the regions of contraction

and expansion of the flow if there are large changes in velocity head or friction slope.

7. Have X3 lines or other appropriate methods been used to constrain flow to the bridge opening until the roadway is overtopped? Are both the upstream and downstream cross sections overtopped at about the same discharge? Constraint elevations should be adjusted until this is achieved.

8. Have the bridge abutments been specified as the right- and left-bank stations? They should be.

Special-bridge method. Questions to consider in checking the input data for the special-bridge method are as follows:

1. Is a pier width indicated for the bridge? If not, the program reverts to the standard step method to compute the water surface elevation for low flow. If a low-flow solution is required, all the questions in the preceding section on the normal-bridge method should be addressed. If piers are present but the pier width is specified as zero to avoid the use of the Yarnell equation for low flow, piers should be defined with GR points for the upstream and downstream bridge cross sections.

2. If the Yarnell equation is to be used for low-flow computations, has the equivalent trapezoidal bridge opening been modeled correctly? This can be determined by making two comparisons: (1) a comparison of the total bridge area specified on the SB line with the trapezoidal area printed in output and (2) a comparison, for the same low-flow water surface elevation, of the channel area computed for the bridge with the channel area computed for the cross section immediately upstream. For a good approximation, the area of the equivalent cross section should be within 10 percent of that for the actual one.

3. Is the total orifice loss coefficient XKOR high enough? The value should be at least 1.5 and may be as high as 3 or higher for long culverts. The value for K_e, representing the intake loss part of the total, may range from 0.1 to 0.9.[1] Hydraulic computations should be used to verify the selection of an appropriate value for XKOR.

4. Is the weir coefficient appropriate? The same coefficient should not be used for all bridge fills. The coefficient for the most efficient broad-crested weir approaches 3.08. A value this high might be appropriate for a highway crossing with no guardrails. For railroad crossings, highways with guardrails, and other partially obstructed locations, a value of 2.5 to 2.8 would be more reasonable. A field inspection should be made of each road crossing subject to weir flow and the results used to help define weir-flow parameters. The weir

length should be reduced for areas of total obstruction to flow, and weir coefficients should be adjusted for partial obstructions.

5. Are the GR lines used only for defining ground points? In the special-bridge method, the roadway elevations are specified with BT lines; they should never be specified with GR lines.

6. Is the same set of GR lines used at both the upstream and downstream faces of the bridge? Poor floodway definition may result if these cross sections are significantly different.

7. Has the effective-area option specified on the X3 line or some other method been used to constrain the discharge next to the upstream and downstream faces of the bridge to the width of the bridge opening until the roadway is overtopped? For nearly every bridge analyzed, whether by the special-bridge method or the normal-bridge method, the bridge abutments should be designated as the right- and left-bank stations. Refer to Chap. 13 for details on the modeling of effective flow at bridges.

8. Do special notes in the output concerning bridge flow make sense? Inconsistent notes may provide a clue to errors. For example, if pressure-flow conditions exist, note "low flow by normal bridge" may indicate that the SB line is being ignored. This would happen if a "1," calling for the special-bridge method, were inadvertently omitted from field 3 of the X2 line.

Floodway determination.   Some questions that ought to be considered concerning output from floodway computations follow. A floodway is defined as the channel and overbank areas required to pass a flood of a specified recurrence interval.

1. Do the floodway boundaries appear reasonable, i.e., with fairly smooth transitions and no large undulations? See the discussion in Chap. 14 on the smoothing of floodway boundaries.

2. Are changes in the water surface elevation resulting from encroachments close to or within allowable limits?

3. Should additional encroachment methods be applied within the reach to improve the floodway?

4. Have any bridge openings been subject to encroachment?

5. Does critical depth occur in the floodway? If critical depth occurs at a cross section, it may preclude using any encroachments at this location. If critical depth occurs at several locations for the 100-year flood with no encroachments, a floodway should probably not be developed for this reach.

6. Do velocities increase excessively? Does a hazardous condition exist

in an overbank area with the floodway in place, but not under natural conditions? A hazardous condition is defined as 3 ft/s and a 3-ft depth.

7. Are encroachment stations as specified in encroachment method 1 or 2 within the channel? No channel encroachment is allowed.

Verification

When an HEC-2 model is being developed, it should be verified with observed data. Computed water surface elevations should be compared with high-water marks from past floods and rating curves at stream gages. If no observed data exists, sensitivity analyses should be performed to determine how critical various parameters, such as *n* values, are to the results.

If reliable observed data exists, it can be used to calibrate the model or at least provide an indication as to whether additional adjustments and improvements are needed. Some of the factors that might be considered in reconciling differences between computed and observed data are as follows.

Manning's *n*. There is usually some leeway in assigning *n* values, and these might be adjusted upward or downward slightly to achieve a better fit of actual profiles with computed profiles.

Discharge values. How reliable are the discharge values obtained from the hydrologic model or from other sources? If differences between the observed and computed profiles are great, on the order of a few feet or more, this should be investigated.

Survey data. The accuracy of the survey data can have a significant effect on the accuracy of the computed profiles. Checking the survey data for accuracy and consistency relative to developing an input file is discussed in Chap. 11.

Bridge method. At some locations, changing the bridge method may improve the profile.

Special conditions. If a high-water mark is unusually high at a bridge, it may have resulted from a snag or debris caught on the piers. A dam failure or a large diversion upstream can also abnormally affect watermarks.

Changed conditions. The replacement of a bridge, channel modification, encroachment, development of adjacent land, and other changes which have occurred since watermarks or records were made would complicate verification.

Faulty data. Questionable data is always a possibility. For example, insufficient rainfall means that the discharge cannot be determined accurately; information from residents regarding high-water marks may be in error; and so on.

Sensitivity of data

Sensitivity analyses of several factors that can affect the accuracy of output and the cost of water surface profile computations have been performed. A comparison of results from computations based on surveyed cross sections with the results from computations based on cross sections derived in other ways revealed differences in water surface elevations of 1 to 4 ft.[4,5] An analysis of the effect of cross-section spacing on computed results produced differences in water surface elevations of 0.3 to 4 ft.[6] In this analysis, several models were tested by removing cross sections and comparing the results before and after removal. Another investigation found that expansion and contraction coefficients for bridge routines in HEC-2 have a relatively small impact except in high-velocity streams.[4]

A study in 1986 focused on two major factors related to water surface profile accuracy: (1) the survey technology employed for determining cross-section geometry and (2) the degree of confidence in Manning's coefficient. Methods for determining cross-section geometry include field surveys with land-surveying instruments, development of aerial spot elevations from aerial stereo models (photogrammetry technology), generation of topographic maps from aerial photography, and hydrographic measurements. Since information on the reliability of estimating Manning's coefficient could not be found, experimental data was developed and used to assess the impact of this factor on profile accuracy.

Findings of the study[7] include the following:

1. Cross sections obtained from aerial spot elevation surveys that conform to mapping industry standards are significantly more accurate than cross sections obtained from topographic maps.

2. Aerial spot elevation surveys are generally more cost-effective than field surveys when more than 15 surveyed cross sections are required.

3. The reliability of the estimation of Manning's n has a major effect on the accuracy of computed water surface profiles. Reliable estimates of n require aerial and field reconnaissance in conjunction with estimation guidelines and tables, such as those in Chow.[3]

4. Significant computational error can result from cross-section spacing that is often considered to be adequate. This error can effectively be eliminated by the addition of interpolated cross sections, provided that original cross sections adequately portray the flow geometry.

On a large project, sensitivity analyses performed at the outset on roughness coefficients and other factors for a short section of the river will provide valuable guidance for conducting the study. Cross-section spacing can be analyzed by observing the effect of changing the number of cross sections and reach lengths on computed results. The effect of n-value changes on profile computations can be determined quite easily with HEC-2 by making several runs with different n values adjusted by ratios specified in input. This procedure can also be used with discharges.

Contraction and expansion coefficients are multiplied by changes in velocity head to determine minor energy losses, designated as OLOSS in output. If an examination of the output indicates that the velocity head is small and the assumed contraction and expansion coefficients are producing small energy losses, the computed profile may be relatively insensitive to changes in these coefficients. On the other hand, if velocities and energy losses are large, the values used for these coefficients will have a significant impact on results and should be carefully evaluated.

Workshop Problem 12.1: Output Analysis—Four Case Studies

Purpose

The purpose of this workshop problem is to reinforce familiarity with HEC-2 program output and the ways to examine the data and evaluate it for reasonableness.

ID	1	2	3	4	5	6	7	8	9	10
T1	OUTPUT ANALYSIS WORKSHOP PROBLEM 1									
T2	Q - 1500 CFS									
T3	EASY CREEK									
J1	0	0	0	0	0	0	0	1500	15.5	0
J2	-1									
J3	38	1	2	4	26	58	13	14	15	
NC	.07	.07	.035	.1	.3	0	0	0	0	0
X1	1	12	70	140	0	0	0	0	0	0
X3	10	0	0	0	0	0	0	0	0	0
GR	18	0	18	40	16	70	8	80	8	90
GR	16	120	18	140	16.5	160	16	180	16.5	200
GR	17	220	18	240	0	0	0	0	0	0
X1	1.5	0	0	0	700	700	700	.9	.8	0
X3	10	0	0	0	0	0	0	0	0	0
X1	2	14	70	110	500	500	500	0	0	0
X3	10									
GR	21.6	0	17.6	0	11.6	14	12.6	40	16.6	60
GR	18.8	70	9.6	90	9.6	100	18.8	110	13.6	145
GR	13.6	190	15.6	200	19.6	230	21.6	240	0	0
X1	3	11	160	210	1500	1500	1500	0	0	0
X3	10									
GR	22	0	20	50	18	80	22	160	12	165
GR	12	200	16	205	20	210	21	230	21	240
GR	26	290	0							
X1	4	0	0	0	1000	1000	1000	0	1.5	0
X3	10									
EJ										
ER										

Figure P12.1 Input file for output analysis problem 1.

Problem statement

The four input files—Figs. P12.1, P12.2, P12.3, and P12.4—have been developed to perform water surface profile computations with HEC-2 at four different locations—Easy Creek, Trail Creek, Rocky Creek, and Baker Creek. The output that is generated by running each of these files with HEC-2 has, in each case, at least one major error that can be identified through careful analysis. A topographic map, useful in the analysis of the Trail Creek output, is shown in Fig. P12.5. Since two of these locations, Rocky Creek and Baker Creek, have bridges, it may be advisable to defer working these parts of the problem until after Chap. 13 is read.

```
ID.....1.......2.......3.......4.......5......6.......7.......8.......9......10

T1    OUTPUT ANALYSIS WORKSHOP PROBLEM 2
T2    X-SECTIONS TAKEN FROM FLOOD QUAD
T3    TRAIL CREEK 100-YR FLOOD 5450 CFS
J1         2                                              624.3
J3    38    1      2       3      4       5      39      43     13      14
J3    15
NC    .10   .10    .07     .10    .30
QT    2     5450   17800
X1    4.0   20     540     610
GR    650   0      645     60     640     140    635     350    630     380
GR    625   410    620     540    615     550    610     560    610     590
GR    615   600    620     610    625     680    630     700    635     740
GR    640   770    645     790    640     840    640     930    650     1000
X1    5.5   20     530     590    1350    1750   1700
GR    680   0      670     60     660     160    650     240    640     300
GR    635   340    630     380    625     460    624     530    615     545
GR    615   575    624     590    625     630    630     710    635     750
GR    640   790    650     900    660     980    670     1060   680     1180
X1    7     30     550     605    1600    1400   1800
GR    680   0      670     160    660     220    650     300    640     370
GR    635   410    628     550    621.5   565    621     575    621.5   585
GR    628   605    628.15  650    628.31  700    628.47  750    628.63  800
GR 628.8   850    628.96  900    629.12  950    629.28  1000   629.45  1050
GR629.61   1100   629.77  1150   629.93  1200   630     1220   635     1280
GR    640   1320   650     1380   660     1450   670     1530   680     1620
X1    7.5   15     330     380    1200    900    1100
GR    655   0      650     40     645     100    640     140    635     220
GR    632   330    626     340    625     355    626     370    632     380
GR    635   410    640     450    645     520    650     580    655     620
X1    8.0   26     385     430    1000    1150   1100
GR    680   0      675     150    670     190    665     230    660     280
GR    655   320    650     340    645     350    640     370    636     385
GR    636   400    629     405    629     415    636     430    636     510
GR    640   590    645     660    650     700    655     730    660     770
GR    665   800    670     840    675     880    680     920    685     960
GR    690   1010
EJ
T1    OUTPUT ANALYSIS WORKSHOP PROBLEM 2
T2    X-SECTIONS TAKEN FROM FLOOD QUAD
T3    TRAIL CREEK SPF 17800 CFS
J1         3                                              631.85
J2    15


ER
```

Figure P12.2   Input file for output analysis problem 2.

```
ID.....1.......2.......3.......4.......5.......6.......7.......8.......9......10
```

T1	OUTPUT ANALYSIS WORKSHOP PROBLEM 3								
T2									
T3	ROCKY CREEK 5000 CFS								
J1	0	2	0	0	.002	0	0	0	30	0
J3	100	105	0	0	0	0	0	0	0	0
QT	2	5000	6000							
NC	.07	.07	.04	.3	.5					
X1	1	13	200	255						
GR	48.5	0	38	70	32.5	162.5	26.5	200	21	200
GR	17.5	217.5	17.5	237.5	21	255	27	255	33	312.5
GR	40	375	42	410	47.5	450				
X1	2	0	0	0	300	300	300			
X3	10	0	0	0	0	0	0	31	31	0
SB	1.05	1.6	2.6	0	53	6	6680	0	17.5	17.5
X1	3	0	0	0	40	40	40			
X2	0	0	1	30	32.5	0	0	0	0	0
X3	10	0	0	0	0	0	0	32.5	32.5	0
BT	7	0	48.5	0	70	38	0	162.5	32.5	0
BT	312.5	33	0	375	40	0	410	42	0	450
BT	47.5	0								
X1	4	0	0	0	75	75	75			
EJ										
T1	OUTPUT ANALYSIS WORKSHOP PROBLEM 3								
T2									
T3	ROCKY CREEK 6000 CFS								
J1	0	3	0	0	.002	0	0	0	31.5	0
J2	15									

ER

Figure P12.3   Input file for output analysis problem 3.

Tasks

1. Execute the program with each of the input files to obtain computed output. Obtain plots of cross sections and profiles to use in the analysis.

2. Review each set of output, using the guidelines and review suggestions presented in this chapter, to determine what the problem is.

3. Determine the changes that are needed for each data set to correct the problem, and revise the input file accordingly.

4. Execute HEC-2 again, using each of the four revised input files, and compare the new output with the original output. Repeat this procedure, if necessary, until reasonable results are obtained.

5. Write a brief description of the problems identified and the corrections made to solve them.

```
ID.....1.......2.......3.......4.......5......6.......7.......8......9......10

T1      OUTPUT ANALYSIS WORKSHOP PROBLEM 4
T2      PERCHED BRIDGE WITH LOW OVERBANKS
T3      BAKER CREEK
J1      0       0       0       0       0       0       0     8000    74.5      0
NC     .08     .08     .03      .1      .3
X1      1      12     200     250
GR    150       0      90      25    87.5     150      95     200    52.5     200
GR     50     215      50     235    52.5     250      95     250      75     335
GR     75     775     150     850
X1      2      12     200     250     200     200     200
GR    150       0      90      25    87.5     150      95     200    52.5     200
GR     50     215      50     235    52.5     250      95     250      75     335
GR     75     775     150     850
X1      3      12     200     250     200     200     200
GR    150       0      90      25    87.5     150      95     200    52.5     200
GR     50     215      50     235    52.5     250      95     250      75     335
GR     75     775     150     850
SB     .9    2.04     2.7       0      20       2    1845       3      50      50
X1      4      12     200     250      20      20      20
X2      0       0       1      90      75
BT      8       0     150     150      25      90      90     150    87.5    87.5
BT    200      95      90     250      95      90     335      75      75     775
BT     75      75     850     150     150
GR    150       0      90      25    87.5     150      95     200    52.5     200
GR     50     215      50     235    52.5     250      95     250      75     335
GR     75     775     150     850
X1      5      12     200     250      50      50      50
GR    150       0      90      25    87.5     150      95     200    52.5     200
GR     50     215      50     235    52.5     250      95     250      75     335
GR     75     775     150     850
EJ


ER
```

Figure P12.4 Input file for output analysis problem 4.

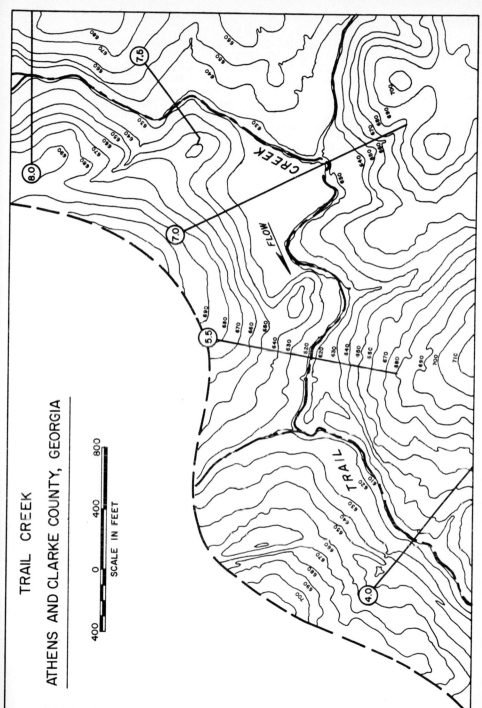

Figure P12.5 Trail Creek cross-section layout for output analysis problem 2.

References

1. Hydrologic Engineering Center, *HEC-2 Water Surface Profiles, Users Manual,* U.S. Army Corps of Engineers, Davis, Calif., 1982.
2. Jacob Davidian, "Computation of Water-Surface Profiles in Open Channels," Chap. A15 in *Techniques of Water-Resources Investigations of the United States Geological Survey,* Book 3: *Applications of Hydraulics,* U.S. Department of the Interior, 1984.
3. Ven T. Chow, *Open Channel Hydraulics,* McGraw-Hill, New York, 1959.
4. James G. Beasley, "An Investigation of the Data Requirements of Ohio for the HEC-2 Water Surface Profile Model," master's thesis, Ohio State University, Columbus, 1973.
5. Hydrologic Engineering Center, "Sensitivity Analysis of Factors That Influence Water Surface Profiles," U.S. Army Corps of Engineers, Davis, Calif., 1970.
6. Barr Engineering Company, "The Effect of Cross Section Data Errors on Water Surface Profile Determination," unpublished report for the Minnesota Department of Natural Resources, Minneapolis, 1972.
7. Hydrologic Engineering Center, *Accuracy of Computed Water Surface Profiles,* U.S. Army Corps of Engineers, Davis, Calif., 1986.

Water Surface Profiles through Bridges

Introduction

General

Since most floodplain studies are done in urban areas with numerous bridges, bridge analyses are a major part of the water surface profile computations. Because of the wide range of bridge types and the complex flow conditions that exist at bridges, bridge-flow analysis presents one of the most difficult problems to deal with.

Energy losses at bridges and culverts consist of the losses in the reaches immediately upstream and downstream from the structure and the losses through the structure itself. In the reach immediately upstream from a bridge, the flow is in a state of transition as it contracts to enter the bridge opening. Immediately downstream from the bridge, the flow is again in transition as it expands while flowing away from the bridge. In both of these reaches the energy losses are computed in HEC-2 with the standard step method. Losses through the structure itself are handled in three different ways, depending upon which method of analysis is used—the normal-bridge method, the special-bridge method, or external hydraulic calculations.

In the normal-bridge method, standard step calculations are carried through the structure. Cross sections located within the bridge are treated much the same as river cross sections where no bridge exists, except that the area of the structure below the water surface is blocked out of the total flow area, and the wetted perimeter is changed to reflect contact of the water with bridge surfaces. Friction factors are modified to reflect the roughness of the bridge opening. This method is particularly applicable to bridges without piers (piers are required for low-flow analysis with the special-bridge method), to bridges under high submergence, and to culverts with low-flow conditions.

In the special-bridge method, changes in water surface elevation through the bridge structure are computed with momentum and other hydraulic equations. This method can be used to analyze most types of bridges, but it is most applicable to bridges with piers where low flow controls, to bridges with pressure-flow or weir-flow conditions, and to bridges where the flow passes through critical depth while going through the structure. With this method, losses through the structure can be computed for low flow, weir flow, pressure flow, or any combination of these conditions.

The results of external calculations of energy losses through a bridge are introduced via the input file.

Nature of flow through bridges

The nature of flow through bridges is illustrated in Fig. 13.1. In this conceptualization,[1] the flow is divided into four regions: accretion, contraction, expansion, and abstraction. The stream is considered symmetrical about its centerline, and only the flow on one side of the centerline is shown.

Upstream from the bridge just far enough for the flow to be beyond the constricting influence of the bridge, the flow lines are parallel. As the flow moves from this point to the bridge, the flow in the overbanks must move to the channel so that all the flow can pass through the bridge opening. In this region of accretion, the flow is considered to be gradually varied. The region of contraction begins close to the upstream face of the bridge, where the flow contracts more severely to enter the bridge opening, and the geometry of the constriction is an important influence. A jet is formed which extends into the region of expansion, where it expands through turbulent diffusion and mixing. The flow in these two regions is rapidly varied, and the energy loss is relatively high.

The region of abstraction, downstream from the bridge, is a counterpart of the region of accretion upstream and has similar gradually varied flow characteristics. In this region, the flow moves laterally across the floodplain and eventually returns to normal flood-flow conditions at some distance downstream.

It is apparent from this conceptualization that flow through bridges does not exactly fit the assumptions of one-dimensional, and gradually varied flow, that are basic to HEC-2. However, the program has been designed to accommodate the complexities of bridge flow and produce reasonable results under most conditions.

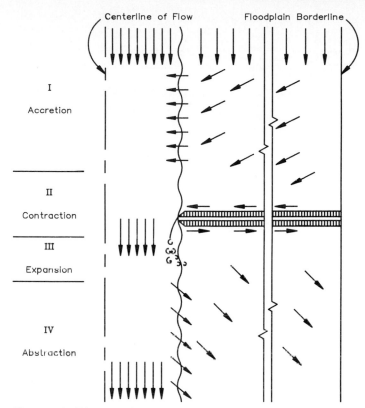

Figure 13.1   Diagram of flow through a bridge constriction.

Classes of low flow through bridges

A low-flow condition exists when all the water flows through the bridge opening and the water surface is at or below the low chord. The profiles shown in Fig. 13.2 illustrate three classes of low flow.

Class A low flow.   Class A low flow exists in a subcritical-flow regime when the water surface profile through the bridge remains above critical depth even though the critical-depth profile rises due to the constriction. The change in water surface elevation caused by the bridge is represented by the distance between the computed water surface elevation and the normal water surface (N.W.S.) profile, which would exist in the absence of the bridge.

Class B low flow.   In Class B low flow, the water surface profile passes through critical depth in the bridge constriction. This may happen both in subcritical flow, as illustrated in Fig. 13.2, and in supercritical

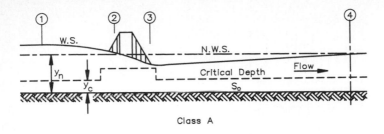

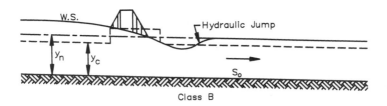

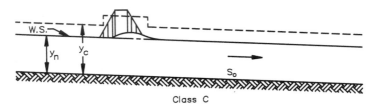

Figure 13.2 Water surface profiles through bridge constrictions for different classes of flow.

flow. In the subcritical case shown, the additional head required to pass the flow through the constriction is represented by the distance between the W.S. and N.W.S. profiles upstream. The water surface profile passes through critical depth, and the flow becomes supercritical for a short distance before returning to the subcritical condition in a hydraulic jump. In the case of supercritical flow, a second type of Class B flow can occur. The water surface profile rises in a hydraulic jump through critical depth, due to the bridge constriction, and the flow becomes subcritical for a short distance before returning to the supercritical condition in a hydraulic drop.

Class C low flow. The Class C profile is a supercritical profile that remains supercritical through the bridge. Although the water surface profile may rise due to the constriction, it does not rise high enough to reach critical depth.

Other flow conditions at bridges

Pressure-flow, weir-flow, and combined-flow conditions may occur at a bridge. Pressure flow occurs when the water surface elevation rises above the low chord of the bridge, submerging the bridge opening. Weir flow may occur in combination with either low flow or pressure flow. If the roadway approaching the bridge is low, it is possible to have weir flow in one or both of the overbank areas concurrent with low flow through the bridge opening. If weir flow occurs at an elevation higher than the low chord of the bridge, combined weir and pressure flow may result. However, water flowing over the bridge or the roadway approaching the bridge is not necessarily weir flow. Submergence and other factors, such as the existence of a relatively smooth flow cross section at a roadway, may not fit the hydraulic conditions for weir flow.

The Normal-Bridge Method

Defining the bridge structure

In the normal-bridge method, the bridge structure can be specified in input in two different ways: (1) by a table of roadway and low-chord coordinate points entered on BT lines and (2) by a set of coordinate points entered on GR lines, which define the essential parts of the structure except for constant low-chord and top-of-road elevations, which are entered on X2 lines.

A simple bridge structure is shown in Fig. 13.3 to illustrate how the channel geometry is modified with the first method. The points represented by small squares define the channel cross section without the bridge. The bridge structure is added with coordinates, represented by small triangles, which give the top-of-bridge elevation and corre-

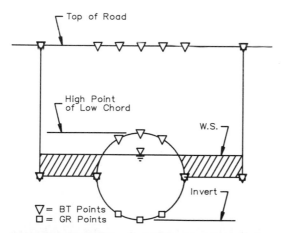

Figure 13.3 Definition of a bridge structure with the normal-bridge method.

sponding low-chord elevation at various stations measured horizontally across the bridge.

Flow in the stream is blocked by the cross-sectional area of the bridge structure defined in this manner. Areas of bridge abutments are blocked out by setting low-chord elevations on BT lines equal to corresponding ground elevations on GR lines at abutment locations.

In the area of the bridge opening, the low-chord elevations on the BT lines define the top of the opening, and the GR points define the bottom. For a curved opening, such as the one shown in Fig. 13.3, the program approximates the curvature of both the low chord and the bottom with straight lines between coordinate points. Increasing the number of points improves the approximation of the curves.

In standard step computations, the program recognizes the blocked-out area of flow depicted by the shaded area and reduces the conveyance accordingly. It also makes adjustments to the wetted perimeter to account for bridge surfaces in contact with the flow.

The roughness coefficient(s) for a bridge opening are specified on an NC, NH, or NV line. If the NV option, which provides for roughness variation in a vertical direction, is used, a table of n values vs. elevation must be specified. For a bridge cross section with a lower-chord roughness that is different from the bottom roughness, a relationship of composite n value vs. elevation must be developed. The program does not have the capability to use two different n values representing different roughness in the lower chord and channel portions directly.

Guidelines for locating bridge cross sections

Six cross sections are needed to model a bridge with this method (Fig. 13.4). For a subcritical regime the first cross section is located downstream from the bridge far enough for the flow to be fully expanded and the flow lines essentially parallel. The second is located close to the downstream side of the bridge. The third and fourth cross sections are located within the bridge, one near each face. The fifth is adjacent to the upstream side of the bridge. And the sixth is located upstream from the bridge just far enough to be beyond the region of accretion so that the flow lines are parallel.

Cross section 1 is the first required for the bridge analysis, but it would not ordinarily be the first cross section in the river model. There would be other cross sections both upstream and downstream from the six that are shown here.

The boundaries of the floodplain are shown with solid lines; the boundaries of effective flow are shown with dashed lines. The bridge abutments block substantial areas from flow in the right and left

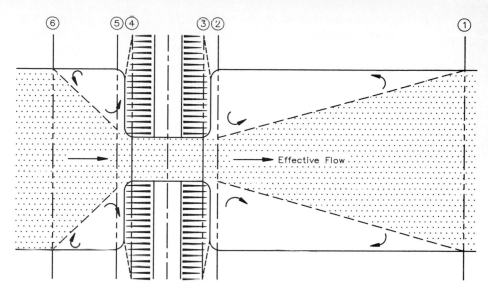

Figure 13.4   Cross-section layout for the normal-bridge method.

overbanks under low-flow conditions, and all the flow must pass through the opening. Thus, the extreme outside points of the opening normally dictate the boundaries of effective flow next to the bridge. And the dashed lines, defining the effective-flow area, converge to this opening width at cross section 2.

The standard step calculations proceed from the full-flow cross section (1) to the contracted cross section (2), just outside of the bridge. Cross section 2 is usually located 1 to 10 ft from the face of the bridge. The distance from cross section 1 to cross section 2 represents the distance required for the flow to expand from the bridge constriction to the full width of the natural cross section. In reality, this distance could vary greatly depending on the incisiveness of the channel, the roughness of the overbanks, the velocity of flow, and other conditions. The angle of expansion, represented by the angle of the effective-flow boundary with the floodplain boundary, may be very small under some conditions. In this case, the distance between cross sections 1 and 2 could be very large.

A rule of thumb developed from field observations sets the expansion angle equivalent to a ratio of 4:1. In other words, cross section 1 is located downstream a distance 4 times the average of the two abutment projections. If the estimated distance is unreasonably long under this rule, it should be arbitrarily shortened, or an intermediate cross section should be located between cross sections 1 and 2. It is a good idea to check the KRATIO for this reach in the output.

Cross section 3 is located just inside of the bridge. It is important that this cross section and cross section 4, just inside of the bridge at the upstream face, be included in the model. A cross section inside of the bridge and one immediately outside of the bridge represent different flow conditions, particularly with respect to roughness and cross-sectional geometry. Between cross sections 3 and 4 the friction loss within the bridge structure is computed. If the distance between these cross sections is excessively long, as might be the case with a long culvert, or there are changes in shape or roughness within the bridge, intermediate cross sections may need to be included between cross sections 3 and 4.

Cross section 5 serves a function similar to that of cross section 2: it redefines the conveyance conditions in the channel just outside of the bridge opening at the upstream face.

From cross section 5, the program steps upstream to the full-flow cross section (6), computing the contraction loss that occurs between these cross sections. The angle of contraction of effective-flow boundaries between these cross sections is larger than the expansion angle of the flow boundaries downstream. Two rules of thumb have been suggested for determining the contraction distance. One sets the distance between cross sections 5 and 6 equal to the average of the two side projections. The other sets the distance equal to the width of the channel opening.

In a multiple profile run, involving several different discharges, the length of the side projections will vary with each discharge. The purpose of the study should be considered in selecting a length which is most significant or representative. For example, in a flood insurance study, the length associated with the 100-year discharge might be used. Since bridges are frequently designed to pass the 100-year flood, a length for the side contraction corresponding to flow at the level of the low chord would be a reasonable approximation in the absence of more definitive hydrologic data.

Example problem: culvert analysis

An example problem from the user's manual[2] is described here to demonstrate the application of the normal-bridge method.

Geometric model. The six cross sections for this example problem are shown in Fig. 13.5. These have the same relative locations in plan view as the cross sections in Fig. 13.4. Cross sections 1 and 6 are the full-flow cross sections, taken downstream and upstream from the culvert, respectively. Cross sections 2 and 5 are located just outside, and cross sections 3 and 4 are located just inside of the ends of the culvert.

Cross sections 2 and 5 represent the limited-flow, or effective-flow,

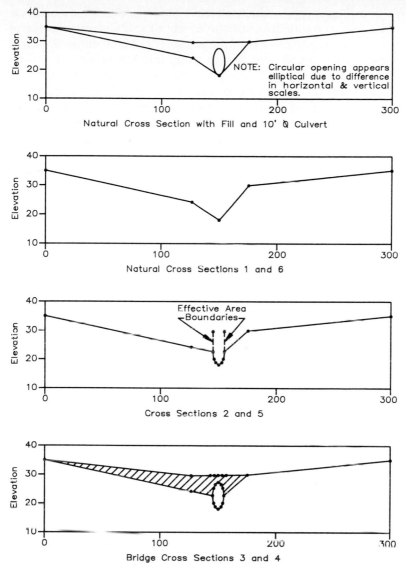

Figure 13.5   Six cross sections for the normal-bridge example problem.

area when all the water is flowing through the culvert. The dashed lines extended vertically from the right- and left-bank stations define the effective-flow boundaries that have been specified for these cross sections on X3 lines. The effective-flow option serves to contain the flow between these vertical lines at the bank stations until the water surface elevation exceeds the elevations assigned to the tops of these effective-flow boundaries. Once these boundary elevations are exceeded, the flow spreads out into the overbanks. Different elevations can be assigned to these two boundaries, and if this is done, flow will spread into one overbank area before going into the other.

Cross sections 3 and 4 are located inside of the bridge, and the structure is defined on BT lines for these cross sections. The BT points in effect define the top of road and block out the area of fill, shown shaded in the figure.

Input file. The input file for this example problem is shown in Fig. 13.6. After the title and job lines, the data for cross section 1, which is a natural-channel cross section, is entered. Since input data for a natural cross section was covered in a previous example, the input for cross sections 1 and 6 will not be discussed here except as related to

```
ID.....1.......2.......3.......4.......5.......6.......7.......8.......9......10

T1 NORMAL BRIDGE METHOD EXAMPLE
T2 USE OF BT DATA TO DEFINE THE BRIDGE STRUCTURE
T3 10-FT CIRCULAR CULVERT
J1                                                      450      25
NC    .08     .08     .03     .3      .5
X1    1       5       125     175     0       0       0
GR    35      0       25      125     18      150     30      175     35      300
X1    2       11      145     155     100     100     100
X3    10                                              29       29
GR    35      0       25      125     23      145     19.5    145.5   18.5    147.5
GR    18      150     18.5    152.5   19.5    154.5   23      155     30      175
GR    35      300
NC                    .012
X1    3                               10      10      10
BT    11      0       35      35      125     30      25      145     30      23
BT  145.5     30      25.5    147.5   30      27.5    150     30      28      152.5
BT    30      27.5    154.5   30      25.5    155     30      23      175     30
BT    30      300     35      35
X1    4                               100     100     100
X2                                                    1
NC                    .03
X1    5                               10      10      10
NC                            .1      .3
X3    10                                              30       30
X1    6       5       125     175     25      25      25
GR    35      0       25      125     18      150     30      175     35      300
EJ


ER
```

Figure 13.6 Input file for the normal-bridge example problem.

the use of revised expansion and contraction coefficients for modeling flow through the bridge. If higher values for expansion and contraction coefficients are to be used to model flow through the bridge, an NC line specifying these higher values should be located in front of the X1 line for cross section 1. Another NC line specifying a change back to the original values for flow upstream from the bridge should be located in front of the X1 line for cross section 6.

Definition of effective flow next to the culvert opening.   Cross section 2, which defines the effective-flow cross section at the downstream face of the culvert, is next. The data for this cross section is shown in Fig. 13.7. A new set of 11 ground points is entered on GR lines to define the bottom half of the culvert. The right- and left-bank stations are redefined to coincide with the right and left sides of the culvert at stations 145 and 155. The reach lengths of 100 feet are based on the 4:1 expansion of flow, as previously discussed.

ID.....1.......2.......3.......4.......5.......6.......7.......8.......9......10									
X1	2	11	145	155	100	100	100		
X3	10							29	29
GR	35	0	25	125	23	145	19.5	145.5	18.5	147.5
GR	18	150	18.5	152.5	19.5	154.5	23	155	30	175
GR	35	300							

Figure 13.7   Data input for cross section 2, including the definition of effective flow on an X3 line.

The effective-area option is specified with the "10" in field 1 of the X3 line. This instructs the program to consider all the flow to be confined within the channel as long as the water surface is lower than the right- or left-bank elevations. One should note that the effective-area option is keyed to the right- and left-bank stations as indicated on the X1 line, so it is important that the bank stations be located properly. The "29" entered in fields 8 and 9 of the X3 line indicates the vertical extension of effective-area control at these two bank stations. This is represented by the vertical dashed lines and points directly above the right- and left-bank stations shown in cross section 2 in Fig. 13.5.

How are the two values of 29 determined for defining effective flow? The minimum top-of-road elevation on the upstream side of the bridge is 30 ft, and as long as the water surface elevation upstream is lower than 30 ft, all the flow will go under the bridge. Above 30 ft, part of the flow will come over the top of the bridge, and flow conditions will change. On the downstream side of the bridge, what elevation of flow will be associated with 30 ft on the upstream side? The difference between upstream and downstream elevations in this instance accounts for energy loss through the bridge, which is not known at this stage of the computations. An elevation somewhat below that on the upstream side probably should be assumed. If the distance between the top of road and low chord is not too great, a reasonable assumption might be the low-chord elevation or an elevation midway between this elevation and the top of road. However, this assumption would not be suitable for culverts with a large amount of fill material between the low-chord elevation and the top of road. An elevation taken a foot or two below the top of road would be a preferred starting assumption.

What would happen if the assumed value were below the computed water surface elevation at this point? Since the effective-area option is an "all-or-nothing-at-all" condition, the effective-area constraints would disappear in this situation, and the flow would fill the overbanks in all areas lower than the water surface elevation. This increased conveyance would be inappropriate unless sufficient water was flowing over the bridge to cause flow in these overbank areas.

The effective-area option can cause a problem in the computations when the computed water surface elevation is near the elevation or elevations specified to control the effective area. In one iteration of the computations, the water surface may be below the specified elevation, and the effective-area constraints are in, but at the next iteration, the water surface elevation is above the specified elevation, and the constraints are out. It may be difficult for the program to obtain a balance in this situation.

There is some latitude in the assumption of effective-area elevations as long as they fit the conditions being modeled and any seeming inconsistencies caused by them can be explained logically. For example, for a condition of flow such that only a small portion would come over the top of the bridge, the effective-area option might be set downstream at an elevation equal to or above the top-of-road elevation. The top width of flow as indicated in output for the upstream side of the bridge would be broad, with flow occurring in both overbanks, but because of the effective-area option, the flow would be very narrow on the downstream side of the bridge. This inconsistency might be explained by reasoning that the small amount of flow over the bridge would be insufficient to cause significant flow in the overbanks downstream.

Definition of the culvert structure. Lines of input for cross sections 3 and 4, located within the bridge, are shown in Fig. 13.8. A change of surface roughness within the bridge is specified on the NC line for cross section 3. Arbitrary reach lengths between cross sections 2 and 3 of 10 ft are entered on the X1 line. Since the configuration of the bottom of the culvert was defined with GR points for cross section 2, the GR data was repeated for cross section 3 by leaving field 2 blank on the X1

ID	1	2	3	4	5	6	7	8	9	10
NC			.012							
X1	3				10	10	10			
BT	11	0	35	35	125	30	25	145	30	23
BT	145.5	30	25.5	147.5	30	27.5	150	30	28	152.5
BT	30	27.5	154.5	30	25.5	155	30	23	175	30
BT	30	300	35	35						
X1	4				100	100	100			
X2							1			

Figure 13.8 Data input for cross sections 3 and 4, defining the bridge structure.

line. Fields 3 and 4 also were left blank, since the right- and left-bank stations for cross section 2 were repeated.

Top-of-road and low-chord points corresponding to GR stations are specified on BT lines included with the data for cross section 3. There must be a GR station at all locations where BT points are specified, but it is not necessary to specify BT points at all GR stations. The purpose of the BT data is to define the area of the bridge structure (in this case, area of fill material) to be subtracted from conveyance and define the configurations of the top of bridge and low chord. In the area of the abutments, a low-chord elevation specified for a BT point should be the same as the ground elevation specified for the corresponding GR point.

At cross section 4, located on the other side of the bridge, but still within the bridge opening, all the data for cross section 3 is repeated except the reach lengths. Reach lengths of 100 ft, equal to the distance through the bridge, are entered in fields 5, 6, and 7 of the X1 line. The GR data and the right- and left-bank stations are repeated by leaving fields 2, 3, and 4 blank on the X1 line. An X2 line with a "1" in field 7 causes the BT data to be repeated.

Cross sections upstream from the culvert. The input data for cross sections 5 and 6, upstream from the bridge, is very similar to that for cross sections 1 and 2, downstream from the bridge. An NC line is used at cross section 5 to change the surface roughness coefficient from the p   ious bridge cross section to an appropriate value for the natural channel. Since cross section 5 is an effective-flow cross section, an X3 line is used to implement this option. However, the effective-flow elevations in fields 8 and 9 are set equal to 30 for this upstream cross section. This elevation is equal to the elevation of the top of road ELTRD. Essentially, all of the preceding discussion of the effective-flow option at downstream cross section 2 applies at cross section 5 as well. The ground points and right- and left-bank stations are redefined at cross section 6 to represent the natural channel under full-flow conditions. The EJ line followed by three blank lines and an ER line end the job and the run, respectively.

Optional method of defining a bridge with X2 and GR lines

In the normal-bridge method, it is possible to define the structure of some bridges without using BT lines. The piers and abutments can be defined with GR points, and a horizontal top of road and horizontal low chord can be specified on X2 lines to complete the structure. This procedure is described with reference to the bridge cross section shown in Fig. 13.9 and the excerpt from the input file shown in Fig. 13.10.

The normal-bridge method of analysis is essentially the same with this option as in the previous example. The only change is in the data for the two cross sections taken inside of the bridge opening. GR points are used in a different manner in this option to define all the essential parts of the cross section except the horizontal lines representing the top of road and the low chord. The 15 GR points used in this example are numbered in Fig. 13.9 in the same sequence in which they are en-

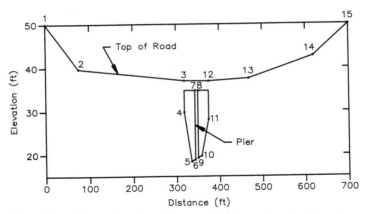

Figure 13.9 Bridge structure defined with GR points and with horizontal lines to define the top of road and low chord at the bridge opening.

```
ID.....1.......2.......3.......4.......5.......6.......7.......8.......9......10

X1    2.1     15     325     375      10      10      10
X2                            35      37
GR     50      0      40      75    37.5     325      30     325      19     345
GR     19     350     35     350      35     352    19.5     352      50     360
GR     28     375     37     375      38     475      43     625      20     700
```

Figure 13.10 Excerpt from an input file corresponding to the definition of the bridge structure in Fig. 13.9.

tered on the GR lines in Fig. 13.10. These points describe the top of road except for a horizontal section over the top of the bridge opening, the sides and bottom of the bridge opening, and the pier. The latter is defined by a line extending from point 6 up the left side of the pier to point 7, across the top of the pier to point 8, and down the right side of the pier to point 9. The missing top of road section and the lower chord are filled in by the elevations specified in fields 4 and 5 of the X2 lines.

Output for the normal-bridge method

When the normal-bridge method is used, additional bridge data is provided with the detailed output for the cross sections at the bridge. For the cross sections at the upstream and downstream faces of the bridge, statement no. 3495 appears if the effective area option is used. This statement indicates that this option is being used and gives the effective elevations, ELLEA and ELREA, for the right- and left-bank stations. For each of the two cross sections located within the bridge, statement no. 3370 gives the number of stations used to describe the bridge deck, NRD, and the elevations of the top of road and low chord, ELTRD and ELLC, respectively.

The Special-Bridge Method

In the special-bridge method, standard step calculations are used downstream and upstream from the bridge the same as in the normal-bridge method, but the changes in energy and water surface elevations through the bridge structure are computed with a separate set of equations. Ordinarily, only four of the six cross sections shown in Fig. 13.4—1, 2, 5, and 6—are required with this method. The particular set of equations selected by the program to compute changes through the structure depends on flow regime, depth of water at the bridge, and other factors. Energy losses are computed for low flow, for pressure flow, or for combinations of one or the other of these with weir flow.

Computational theory and methodology

Determination of low-flow classification. The procedure used for low-flow computations varies depending on the presence or absence of bridge piers. If there are no piers, the low-flow solution reverts to the standard step calculations of the normal-bridge method. This occurs because the determination of bridge loss by equations used in the special-bridge method for low flow is based upon the obstruction to flow caused by piers. Reverting to the normal-bridge method under these conditions may result in erroneous computations for low flow. Cross sections defining bridge geometry and friction coefficients are missing, as explained in the preceding section.

In the special-bridge method, if piers are present, the program computes a momentum balance for cross sections just outside and inside of the bridge to determine the class of flow.

$$m_1 - m_{p1} + \frac{Q^2}{g(A_1)^2} \left(A_1 - \frac{C_D}{2} A_{p1} \right) =$$

$$m_2 + \frac{Q^2}{gA_2} = m_3 - m_{p3} + \frac{Q^2}{gA_3} \qquad (13.1)$$

where A_1, A_3 = flow areas at upstream and downstream sections, respectively

A_2 = flow area (gross area – area of piers) at a section within constricted reach

A_{p1}, A_{p3} = obstructed areas at upstream and downstream sections, respectively

$\bar{y}_1, \bar{y}_2, \bar{y}_3$ = vertical distance from water surface to center of gravity of A_1, A_2, and A_3, respectively

m_1, m_2, m_3 = $A_1\bar{y}_1$, $A_2\bar{y}_2$, and $A_3\bar{y}_3$, respectively

m_{p1}, m_{p3} = $A_{p1}\bar{y}_{p1}$ and $A_{p3}\bar{y}_{p3}$, respectively

C_D = drag coefficient

$\bar{y}_{p1}, \bar{y}_{p3}$ = vertical distance from water surface to center of gravity of A_{p1} and A_{p3}, respectively

Q = discharge

g = gravitational acceleration

The three equalities expressed in Eq. (13.1)[3] represent the momentum flux in the bridge constriction based on channel characteristics and flow depths upstream and downstream from the constricted section and within the constriction itself. By computing the momentum flux based on critical depth inside of the constriction and comparing it with the computed momentum flux based on the downstream depth, the program can determine the class of flow for a subcritical-flow regime. The program can make a similar determination for a supercritical regime by comparing the critical momentum flux within the constriction with the momentum flux based on the upstream depth. Since the momentum computations use a trapezoidal cross section, a trapezoidal approximation of the bridge opening must be specified in input. This is done on the SB line.

Class A low flow. When Class A low flow occurs, the depth of flow remains above critical depth through the bridge, and the Yarnell equation [Eq. (13.2)] is used to determine the change in water surface elevation.[4] As in the momentum calculations, a trapezoidal approximation of the bridge opening is used.

$$H_3 = 2K\,(K + 10\omega - 0.6)\,(\alpha + 15\alpha^4)(\frac{V_3^2}{2g}) \qquad (13.2)$$

where H_3 = drop in water surface elevation from upstream to downstream sides of the bridge

K = pier shape coefficient

ω = ratio of velocity head to depth downstream from the bridge

α = obstructed area ÷ total unobstructed area

V_3 = velocity downstream from the bridge

Pier shape coefficients for use in Eq. (13.2) are shown in Table 13.1. Since the Yarnell equation was derived from empirical data obtained from hydraulic models of piers, caution should be exercised in using pier coefficients for obstructions other than piers. For example, defining areas of fill between circular culverts as piers is not recommended.

TABLE 13.1   Pier Shape Coefficients

Pier shape		K
Square nose and tail		1.25
90° triangular nose and tail		1.05
Twin-cylinder piers without diaphragm		1.05
Twin-cylinder piers with connecting diaphragm		0.95
Semicircular nose and tail		0.90

The drop in water surface elevation H_3, computed with Eq. (13.2), is added to the computed water surface elevation at the downstream face of the bridge to obtain the computed water surface elevation at the upstream face. With the upstream water surface elevation known, the program computes the velocity head and energy elevation at this location.

Class B low flow.   When Class B low flow occurs in a subcritical-profile computation, critical depth is determined in the bridge constriction. A new downstream depth below critical and a new upstream depth above critical are computed, corresponding to the momentum flux at these locations. A note is written to the output, DOWNSTREAM

ELEV IS X, NOT Y, HYDRAULIC JUMP OCCURS DOWNSTREAM. When this occurs, a computation for a supercritical profile (M3 curve) could be started at the downstream cross section with a known water surface elevation equal to X. Depending on how far the supercritical-flow regime extends downstream, this might be useful in defining the profile and in narrowing the uncertainty in locating the hydraulic jump.

When Class B low flow occurs in a supercritical-profile computation, the bridge acts as a control and causes the upstream water surface elevation to rise above critical depth. The program computes upstream and downstream water surface elevations with Eq. (13.1) and writes a note to the output, UPSTREAM ELEVATION IS X, NOT Y, NEW BACKWATER REQUIRED, indicating that a subcritical profile (S1 curve) could be computed upstream from the bridge, starting at elevation X.

Class C low flow.   For Class C low flow, in which the regime remains supercritical through the bridge, the downstream depth and the depth in the bridge are determined with Eq. (13.1) on the basis of the computed upstream depth.

Pressure flow.   The computation of pressure flow in HEC-2 is done with a modified orifice equation [Eq. (13.3)], which is derived by applying the energy equation between a point immediately downstream from the bridge and one immediately upstream from the bridge.

$$Q = A \sqrt{\frac{2gH}{K}} \qquad\qquad (13.3)$$

where H = difference between the energy gradient elevation upstream and the tailwater elevation downstream
   K = total loss coefficient
   A = net area of the orifice
   g = gravitational acceleration
   Q = total orifice flow

The total loss coefficient K, specified as the variable XKOR on the SB line, is equal to 1.0 plus the sum of loss coefficients for intake, intermediate piers, surface roughness, and other losses. Tabulated values for the coefficient C in the commonly used orifice equation $Q = CA(2gH)^{0.5}$ can be used to determine K with the conversion relationship $K = 1/C^2$. However, care should be taken to make certain that the definition of H, the head on the orifice, is compatible.

Values for C have been determined by experiment.[5] The value of C for conditions of full submergence is in the range of 0.7 to 0.9. A value of 0.8 is recommended for a typical two- to four-lane concrete girder bridge. In the absence of calibration data, a value of C of 0.8 ($K = 1.56$) would be applicable to most bridges and short culverts. For long culverts, the coefficient K can be computed as a sum of loss coefficients [Eq. (13.4)].

$$XKOR = k_e + k_f + 1 \qquad (13.4)$$

where k_e is the entrance loss coefficient and k_f is the friction loss coefficient.

Typical values of the coefficients are shown below:

Description	k
Intake (k_e)	0.1 to 0.9
Intermediate piers	0.05
Friction (Manning's equation)	k_f
	$XKOR = \Sigma k + 1$

$k_f = 29n^2 R^{4/0}$

The *Handbook of Hydraulics*[6] gives entrance loss coefficients that are applied to velocity heads for pipe flow. It specifies a coefficient of 0.1 for a flush orifice inlet and a coefficient of 0.15 for a projecting concrete pipe inlet. It gives loss coefficients as high as 0.9 for projecting corrugated metal pipes. All of these coefficients are applied to the velocity head for the pipe.

For multiple culverts with full flow, a single equivalent coefficient can be computed [Eq. (13.5)].

$$Q = \sqrt{2gH} \; AT \; \sqrt{1/K_{equiv}} \qquad (13.5)$$

where $K_{equiv} = \dfrac{AT^2}{\left(\displaystyle\sum_{i=1}^{n} \sqrt{A_i^2/K_i} \right)^2}$

AT = total area
A_i = area of individual culvert
K_i = coefficient for individual culvert
n = number of culverts

In computing the head on the orifice H, the program goes through a trial-and-error procedure using the orifice equation. An upstream energy elevation is assumed, the discharge through the bridge is calculated, and this calculated discharge is compared with the given discharge in the stream. The procedure is repeated if necessary until a difference of less than 1 percent is obtained.

Weir flow. Flow over the bridge and the roadway approaching the bridge is calculated with the standard weir equation [Eq. (13.6)].

$$Q = CLH^{3/2} \qquad (13.6)$$

where C = coefficient of discharge
  L = effective length of inundated roadway
  H = difference between the energy grade line upstream and the roadway crest elevation
  Q = total flow over the weir

The coefficient of discharge C, which is entered on the SB line as the variable COFQ, ranges from 2.5 to 3.1 (1.39 to 1.72 for metric units) for broad-crested weirs under free-flow conditions.[2] The value of C should be increased with increasing head on the weir crest. It should be decreased when obstructions, such as debris caught on the bridge structure and curbs, increase the resistance to flow. For flow with submergence, the program automatically reduces the coefficient C on the basis of criteria developed by the Federal Highway Administration[5] (Fig. 13.11). Coefficients of discharge C_f and C_s—for free flow and flow with submergence, respectively—are obtained from relationships of energy head H to road width l and tailwater submergence D. For values of H/l less than 0.15, values for C_f are taken from Fig. 13.11a. For larger values of H/l, C_f is taken from Fig. 13.11b. If D/H is larger than 0.7, the submergence factor C_s/C_f is obtained from Fig. 13.11c.

Other sources of weir coefficients include the *Handbook of Hydraulics*,[6] which contains tables of coefficients for broad-crested weirs, with the coefficients varying with measured head H and breadth of weir. For rectangular weirs with a breadth of 15 ft and an H of 1 ft or more, the coefficient is 2.63. For trapezoidal weirs, the coefficients are larger, with typical values ranging from 2.7 to 3.08.

The HEC-2 users' manual[2] recommends the assumption of a coefficient of 2.6 for rectangular-weir flow over the bridge deck and a value of 3.0 for weir flow over the roadway approaches. If weir flow occurs over a combination of bridge deck and roadway, an average coefficient weighted by weir length is recommended.

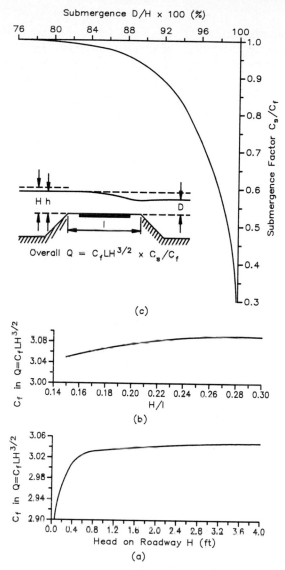

Figure 13.11 Discharge coefficients for flow over roadway embankments.

The program computes the weir loss through an iteration procedure, similar to that described for orifice flow, based on the matching of computed and given Q values until the difference is no more than 1 percent.

Combination flow

Combinations. Various combinations of complex flow conditions can be described with reference to the bridge cross section shown in Fig. 13.12.[7] HEC-2 is capable of computing water surface profiles through the complete range of flow conditions indicated by regions A through D. When the water level is in low-flow region D and the program determines that the flow is Class A, the Yarnell equation is used to determine the change in water surface profile through the bridge.

When the water level is in region C, above the elevation of the low point of the approach roadway (elevation 3) but below the low chord of the bridge (elevation 2), a combination of low flow and weir flow exists. Part of the flow is going through the bridge, and part is going over the roadway on each side of the bridge. The program has to perform a dual trial-and-error computation, using the Yarnell equation for flow through the bridge and the weir equation for flow over the roadway. Since the weir in this case has a complex shape, it is necessary to define the roadway, which is the crest of the weir, with a bridge table (on BT lines).

A combination of pressure flow and weir flow exists when the water level is in regions B and A. In the computation of pressure flow through the bridge with the orifice equation, the trapezoidal approximation of the bridge opening is not used; instead, the total area of the opening as specified on the SB line is used.

When low flow or pressure flow is combined with weir flow, a trial-and-error procedure is used to determine the amount of flow in each component. The procedure involves assuming an energy elevation for each component and computing the total discharge, comparing this total with the specified discharge, and repeating the procedure if necessary until a difference of less than 1 percent is achieved.

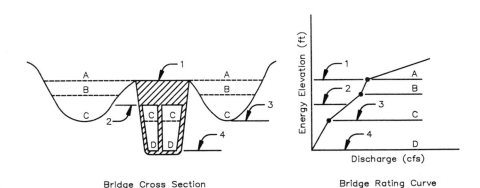

Bridge Cross Section   Bridge Rating Curve

Figure 13.12 Illustration of various complex flow conditions at a bridge with low overbanks.

Computational logic. One of the first steps in the computational logic for combined-flow analysis is to determine which type of flow controls. The water surface elevation on the other side of the bridge is computed, under the assumption of low-flow conditions for the total flow in the channel. The height of the trapezoidal opening is unrestricted in this preliminary computation, which may result in a water surface elevation well above the low chord of the bridge. If the bridge does not have piers, the special-bridge equations for low flow cannot be used for this computation, and the water surface elevation is assumed to be the same as at the previous cross section.

The program checks for potential weir flow by comparing this computed or assumed water surface elevation with the minimum top-of-road elevation ELTRD. If the comparison indicates possible weir flow, the energy elevation is computed—the velocity head from the previous cross section is added to the water surface elevation. This low-flow energy elevation EGLWC is then compared with the maximum elevation of the low chord ELLC. If the EGLWC is higher than the ELLC, an energy elevation EGPRS is computed under the assumption that all the flow in the channel is pressure flow. If the EGLWC is lower than the ELLC, the program concludes that low flow exists and checks again to determine if weir flow also exists. If there is weir flow, a trial-and-error solution will be made for a combination of low flow and weir flow. If weir flow is not indicated, a low-flow solution will be computed.

When the EGPRS is calculated, as a result of the low-flow energy elevation being greater than the low-chord elevation, it is compared with the EGLWC to determine whether pressure flow or low flow controls. However, when the minimum elevation of the top of road ELTRD is less than the maximum elevation of the low chord ELLC, indicating possible weir flow, a different comparison is made. The low-flow energy elevation EGLWC is compared with the estimated maximum energy elevation for low flow, which is computed as 1.5 times depth plus invert elevation. The estimated maximum energy elevation is used in the comparison rather than EGPRS because the low road elevation would cause weir flow to exist prior to the occurrence of pressure flow. Depth is defined as the difference between the low-chord elevation and the invert elevation.

At critical depth in a rectangular flow cross section, 1.5 times the depth is the specific energy (Fig. 13.13a). If critical depth occurred just at the maximum low chord elevation, it would produce the maximum possible energy elevation for low flow, as shown in Fig. 13.13b. An energy elevation greater than this value would indicate pressure flow. For energy elevations between the low chord and this maximum for low flow, the program computes the energy elevations for combined low and weir flow and combined pressure and weir flow and compares them. The higher of the two controls.

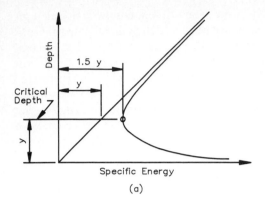

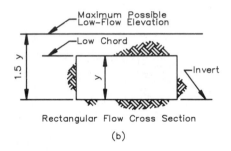

Figure 13.13 Relationship of depth to specific energy for a rectangular flow cross section. (a) Specific energy curve; (b) rectangular cross section.

Location of cross sections for special-bridge models

A plan view of a typical layout for cross-section spacing in the special-bridge method is shown in Fig. 13.14. Four cross sections are shown, in contrast to the six used in the normal-bridge method. The two that are missing are the two that are located within the bridge structure itself. These are not needed because of the way the water surface profile is determined. Standard step calculations are used to compute the water surface profile from cross section 1 to cross section 2. Standard step calculations are not used between cross sections 2 and 3. The change in water surface elevation through the structure is computed with semiempirical hydraulic equations, and this change is added to the computed water surface elevation at cross section 2 to obtain the computed water surface elevation at cross section 3. Standard step calculations are then resumed at cross section 3 and continued to cross section 4 and beyond.

The four cross sections in the special-bridge method, two upstream and two downstream, are essentially the same as the four corresponding cross sections in the normal-bridge method. One difference in the input that should be noted is the additional data related to the special-bridge method that is required for cross section 3, next to the upstream face of the bridge. This will be explained in the example which follows.

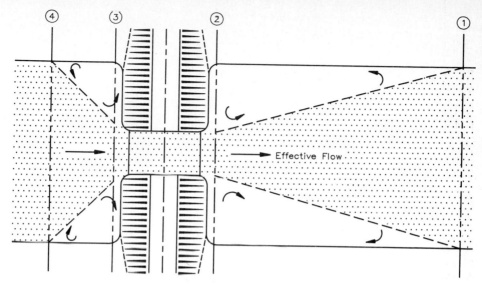

Figure 13.14 Cross-section layout for the special-bridge method.

Example problem: special-bridge method

The purpose of this example is to demonstrate how an input file is developed for the cross sections shown in Fig. 13.15. To simplify the discussion, the natural cross sections are the same at all locations. This common cross section is shown in Fig. 13.15a. A cross section through the bridge structure is shown in Fig. 13.15b. The input file for this example is shown in Fig. 13.16.

The data included in the first part of the input file, down through the data for cross section 2, is similar to the data in the normal-bridge example discussed previously. The discussion here will be confined to the SB line and the lines for cross section 3, which have data that is different in the special-bridge method. These lines of input are shown in Fig. 13.17.

The SB line is used to define the hydraulic characteristics of the bridge. The pier shape coefficient XK for the Yarnell equation used in low-flow calculations is specified in field 1 of the SB line. The orifice equation coefficient XKOR and the weir equation coefficient COFQ are specified in fields 2 and 3, respectively. If a horizontal fixed-length weir surface is to be used in lieu of a more complex surface, which would be defined with BT points, the weir length RDLEN is specified in field 4. In this example, field 4 is left blank, indicating that BT lines will be used.

Variables specified in the remaining fields on the SB line, fields 5 through 10, define the trapezoidal approximation of the bridge opening used in low-flow computations. To obtain good accuracy with the momen-

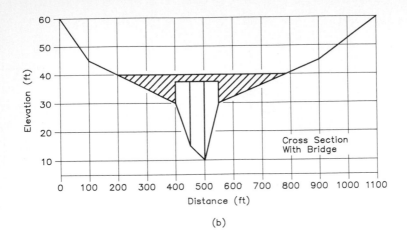

(b)

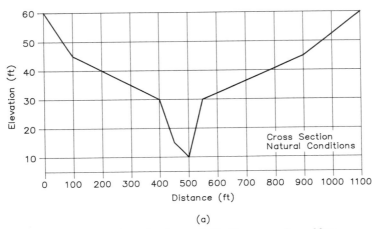

(a)

Figure 13.15   Cross sections for the special-bridge example problem.

tum and Yarnell equations used in low-flow computations, it is necessary to configure the trapezoid so that its area of flow very closely approximates the area of flow in the actual bridge opening over the full range of low-flow depths. In other words, at any depth under low-flow conditions, the area of flow in the trapezoid should be very close in magnitude to the area of flow in the actual bridge opening. If it is impractical to obtain a close approximation over the full range of depths, the trapezoid should be configured to obtain a good approximation at the most significant or frequently used depths in the analysis.

A common way to design the trapezoid is to sketch it on a drawing of the bridge opening, superimposing the bottom and sides to match those of the actual opening as closely as possible. The invert elevation can be raised or lowered slightly if necessary. (See the enlarged cross-

ID.....1.......2.......3.......4.......5.......6.......7......8.......9......10

ID	1	2	3	4	5	6	7	8	9	10
T1 EASY CREEK										
T2 WATER SURFACE PROFILE										
T3 SPECIAL BRIDGE MODEL										
J1		2						32		
NC	.08	.08	.04	.3	.5					
QT	1	10000								
X1	1	8	400	550						
GR	60		45	100	30	400	15	450	10	500
GR	30	550	45	900	60	1100				
X1	2				500	500	500			
X3	10							38	38	
SB	1.05	1.5	2.5		60	4	2700	2	12.5	12.5
X1	3				100	100	100			
X2			1	37	40					
X3	10							40	40	
BT	6		60		100	45		200	40	
BT	780	40		900	45		1100	60		
X1	4				150	150	150			
EJ										
ER										

Figure 13.16 Input file for the special-bridge example problem.

ID.....1.......2.......3.......4.......5.......6.......7......8.......9......10

ID	1	2	3	4	5	6	7	8	9	10
SB	1.05	1.5	2.5		60	4	2700	2	12.5	12.5
X1	3				100	100	100			
X2			1	37	40					
X3	10							40	40	
BT	6		60		100	45		200	40	
BT	780	40		900	45		1100	60		

Figure 13.17 Data input for cross section 3, including the bridge entered on the SB line.

section of the bridge structure in Fig. 13.18.) After a suitable trapezoid is determined, the dimensions are measured, and the appropriate entries are made on the SB line. The required variables are entered in fields 5 through 10 as follows: the bottom width in 5, the total width of piers in 6, the net area of the opening below the low chord in 7, the vertical side slopes in 8, the channel invert elevation at the upstream side of the bridge in 9, and the channel invert elevation at the downstream side in 10. These entries are 50, 4, 2800, 2, 13, and 13, respectively.

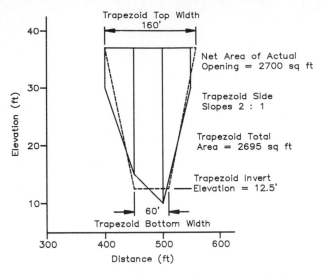

Figure 13.18 Definition of an approximate trapezoidal opening for the special-bridge computation.

The rest of the data for defining the bridge structure is included with the data for cross section 3, immediately upstream from the bridge. The X2 line has a "1" in field 3, indicating to the program that the special-bridge method is to be used. The low-chord elevation, 37, in field 4 is a control point at which the program begins checking for pressure flow, and the top-of-road elevation, 40, in field 5 is a control point for weir flow. These are key decision criteria in the program, and as such, they should always be specified on the X2 line, even though top-of-road elevations may also be specified on BT lines.

The BT lines serve a different function in the special-bridge method than they do in the normal-bridge method. BT points are used to define the top-of-road profile for weir-flow computations. Since the low chord is not defined by the BT points, every third field on the BT line that normally would contain a low-chord point is left blank. The BT points do not have to coincide with GR points in this application.

The fact that the BT data can be specified quite differently in the special-bridge method as compared with the normal-bridge method has significant implications when the special-bridge computation reverts to a normal-bridge computation for low flow in a bridge without piers. The geometry and roughness coefficients are not likely to be appropriately defined for normal-bridge analysis. This situation should be avoided, or it should be corrected when it occurs.

SPECIAL BRIDGE

SB	XK	XKOR	COFQ	RDLEN	BWC	BWP	BAREA	SS	ELCHU	ELCHD
	1.05	1.50	2.50	.00	60.00	4.00	2700.00	2.00	12.50	12.50

*SECNO 3.000
CLASS A LOW FLOW

3420 BRIDGE W.S.=   32.25 BRIDGE VELOCITY=   5.30   CALCULATED CHANNEL AREA=   1886.

EGPRS	EGLWC	H3	QWEIR	QLOW	BAREA	TRAPEZOID AREA	ELLC	ELTRD
.00	32.65	.02	0.	10000.	2700.	2573.	37.00	40.00

3495 OVERBANK AREA ASSUMED NONEFFECTIVE,ELLEA=   40.00 ELREA=   40.00

3.00	22.30	32.30	.00	.00	32.65	.35	.02	.00	30.00
10000.	0.	10000.	0.	0.	2095.	0.	29.	3.	30.00
.03	.00	4.77	.00	.000	.040	.000	.000	10.90	400.00
.000518	100.	100.	100.	0	0	0	.00	150.00	550.00

Figure 13.19   Detailed output for a special-bridge cross section.

Output for the special-bridge method

When the special-bridge method is used, several lines of bridge data are included with the detailed output for the cross section associated with the SB line of input. For example, detailed output for cross section 3 in the special-bridge problem just described is shown in Fig. 13.19. The variables specified on the SB line of input are listed first. Then, after the cross-section number, the class of flow is indicated. On the next line, a special note defining the flow through the bridge is given. In this case, statement no. 3420 gives the water surface elevation under the bridge, the velocity through the bridge, and the area of flow through the bridge. Variables computed for the bridge flow are shown on the next line. These variables, some of which are described in the preceding section on computational logic, are as follows:

EGPRS   The energy grade line computed under the assumption that flow is pressure flow.

EGLWC   The energy grade line elevation computed under the assumption that low flow exists.

H3   Drop in water surface elevation from upstream to downstream sides of the bridge, computed with the Yarnell equation, under the assumption that Class A low flow exists.

QWEIR	Total weir flow at the bridge.
QLOW	Total low flow at the bridge.
BAREA	Net area of the bridge opening below the low chord, entered on the SB line.
TRAPE-ZOID AREA	Net area of the bridge opening up to the low chord as defined by variables on the SB line representing the bottom width of the bridge opening (BWC), the total width of piers (BWP), and the side slopes (SS). Should be close to the value of BAREA.
ELLC	Elevation of the low chord. Equals ELLC specified on the X2 line, if used. Otherwise, it equals the maximum low-chord elevation in the BT data.
ELTRD	Elevation of the top of roadway. Equals ELTRD specified on the X2 line, if used. Otherwise, it equals the minimum top-of-road elevation in the BT data.

Statement no. 3495, which precedes the detailed cross-section output, indicates that the effective-area option is used and gives the elevations used at the left- and right-bank stations, ELLEA and ELREA, respectively. A statement no. 3495, giving similar information, also is included with detailed output (not shown) for the cross section at the downstream face of the bridge.

Externally Computed Bridge Losses

Total bridge loss between cross sections 1 and 4 in Fig. 13.14 can be computed externally and introduced in the data for cross section 4 by two different methods. One method is to specify a computed change in water surface elevation BLOSS on the X2 line. This method is not suitable for a multiple profile run because only a single value can be specified, which would apply to all profiles. The other method, in which the loss is input with an X5 line, is suitable for a multiple profile run. Several different values for the loss can be specified on the X5 line, and the program can select one of these for each profile. The value to be selected for a particular profile is indicated on the job line, J1, in field 2. The X5 line can introduce a change in water surface elevation between cross section 1 and 4 or a new elevation at cross section 4. An example of an input file containing an X5 line for a three-profile run is shown in Fig. 13.20.

```
ID.....1.......2.......3.......4.......5.......6.............,.......8.......9......10
T1    SPECIAL BRIDGE PROBLEM
T2    EXTERNALLY COMPUTED CHANGE IN WATER SURFACE ELEVATION THROUGH BRIDGE
T3    CHANGE IS SPECIFIED ON X5 LINE OF INPUT
J1            2                                                       30
NC  .08     .08     .05     .3     .5
QT   3     2000    4500    6000
X1   1      10      325     375     0       0       0
GR  50      0       40      75     35      250      30     325      19     345
GR  20     360      28      375    38      475      43     625      50     700
X1   4                              415     415     415
X5  -3     1.07    2.20    2.56
EJ
T1
T2    SECOND PROFILE
T3
J1            3                                                       34
J2   2
T1
T2    THIRD PROFILE
T3
J1            4                                                       36
J2  15



ER
```

Figure 13.20   Example of X5-line input for externally computed energy loss at a bridge.

Selection of a Method for Bridge Analysis

This section provides some guidelines for choosing an appropriate method for bridge analysis.

General

The normal-bridge method is most applicable when friction losses are the predominate consideration or bridge conditions make it impractical to use the special-bridge method. The special-bridge method is most applicable for conditions of weir flow, pressure flow, low flow, or some combination of these that can best be modeled with the hydraulic equations available in this method. If critical depth occurs through the bridge, only the special-bridge method will define the profile. Because of the wide variation in bridge conditions, the selection and application of a method, or methods, requires imagination and good engineering judgment.

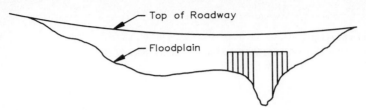

Figure 13.21   Bridge on fill.

Bridge on fill

If the approach roads across the floodplain are on fill measuring 3 ft or more, as shown in Fig. 13.21, and weir flow is a possibility, use of the special-bridge method is indicated.

Low-profile bridge

If the approach roads and bridge are not on fill and the flood would flow through and around the bridge as it would in a valley cross section (Fig. 13.22), friction loss would dominate. Use of the normal-bridge method is indicated.

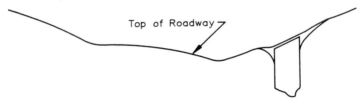

Figure 13.22   Low-profile bridge.

High bridge

If the top of road and the low-chord elevations are so high as to preclude flood flows from reaching the low chord, either method could be used. The selection would depend on other factors, such as the number of openings. For multiple openings (Fig. 13.23), the normal-bridge method would be preferred. (See the discussion in the next section.)

Complex and multiple bridge openings

The special-bridge method requires the specification of a trapezoidal approximation of the bridge opening for low-flow computations. If the opening cannot be reasonably approximated by a trapezoid, the

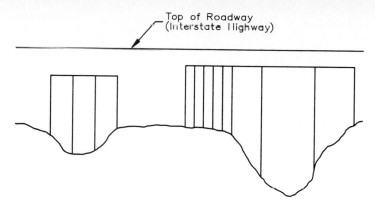

Figure 13.23   High bridge.

normal-bridge method may be the best approach for an analysis of low flow. However, for pressure-flow computations, the special-bridge method can be used because it utilizes the net area of the opening rather than the trapezoidal area.

Multiple culverts, bridges with side relief openings, and separate bridges over a divided channel provide examples of multiple bridge openings. For low-flow conditions, the normal-bridge method generally is most applicable. The special-bridge method cannot be used to model more than one trapezoidal opening, and modeling two or more openings as one having wide piers is generally unsatisfactory. The semiempirical equations used in this method have not been calibrated for such flow conditions.

For pressure flow, the special-bridge method is applicable, but only one low-chord elevation can be used. If the maximum low-chord elevation is the same for all openings or the water surface is above all of them, the orifice equation applies. An equivalent orifice coefficient for multiple culverts is described in the section "Pressure flow."

If low flow exists in some culverts concurrently with pressure flow in others of a group, the special-bridge method cannot provide a direct solution. As an alternative, the openings can be modeled separately, and an external procedure for analyzing "divided flow"[8] can be used. A normal-bridge solution can be obtained directly if two assumptions are reasonable: (1) a distribution of flow based on conveyance and (2) a single water surface elevation for the entire bridge section. High n values can be used to effectively model some bridges with multiple openings, such as the one shown in Fig. 13.24. It is impractical to use the special-bridge method to model low flow through this bridge be-

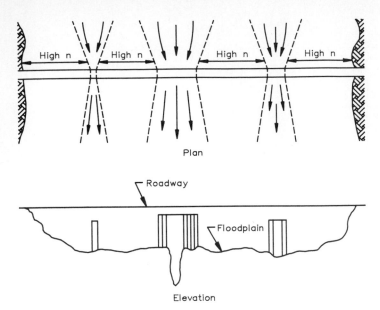

Figure 13.24 Use of high n values to model bridge with multiple openings.

cause of the difficulty of approximating a single trapezoidal opening. Use of the effective-area option specified with an X3 line also is impractical because of the impossibility of assigning right- and left-bank stations that make sense.

One possible approach for these conditions is to use NH lines with extremely high n values to block out flow in noneffective areas outside of the bridge openings. The piers are defined on GR lines for the cross sections at the upstream and downstream sides of the bridge, and the width of obstruction BWP is set equal to zero on the SB line. A value of zero for BWP triggers the use of the normal-bridge method for low-flow computations. The effect of using high n values can be seen in the plan view, Fig. 13.24. This approach has another advantage if storage is needed for modified Puls routing: the high n values block out the flow but retain the storage volume.

If weir flow, which requires the use of the special-bridge method, is anticipated, it probably would be appropriate to redefine the bridge model so that it is consistent with special-bridge requirements (for example, change high n values to more realistic values) and make a separate run. When flow comes over the top of the bridge, conveyance will occur in the areas considered noneffective during low flow.

Culverts

The modeling of culverts can be one of the most difficult tasks in water surface profile analysis. In some cases, it is appropriate to use hydraulic calculations[9] through a culvert or group of culverts and enter the results in the HEC-2 input with an X5 line. The application of HEC-2 to multiple culverts is discussed in the preceding section on multiple bridge openings. Some ideas regarding application to long culverts are presented here.

The normal-bridge method is the most suitable approach for long culverts under low-flow conditions. Cross sections can be taken through the culvert as needed to model friction loss and changes in grade and slope. Since the orifice coefficient can be adjusted to account for friction loss, the special-bridge method is suitable for pressure flow through long culverts.

Special culvert analysis capability is being developed for a new version of HEC-2 to be released in 1989. This new capability is based on equations and charts in highway design publications.[10,11]

Bridges that are a minor obstruction to flow

If a bridge, including its abutments, does not encroach significantly on the flow, it can be analyzed with the normal-bridge method. Although the use of six cross sections is ordinarily recommended for this method, only one cross section through the bridge may be sufficient to include its effect in the model.

Bridges under high submergence

Weir equations are normally based on the assumption of free flow. Corrections for submergence due to tailwater conditions are included in weir calculations of the special-bridge method; however, these were developed for an ogee-shaped weir and are not reliable for weirs with other shapes, especially under high submergence. The normal-bridge method is preferable under these conditions.

Dams and weirs

Flow over weirs and uncontrolled dams can be modeled with the special-bridge method. The same basic data as for a bridge is required. Since the special-bridge method goes through a sequence of assuming low flow and pressure flow prior to determining if weir flow exists, it is necessary to specify on the SB line small values defining the trapezoid and orifice area. The small areas so defined will cause the program to solve for a combination of pressure flow and weir flow. With a very small orifice area, the pressure flow will be negligible, and a weirflow solution will essentially be obtained.

Perched bridges

A perched bridge is defined as a bridge which has approach roads on each side at floodplain level, and only in the immediate vicinity of the bridge does the road rise above the floodplain to span the watercourse. In a typical flood-flow situation with this type of bridge, there is combined low flow under the bridge and weir flow around one or both sides of the bridge. Because the approach roads in the floodplain may not be much higher than the surrounding ground, the assumption of weir flow often is not justified. Standard step calculations in the normal-bridge method are generally preferred for this type of bridge, especially when a large percentage of the total discharge is carried in the overbank areas.

Low-water bridges

A low-water bridge is designed to carry only low flows under the bridge. Flood flows are typically carried over the bridge and road as a combination of pressure flow and weir flow. This implies that the special-bridge method should be used, but the application of the submergence correction by the program may introduce significant error if there is high submergence. When the tailwater is expected to be high, it may be preferable to use the normal-bridge method. Also, if a large part of the flow is going over the top of the bridge, it may be reasonable to define a simple cross section without an opening for the bridge. Omitting the pressure flow may have an insignificant effect on the results.

Skewed bridges

The horizontal dimensions of a skewed bridge, measured along the longitudinal axis of the bridge, are adjusted to define an equivalent cross section perpendicular to flow (Fig. 13.25). In the normal-bridge method, the dimensions of the bridge are multiplied by the cosine of the skew angle. The cosine is specified as the variable PXSECR on the X1 line to adjust the GR-line data and as the variable BSQ on the X2 line to adjust the BT-line data. If the special-bridge method is used, values for the SB line must be adjusted manually. There is no internal method in the program for adjusting this data.

Skewed crossings with angles up to 20° produced no objectionable flow patterns in tests of low-flow models.[5] For higher angles, flow efficiency decreased, but for small flow contractions, angles up to 30° were considered acceptable.

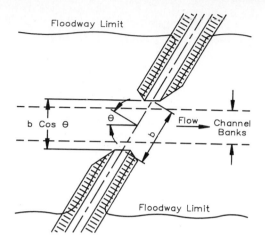

Figure 13.25   Skewed bridge.

Parallel bridges

With increased construction of divided highways, the analysis of flow through parallel bridges has become common. Frequently, parallel bridges are identical, and model studies[5] indicate that the loss for two bridges ranges from 1.3 to 1.55 times the loss for one bridge, over the range of bridge spacings tested. Presumably, the loss for two would be 2.0 times the loss for one if the bridges were far enough apart. Computing a single-bridge loss and then adjusting it with appropriate criteria[5] is an expedient approach to parallel-bridge analysis. If two bridges are modeled together, incorporation of losses from the expansion and contraction of flow between the bridges is an important element.

Workshop Problem 13.1: Special-Bridge Method

Purpose

The purpose of this workshop problem is to provide an understanding of the special-bridge method and the development of an input file for a model using this method.

Problem statement

A plan, thalweg profile, and bridge cross sections are shown in Fig. P13.1. The channel cross section at the bridge is representative of the entire reach. The bridge has two square piers, each 2 ft wide. The remaining area of the bridge opening measures 2250 ft^2. The channel n value is 0.04, and the overbank n values are both 0.08. The contraction and expansion coefficients are 0.3 and 0.5, respectively.

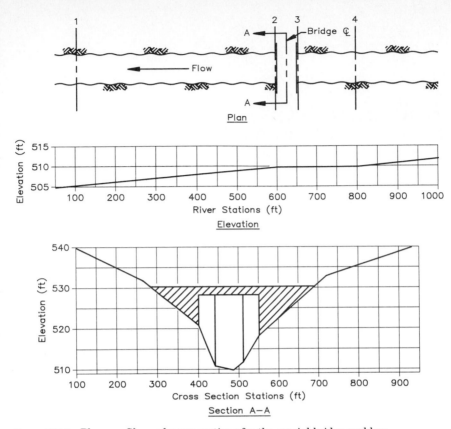

Figure P13.1   Plan, profile, and cross sections for the special-bridge problem.

Tasks

Complete the data input file shown in Fig. P13.2 for a multiple profile subcritical run to compute water surface profiles through the bridge. Start with a cross section at station 100, using the following discharges and water surface elevations:

Discharge, ft³/s	WSEL, ft
8,000	525
10,000	530
25,000	532

After the input file is completed and checked, execute HEC-2 to obtain the standard output. What are the five computed components of loss through the bridge for the flow of 8000 ft³/s?

ID.....1.......2.......3.......4.......5.......6. ....,,7.......8.......9......10

T1	WATER SURFACE PROFILE COMPUTATIONS								
T2	BRIDGE LOSSES								
T3	SPECIAL BRIDGE METHOD								
J1		2.						525.	
J3	100	105							
NC	.080	.080	.040	.300	.500				
QT	3.	8000.	10000.	25000.					
X1	1.	10.	400.	550.				-4.5	
GR	540.	100.	532.	260.	521.	400.	511.	440.	510.	485.
GR	512.	510.	518.	550.	526.	650.	533.	720.	540.	930.
X1									
X3									
SB									
X1									
X2									
X3									
BT									
BT									
X1									
EJ									
T1									
T2									
T3									
J1									
J2									
T1									
T2									
T3									
J1									
J2									

ER

Figure P13.2   Partially complete data input file for Workshop Problem 13.1.

Workshop Problem 13.2: Normal-Bridge Method

Purpose

The purpose of this workshop problem is to provide an understanding of the normal-bridge method and the development of an input file for a model using this method.

Problem statement

Use the normal-bridge method to analyze the same bridge described in Workshop Problem 13.1.

Tasks

Prepare a data input file for a single profile run, using a discharge of 8000 ft³/s. Copy the data file for Workshop Problem 13.1 into a file with another name, and revise it in accordance with the requirements for the normal-bridge method. Use an n value of 0.015 through the bridge opening, including the piers. Enter the data for defining the piers on X4 lines. After the file is revised and checked, execute HEC-2 to obtain the standard output.

Compare the results of this run with the results for the flow of 8000 ft³/s in the special-bridge analysis (Workshop Problem 13.1). Are the differences reasonable?

References

1. Emmett M. Laursen, "Bridge Backwater in Wide Valleys," *Journal of the Hydraulics Division,* ASCE, vol. 96, no. HY4, April 1970.
2. Hydrologic Engineering Center, *HEC-2 Water Surface Profiles, User's Manual,* U.S. Army Corps of Engineers, Davis, Calif., September 1982.
3. Koch-Carstanjen, "Von Der Bewegung Des Wassers Und Den Dabei Auftretenden Kraften," Springer, Berlin, 1926. (A partial translation appears in App. I of *Report on Engineering Aspects of Flood of 1938,* U.S. Army Engineer Office, Los Angeles, May 1939.)
4. D.L. Yarnell, *Bridge Piers as Channel Obstructions,* Technical Bulletin 442, U.S. Department of Agriculture, November 1934.
5. Joseph N. Bradley, "Hydraulics of Bridge Waterways," *Hydraulic Design Series No. 1,* 2d ed., Bureau of Public Roads (now the Federal Highway Administration), 1970, rev. March 1978.
6. Horace W. King and Ernest F. Brater, *Handbook of Hydraulics,* McGraw-Hill, New York, 1963.
7. Bill S. Eichert and John Peters, "Computer Determination of Flow through Bridges," *Journal of the Hydraulics Division,* ASCE, vol. 96, no. HY7, July 1970.
8. Ven T. Chow, *Open Channel Hydraulics,* McGraw-Hill, New York, 1959.
9. G. L. Bodhaine, "Measurement of Peak Discharge at Culverts by Indirect Methods," Chap. A3 in Book 3 of the manual series *Techniques of Water-Resources Investigations of the United States Geological Survey,* U.S. Department of the Interior, 1982.
10. Bureau of Public Roads, *Hydraulic Charts for the Selection of Highway Culverts,* Hydraulic Engineering Circular 5, U.S. Department of Commerce, December 1965.
11. Federal Highway Administration, *Hydraulic Design of Highway Culverts,* U.S. Department of Transportation, 1985.

Floodway and Channel Improvement Analyses

Floodway Determination

Introduction

The evaluation of the impact of encroachments on water surface profiles is of considerable interest to planners, land developers, and engineers. It is also a significant factor in delineating floodways in flood insurance studies.

A floodway reserves an unobstructed portion of a stream channel and floodplain for the passage of floodwaters while providing for appropriate use of adjacent lands. The purpose of a flood insurance study is to delineate floodway boundaries so that a base flood of a specified frequency can be accommodated within acceptable limitations. The size of the "base flood" and the "acceptable limitations" are institutional issues determined by statutes and regulations. Delineating a floodway requires an analysis of the effects of eliminating areas of flow in the overbanks on computed water surface elevations. The flow areas which can be eliminated without violating floodway criteria can be considered for development.

Guidelines for establishing a floodway

Guidelines of the U.S. government, established for a consistent approach to floodway delineation, are as follows:[1,2]

1. Floodway determinations are based on the discharge of a 1 percent chance exceedance flood.

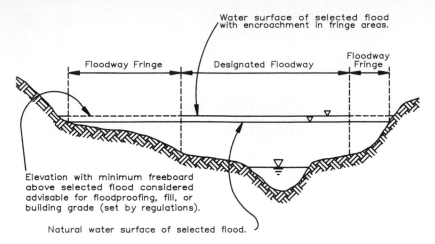

Figure 14.1 Cross section of a floodway and floodway fringes.

2. The floodplain is divided into zones consisting of the designated floodway and floodway fringes (Fig. 14.1).

3. The floodway is designed to pass the base flood without raising the water surface elevation more than 1 ft above what it would be for the existing floodplain.

4. The floodway fringe is the area between the boundaries of the floodway and existing floodplain. For hydraulic computations, it is generally assumed that all conveyance in the floodway fringe will be eliminated.

5. The floodway is determined by means of an equal reduction of conveyance on the two sides of the stream unless provisions for deviation are applied.

6. The hydrology and hydraulics for floodway determination are based on existing conditions. Specific guidelines are provided for considering future flood control works.

7. The adoption of a final floodway is based on coordination with state and local officials.

These guidelines are applicable except in states that have more restrictive requirements specified in legally enforceable statutes.

Modeling procedures

The following procedures are used in the application of the HEC-2 water surface profiles program to floodway determination.[1]

1. Develop a model of the study reach. Calibrate the model to replicate water surface profiles from historical floods if high-water marks or gaged data is available.

2. Compute water surface profiles for existing conditions, using a range of discharges, including the 1 and 0.2 percent chance floods.

3. Analyze the water surface profiles for the study reach to ensure that the model is producing reasonable results. The flow distribution option in HEC-2 is useful in this step, as will be discussed in a subsequent section.

4. Compute the water surface profile of the base flood (1 percent chance event) for encroachment conditions, using one or more of the encroachment options in HEC-2.

5. Compare the results of computations for existing and encroachment conditions. Make adjustments in the model for excessive changes in computed water surface elevation, significant increases in velocity, unreasonable depths of water in the floodway fringe, undulating flow boundaries, and results inconsistent with local needs.

6. Rerun the model after adjustments are made in the encroachments, and again examine the results. Repeat this process of adjusting the encroachments (perform several iterations, if necessary) until acceptable results are obtained.

Undulating flow boundaries

The program computes water surface elevations and other variables, including encroachment characteristics, for one cross section at a time. This may result in top widths between encroachments varying from cross section to cross section. The boundaries drawn between encroachment limits along each side of the channel may tend to undulate, as shown in Fig. 14.2. These floodway boundaries are impractical for implementation. Developing the model so that smooth transitions for flood-flow boundaries are produced under existing conditions will facilitate achieving acceptable boundaries in the floodway run.

Boundaries that are redrawn so that the undulations are smoothed out, as shown by dashed lines in Fig. 14.2, are more practical. Smoothing the boundaries in this way will result in changes in computed water surface elevations at the cross sections, so care should be exercised to make sure that the target change is not exceeded significantly. It is not unreasonable for the change in water surface elevation to vary between cross sections as long as the change stays below the target amount. The objective of delineating a designated floodway is to ob-

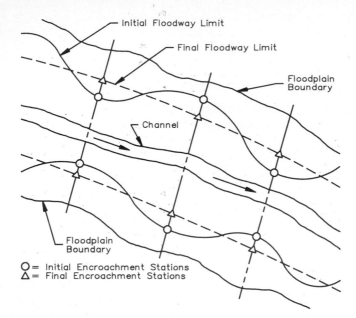

Figure 14.2 Initial and final floodway limits: correcting the problem of undulating boundaries.

tain consistent boundaries with changes in water surface elevations at cross sections that do not exceed the target but may be below it.

Analysis of existing conditions

The first profile computed in a floodway determination is for the base flood under existing conditions. The existing channel geometry and any structures involved are described in input in the same basic format as for a typical water surface profiles model not used for encroachment analysis.

Requesting the flow distribution option for some or all of the cross sections in this run is recommended to obtain useful information for analyzing the floodway. With this option, the program will print out the lateral distribution of percentage of discharge, area, and velocity for up to 13 elements of the overbank areas of a cross section defined by GR points. The 1988 version of HEC-2 also provides the hydraulic depth in each element, and it provides for flow distribution within the channel if variation in channel n values causes the channel to be divided into elements for computation. An example of flow distribution output is shown in Fig. 14.3.

The part of the flow distribution output that is generally most use-

```
FLOW DISTRIBUTION FOR SECNO=       .34          CWSEL=     394.69

STA=    38.      50.    275.     486.    1720.   1780.    1799.
 PER Q=      .1   92.8     2.4     2.3     2.1     .2
 AREA=     27.5  4356.1   495.7   484.7   281.6   44.1
 VEL=       2.4   11.3     2.5     2.5     4.0     2.5
```

Figure 14.3   Example of flow distribution printout.

ful in floodway determination is the breakdown in percentage of discharge. Since conveyance is directly proportional to discharge, the distribution percentages of conveyance and discharge are identical. Several of the encroachment options are based upon conveyance reduction, so this information provides a basis for analyzing the effects of various encroachments.

Flow distribution also is useful in determining ineffective-flow areas. Subdivisions with very low or insignificant velocities can be readily identified.

The flow distribution option can be specified in input for all cross sections or for individual cross sections. Setting the variable ITRACE equal to 15 in field 10 of the J2 line will provide flow distribution for all cross sections. Setting ITRACE equal to 15 on an X2 line provides flow distribution for a single cross section only.

Determination of a floodway with HEC-2 encroachment methods

There are six methods in HEC-2 for analyzing encroachments in floodway studies. Stations and elevations of left- and right-bank encroachments can be specified for individual cross sections with method 1. A floodway with a fixed top width and with left and right encroachments that are equidistant from the centerline of the channel can be specified with method 2. Encroachments can be specified as a percentage reduction in the natural discharge-carrying capacity of each cross section with method 3. Method 4 determines encroachments so that each modified cross section will have the same discharge-carrying capacity as the natural cross section, but at a specified higher water surface elevation. Method 5 is an optimization solution of method 4; it optimizes the difference in water surface elevations between the natural and encroachment conditions. Method 6 also is an optimization solution of method 4; it optimizes the difference in the energy grade line elevations between the natural and encroachment conditions.

Encroachment method 1.   In this method, diagramed in Fig. 14.4, encroachment stations are specified on an ET or X3 line. These stations

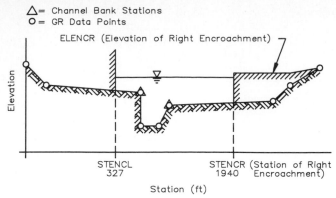

Figure 14.4   Method 1 encroachments.

do not have to coincide with GR points; they can be located anywhere within the defined cross section. For most applications, encroachments are specified on an ET line, which can accommodate several different specifications for a multiple profile run. The field containing the value to be utilized for a particular profile is indicated in field 2 of the J1 line, similar to the way different discharge values are indicated on the QT line.

Method 1 encroachments can also be specified by stations and elevations on an X3 line. The specification of elevations with this approach is a distinguishing feature. The ET-line encroachments for method 1 are, in effect, vertical walls extending upward to 100,000 ft. No limiting elevations can be specified. On the other hand, the use of the X3 line is limited to a single profile run. Only one set of values is specified on the X3 line, and these values would apply to all profiles in a multiple profile run if the X3 line were used.

An example of input data on an ET line for the encroachments in Fig. 14.4 is shown in Fig. 14.5. The "3" in field 2 of the J1 line indicates the field on the ET line in which the encroachment is specified. In this case, field 3 contains the value 9.1, which calls for encroachment method 1 and the location of the left encroachment station in field 9 of the ET line. The value to the right of the decimal point indicates the method; the value to the left indicates the field on the ET

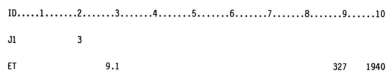

Figure 14.5   Example input for method 1.

line where the left encroachment station is found. The right encroachment station is in the next field to the right, field 10.

Method 1 is the same as other methods with respect to n values. The n values assigned to the ground or road surfaces where the encroachment stations are located are adopted by the program for the surfaces of the encroachments.

Method 1 differs basically from some of the other encroachment methods. Method 1 encroachments are not limited to the overbank areas; they can be located within the channel as well. They are not repeated; they must be specified on an X3 or ET line for each cross section to which they apply. Method 1 is the only method which can be used to turn off the encroachments of other methods at bridges and other locations where it is desired to have no encroachment. This is accomplished by specifying method 1 encroachments at widely separated stations so that the encroachments are far removed from the flow.

Encroachment method 4.   This is probably the most widely used encroachment method in HEC-2 (see Fig. 14.6). It establishes encroachment limits at a cross section on the basis of a target incremental increase in the "natural" water surface elevation. The program first computes the natural water surface elevation. It increases this elevation by the target amount and computes the corresponding percentage increase in conveyance for the cross section. Then it sets encroachment limits by eliminating the equivalent of the increased conveyance from the two extremities of the overbank areas in equal amounts. This method will not extend encroachments into the channel, so if there is insufficient conveyance in one of the overbanks, the program will try to make up the deficit from the other. Once the encroachment stations are set, the program computes the water surface elevation for the encroached cross section. It repeats this procedure for all cross sections, one at a time.

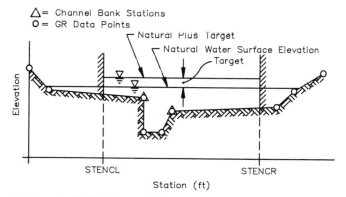

Figure 14.6   Method 4 encroachments.

```
ID.....1.......2.......3.......4.......5.......6.......7.......8.......9......10

J1          5

ET                                  10.4
```

Figure 14.7 Example input for method 4.

This encroachment option is specified on the J1 line and the ET line as shown in Fig. 14.7. The value "10.4" in field 5 of the ET line indicates a target change in water surface elevation of 1 ft and the selection of method 4. The number on the left side of the decimal point is a different indicator in this method as compared with the number to the left in method 1. Here, the number represents the target change in water surface elevation above the "natural" elevation, expressed in tenths of a foot. Thus, the "10" indicates a target change of 1 ft.

There is an alternative scheme for computing encroachments with method 4. In lieu of the conventional method of eliminating conveyance in equal amounts from each overbank, this scheme eliminates conveyance from the overbanks in proportion to the distribution of existing conveyance between the overbanks. This alternative is specified on the ET line with a minus sign before the variable indicating the target change and the method—for example, -10.4.

Some users are confused by the results of method 4 because of the expectation that the encroachments it generates will produce a water surface profile exactly equal to the natural profile increased by the target change. Method 4 does not produce this result. In a single computation it determines the encroachments on the basis of a reduction in conveyance in the overbanks. After these encroachments have been established, the program incorporates these cross-sectional changes in standard step computations of the water surface profile. The resulting profile may be slightly above or below the level targeted. A redistribution of flow and velocity occurs when a cross section is modified. This change plus the backwater effect of downstream encroachments can produce a result that is different from the specified incremental change in water surface elevation.

Encroachment method 5. This method is similar in many ways to method 4. It operates on a target increase in water surface elevation and eliminates conveyance in the overbank areas to offset the increase in conveyance that would otherwise result. It has an optimization feature which method 4 does not have. Instead of performing only one computation to determine the encroachments that are needed to produce the target increase, it will go through several iterations to achieve the desired target.

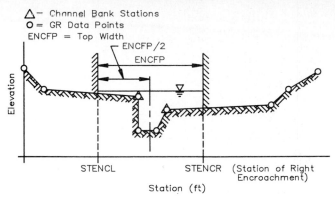

$\triangle$ - Channel Bank Stations
$\bigcirc$ = GR Data Points
ENCFP = Top Width

Figure 14.8 Method 2 encroachments.

Data input for method 5 is similar to that for method 4. The number or numbers entered on the ET line represent the target and the method.

Encroachment method 6. This is an optimization method that is similar to method 5 except that it optimizes a target difference in energy grade line elevation between natural and encroachment conditions. This relates to Federal Insurance Administration criteria for supercritical flow or high-velocity flow for which normal encroachment analysis techniques are not practical. In such cases, a target change in energy grade line may be used in the determination of a designated floodway. The target and the method are indicated on the ET line in a manner similar to that for method 4.

Encroachment method 2. This method is designed to analyze a floodway that has a constant top width and is centered on the channel centerline, as shown in Fig. 14.8. The channel centerline is located midway between the right- and left-bank stations. The top width ENCFP can be specified on an ET or X3 line, and it will be used for the current cross section and all subsequent cross sections until changed by another ET or X3 line. Similar to method 1 encroachments, the encroachments in this method are not limited to the overbank areas; they may fall within the channel as well. An example of data input for method 2 is shown in Fig. 14.9. The number on the left side of the decimal point in the variable entered on the ET line represents the fixed top width in feet.

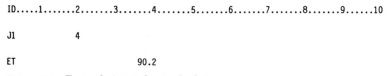

```
ID.....1.......2.......3.......4.......5.......6.......7.......8.......9......10

J1              4

ET                              90.2
```

Figure 14.9 Example input for method 2.

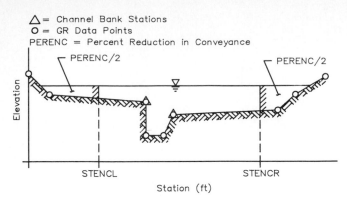

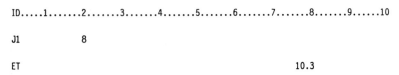

Figure 14.10   Method 3 encroachments.

Encroachment method 3.   Method 3, like method 4, calculates a reduction in conveyance in the two overbank areas; but in this method, the amount of reduction must be specified in input. One-half of the specified conveyance reduction is eliminated from each side of the cross section (see Fig. 14.10). If one-half of the reduction exceeds the overbank conveyance available on one side, the deficit is made up from the other side. If the conveyance reduction cannot be accommodated by the combined overbank areas, the encroachment stations are limited to the right- and left-bank stations. Method 3 encroachments cannot be located within the channel.

An example of data input for the ET line is presented in Fig. 14.11. In field 8, the "10" on the left side of the decimal point represents a 10 percent total reduction in conveyance in the natural cross section. A proportional conveyance reduction can also be specified in this method: a minus sign is added before the value or values entered on the ET line.

```
ID.....1.......2.......3.......4.......5.......6.......7.......8.......9......10

J1          8

ET                                                      10.3
```
Figure 14.11   Example input for method 3.

Bridge encroachments.   Encroachments can be specified at bridges with any of the six methods used in HEC-2: 0.01 is added to the data on the ET line. For example, 9.11, 100.21, 10.31, 10.41, 10.51, and 10.61 would request bridge encroachments for methods 1 through 6, respectively. In methods 1 and 2, bridge encroachments are applied directly to the bridge. In methods 3 through 6, the encroachments determined for the cross section at the downstream face of the bridge are applied to the bridge.

Example problem: floodway determination

Data input file. The HEC-2 input file in Fig. 14.12 represents a water surface profile model that has been calibrated for existing conditions on a reach of Cow Creek.[3] It contains data input for a first-trial floodway computation consisting of three profiles for a peak discharge of 53,000 ft³/s (1 percent probability flood). The three profiles are for the following conditions: (1) the existing floodplain, (2) encroachments added with method 4 on the basis of an 0.8-ft target increase in water level, and (3) encroachments added with method 4 on the basis of a 1.0-ft target increase.

Only a few extra lines are required in a basic water surface profile model to analyze encroachments. For the first trial, the two levels of method 4 encroachments are specified on a single ET line. The target increases of 0.8 ft and 1.0 ft are specified in columns 3 and 4 with "8.4" and "10.4," respectively. Although a maximum 1-ft increase in water surface elevation is desired in the floodway, analysis of a series of profiles generally provides a better basis for selecting the limits of encroachment. To keep this example brief, only the two target levels are used in this first-trial run.

The flow distribution option is specified with "15" for the variable ITRACE in the tenth field of the J2 line. Two predefined summary tables that are particularly useful in floodway computations, tables 110 and 200, are specified on the J3 line. The tables generated for this initial run are shown in Figs. 14.16 and 14.17a and b.

The same 1 percent probability peak discharge used for each profile, 53,000 ft³/s, is entered on the QT line in fields 2, 3, and 4, corresponding to the same fields on the ET line where the encroachment specifications are given. The J1 line for each profile computation indicates (in field 2) which field of the QT and ET lines is to be used for that particular profile. The J1 line also indicates (in field 9) the starting water surface elevation.

In this example, a known water surface elevation of 392 ft is shown on the J1 line for the first profile (existing conditions). On the J1 lines for the two encroachment profiles, 392.8 ft and 393.0 ft are used for starting water surface elevations. These were obtained by adding the target changes of 0.8 ft and 1.0 ft, respectively, to the known elevation for existing conditions. It is assumed that encroachments downstream cause a rise in the water surface elevation comparable to the target

```
ID.....1......2.......3.......4.......5.......6.......7.......8.......9......10
```

Card	1	2	3	4	5	6	7	8	9	10
T1	FLOODWAY DETERMINATION EXAMPLE									
T2	COW CREEK NEAR PALO CEDRO									
T3	EXISTING CONDITIONS									
J1		2							392	
J2	1		-1							15
J3	110	200								
NC	.05	.05	.045	.1	.3					
QT	3	53000	53000	53000						
ET			8.4	10.4						
X1	.08	21	300	530						
GR	410	0	400	40	390	300	380	330	372	350
GR	368	360	368	480	372	490	380	520	390	530
GR	395	700	390	820	380	870	375	890	380	940
GR	390	960	395	1000	390	1100	390	1160	400	1200
GR	410	1500								
X1	.21	18	150	390	750	750	800			
GR	410	0	400	130	390	150	380	175	373	200
GR	369	210	369	300	373	320	380	380	390	390
GR	395	600	390	880	385	900	390	980	400	1040
GR	403	1100	395	1400	400	1600				
X1	.34	16	50	275	650	690	670			
GR	410	0	390	50	373.5	100	370	110	370	210
GR	373.5	220	380	250	390	275	395	500	395	1500
GR	390	1720	390	1780	400	1820	403	2200	398	2900
GR	400	3300								
X1	.55	13	20	350	1100	1100	1100			
GR	410	0	390	20	374.5	90	371.5	100	371.5	200
GR	374.5	210	380	280	390	350	395	600	395	1200
GR	400	2200	405	2350	400	2900				
X1	.86	13	120	480	1550	1400	1500			
GR	420	0	400	120	376	200	372	210	372	370
GR	376	380	400	480	405	550	400	600	395	800
GR	395	1700	400	2050	410	2400				
X1	.94	14	75	255	400	400	400		2	
GR	410	0	408	50	383	75	380	110	367	120
GR	367	210	384	255	390	275	392	276	397	300
GR	400	680	398	800	400	1050	408	2000		
ET					8.4					
X1	1.06	10	190	450	700	600	650			
GR	410	0	400	75	390	190	377	240	373	250
GR	373	400	390	450	395	500	400	1050	407	1600
X1	1.27	14	550	820	1050	1050	1050			
GR	410	0	400	280	390	550	380	590	378	610
GR	374	620	374	710	378	720	380	775	390	820
GR	400	920	405	1200	400	1500	408	3000		

Figure 14.12 Input file for an HEC-2 model of existing conditions on Cow Creek.

```
EJ
T1      FLOODWAY DETERMINATION EXAMPLE
T2      METHOD 4 (0.8 FT TARGET INCREASE)
T3      COW CREEK
J1            3                                      392.8
J2    2              -1
T1      FLOODWAY DETERMINATION EXAMPLE

T2      METHOD 4 (1.0 FT TARGET INCREASE)
T3      COW CREEK
J1            4                                      393.0
J2   15              -1



ER
```

Figure 14.12   (*Continued*)

increases. If the starting elevation is fixed by a lake or some other control, then the encroachment profiles start at the same elevation as that for the profile for existing conditions.

Output for the first trial. An example of detailed cross-section output, including flow distribution data, is shown in Fig. 14.13. Detailed data for cross section 0.21, a divided-flow cross section, is shown in the figure. A cross-section plot, generated with PLOT2 by means of a dot matrix printer and the "print screen" command, is shown in Fig. 14.14. The flow distribution information reveals that 93 percent of the flow is between the right- and left-bank stations and that the balance of flow is in the right-overbank area.

An example of detailed output for this same cross section in the second profile, with encroachments included, is shown in Fig. 14.15. The

```
*SECNO  .210

3265 DIVIDED FLOW

        .21   24.28  393.28     .00     .00  394.97   1.69   1.57     .12  390.00
     53000.     19.  49315.   3666.     11.   4570.  1137.   108.     12.  390.00
        .02    1.80   10.79    3.22    .050    .045   .050   .000  369.00  143.44
    .002201    750.    800.    750.       2       0      0    .00  687.60  999.68

FLOW DISTRIBUTION FOR SECNO=      .21          CWSEL=   393.28

STA=   143.    150.    390.    528.    880.    900.    980.   1000.
 PER Q=     .0    93.0      .8     1.1     1.0     3.9      .1
 AREA=   10.8  4569.5   225.8   301.1   115.6   462.3    32.3
  VEL=    1.8    10.8     1.9     1.9     4.4     4.5     1.9
```

Figure 14.13 Detailed output for cross section 0.21 with existing conditions (profile 1).

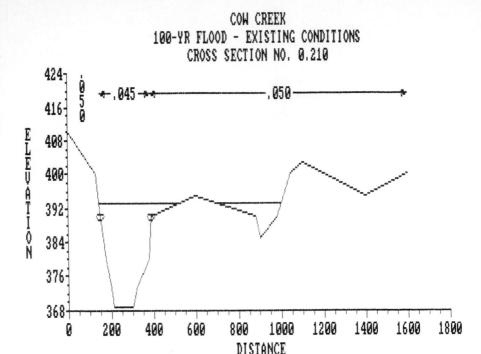

Figure 14.14 Plot of cross section 0.21 (made with a dot matrix printer and the "print screen" command).

first set of information, identified with the number 2800, is conveyance data computed by the program. Computed conveyance and water surface elevations for existing conditions and encroachment conditions are shown on the first line. The ratio of .0000 indicates that the two conveyances are equal. The second line shows conveyance and distribution of conveyance in the existing cross section (no encroachment) at the target water surface elevation (394.08 ft). This reflects an

```
*SECNO  .210
 2800 NAT Q1= 11296.01  WSEL=    393.28  ENC Q1= 11296.01  WSEL=    394.08 RATIO=   .0000
 NAT Q1=  12361.  RATIOS LOB,CH,ROB=   .0006   .9107   .0887  WSEL=    394.08

 3301 HV CHANGED MORE THAN HVINS

 3470 ENCROACHMENT STATIONS=    150.0    404.8  TYPE=    4  TARGET=      .086
      .21    25.04    394.04    .00   393.28   395.96    1.91    1.61    .16   390.00
   53000.       0.   52828.   172.       0.    4753.      57.      98.     7.   390.00
      .02      .00    11.11   3.00  ⟍  .050    .045     .050    .000  369.00   150.00
  .002264     750.     800.   750.       2       0        0      .00  254.81   404.81
```

Figure 14.15 Detailed output for cross section 0.21 with encroachments (profile 2).

intermediate computational step in which the elevation is raised by the target amount and the conveyance is recomputed.

The second set of encroachment information given for cross section 0.21 is identified with the number 3470. Computed encroachment stations, the encroachment method used, and the ratio of conveyance reduction (TARGET) are shown on the line immediately preceding the standard cross-section output.

The sequence of computations is reflected in this output. First, the water surface elevation is increased over the level for existing conditions, and a new value for conveyance is computed. The resulting increase in conveyance is eliminated by encroachment, and new encroachment stations are determined. Next, the flow area outside of the encroachment stations is eliminated, and the standard step method is used to compute the water surface elevation for the modified cross section.

Summary output for the first trial.   The predefined summary tables, 110 and 200, requested by the J3 line of input are shown in Figs. 14.16 and 14.17a and b.

Summary table 110 shows not only computed water surface elevations but also the changes in elevation (DIFKWS) between the base profile (existing conditions) and the encroachment profiles. The DIFKWS values make it easy to determine if the changes in elevation exceed the target change desired. The variables TOPWID, QLOB, QCH, and QROB indicate the consistency of flow distribution from cross section to cross section. The values for the variable PERENC indicate whether conveyance reductions appear reasonable. The encroachment stations STENCL and STENCR are conveniently located, for reference, next to the corresponding bank stations STCHL and STCHR.

Summary table 200 presents the floodway results in a form that is compatible with the requirements of a Federal Emergency Management Agency (FEMA) floodway study report.

Evaluation of the results.   In a review of the results of the first trial, the entire run should be analyzed for reasonableness. The impact on top widths and velocities, as well as changes in water surface elevation, is important. Since changes in water surface elevation frequently exceed the target increases specified, these may need to be adjusted. The problem often stems from the backwater effects of excessive encroachment downstream. In the example, the water surface elevations for cross sections 0.55 to 1.27 exceed the 1-ft target for the third profile (Fig. 14.17b). The water surface elevations computed in the second profile (Fig. 14.17a) are all below the target, but the first three are probably lower than necessary.

SUMMARY PRINTOUT TABLE   110

SECNO	CWSEL	DIFKWS	EG	TOPWID	QLOB	QCH	QROB	PERENC	STENCL	STCHL	STCHR	STENCR
.080	392.00	.00	393.28	662.00	65.35	42754.07	10180.58	.00	.00	300.00	530.00	.00
.080	392.80	.80	394.19	489.24	.00	45707.13	7292.87	.08	300.00	300.00	530.00	916.84
.080	393.00	1.00	394.43	490.91	.00	46604.09	6395.91	.10	300.00	300.00	530.00	906.92
.210	393.28	.00	394.97	687.60	19.36	49314.84	3665.80	.00	.00	150.00	390.00	.00
.210	394.04	.76	395.96	254.81	.00	52827.90	172.10	.09	150.00	150.00	390.00	404.82
.210	394.29	1.01	396.17	240.00	.00	53000.00	.00	.09	150.00	150.00	390.00	390.00
.340	394.69	.00	396.54	733.24	66.50	49202.76	3730.75	.00	.00	50.00	275.00	.00
.340	395.49	.80	397.57	241.88	.00	52666.71	333.29	.10	50.00	50.00	275.00	291.88
.340	395.72	1.03	397.79	225.00	.00	53000.00	.00	.12	50.00	50.00	275.00	275.00
.550	397.47	.00	398.19	1681.14	48.21	46598.15	6353.65	.00	.00	20.00	350.00	.00
.550	398.46	.99	399.25	477.69	.00	49680.37	3319.64	.11	20.00	20.00	350.00	497.69
.550	398.66	1.19	399.48	423.63	.00	50560.73	2439.27	.13	20.00	20.00	350.00	443.64
.860	398.90	.00	399.48	1681.62	.00	44597.52	8402.48	.00	.00	120.00	480.00	.00
.860	399.92	1.02	400.52	1025.97	.00	46647.82	6152.18	.10	120.00	120.00	480.00	1269.91
.860	400.13	1.23	400.74	938.72	.00	47535.79	5464.22	.12	120.00	120.00	480.00	1175.52
.940	398.27	.00	400.43	234.80	330.48	51496.32	1173.19	.00	.00	75.00	255.00	.00
.940	399.32	1.04	401.47	180.00	.00	53000.00	.00	.03	75.00	75.00	255.00	255.00
.940	399.56	1.29	401.67	180.00	.00	53000.00	.00	.03	75.00	75.00	255.00	255.00
1.060	400.54	.00	401.35	1021.68	1667.44	47040.20	4292.36	.00	.00	190.00	450.00	.00
1.060	401.55	1.01	402.41	423.10	.00	49486.50	3513.50	.08	190.00	190.00	450.00	613.10
1.060	401.72	1.18	402.61	367.77	.00	50256.78	2743.22	.10	190.00	190.00	450.00	557.78
1.270	401.43	.00	402.21	1113.86	5191.62	45684.68	2123.70	.00	.00	550.00	820.00	.00
1.270	402.42	.99	403.27	381.23	3996.25	48502.18	501.57	.08	453.07	550.00	820.00	834.30
1.270	402.59	1.16	403.47	358.71	3620.11	49234.06	145.83	.10	466.88	550.00	820.00	825.60

Figure 14.16   Summary table 110.

438

FLOODWAY DATA   METHOD 4 (TARGET .8 FT)
PROFILE NO. 2

STATION	WIDTH	FLOODWAY SECTION AREA	MEAN VELOCITY	WATER SURFACE ELEVATION WITH FLOODWAY	WITHOUT FLOODWAY	DIFFERENCE
.080	617.	5949.	8.9	392.8	392.0	.8
.210	255.	4810.	11.0	394.1	393.3	.8
.340	242.	4625.	11.5	395.5	394.7	.8
.550	478.	7797.	6.8	398.5	397.5	1.0
.860	1150.	9924.	5.3	399.9	398.9	1.0
.940	180.	4503.	11.8	399.3	398.3	1.0
1.060	423.	7590.	7.0	401.5	400.5	1.0
1.270	381.	7519.	7.0	402.4	401.4	1.0

(a)

FLOODWAY DATA   METHOD 4 (TARGET 1.0 FT)
PROFILE NO. 3

STATION	WIDTH	FLOODWAY SECTION AREA	MEAN VELOCITY	WATER SURFACE ELEVATION WITH FLOODWAY	WITHOUT FLOODWAY	DIFFERENCE
.080	607.	5891.	9.0	393.0	392.0	1.0
.210	240.	4812.	11.0	394.3	393.3	1.0
.340	225.	4587.	11.6	395.7	394.7	1.0
.550	424.	7555.	7.0	398.7	397.5	1.2
.860	1056.	9663.	5.5	400.1	398.9	1.2
.940	180.	4547.	11.7	399.6	398.3	1.3
1.060	368.	7331.	7.2	401.7	400.5	1.2
1.270	359.	7357.	7.2	402.6	401.4	1.2

(b)

Figure 14.17 Summary table 200. (a) Method 4 with 0.8-ft target change; (b) method 4 with 1.0-ft target change.

Additional trials. The next step is to attempt to achieve the target levels at all cross sections by adjusting the encroachments in another trial run. It appears from the results in Fig. 14.17a and b that this can be accomplished by using a target of 1.0 ft for the first three cross sections and a target of 0.8 ft for the remaining five. Although the results at cross sections 0.21 and 0.34 are within the target for the third profile, the encroachments at these two locations probably contributed to the excessive change at cross section 0.55. To compensate for this effect, one can relax slightly the encroachments at these two locations

FLOODWAY DATA   COMBINATION OF METHODS 4 AND 1
PROFILE NO. 2

| STATION | WIDTH | ------- FLOODWAY ------- | | WATER SURFACE ELEVATION | | |
		SECTION AREA	MEAN VELOCITY	WITH FLOODWAY	WITHOUT FLOODWAY	DIFFERENCE
.080	607.	5891.	9.0	393.0	392.0	1.0
.210	300.	5030.	10.5	394.3	393.3	1.0
.340	300.	4944.	10.7	395.7	394.7	1.0
.550	478.	7788.	6.8	398.5	397.5	1.0
.860	1150.	9909.	5.3	399.9	398.9	1.0
.940	180.	4500.	11.8	399.3	398.3	1.0
1.060	423.	7586.	7.0	401.5	400.5	1.0
1.270	381.	7515.	7.1	402.4	401.4	1.0

Figure 14.18 Summary table 200 with encroachments adjusted by means of a combination of method 4 and method 1.

by specifying wider encroachment stations with method 1. Thus, for the next trial floodway run (the input file is not shown), the following encroachments are specified on appropriate ET lines: method 4 with a 1.0-ft target for cross section 0.08, method 1 with new encroachment stations for cross sections 0.21 and 0.34, and method 4 with a target of 0.8 ft for the remaining cross sections upstream. The results of these adjustments are shown in Fig. 14.18.

Although the second trial produced changes in water surface elevation within the 1-ft target at all cross sections, other considerations, such as eliminating undulating boundaries, enter into the location of the floodway boundaries. After other adjustments are made, it may be appropriate to run a profile with the final encroachment stations specified by method 1.

Problems encountered in floodway determination

Determining the maximum limits of encroachment in a floodway can be a complicated problem, requiring several trial simulations. Modeling problems related to floodway determination have been identified as follows:[4]

1. Low-gradient streams—usually with floods of low velocity and long duration over a wide area

2. Flood overflow situations—including overflow at drainage divides and on leveed streams

3. Alluvial streams with movable boundaries

4. High-velocity streams—flowing at subcritical and supercritical velocities

5. Developed floodplains—with development in the potential floodway zone

A discussion of these problems in detail is beyond the scope of this book. The problems and suggested solutions are only highlighted here; for additional information, refer to the reference[4] cited.

Low-gradient streams. A significant modeling problem for wide, flat floodplains is distinguishing between conveyance and storage zones. Definition of cross sections is difficult because with the flatness of the floodplain, the limits of the cross sections are not apparent. Defining the limits of effective flow is important because once an area is defined as a storage zone, it is not included as conveyance in water surface profile computations and becomes part of the floodway fringe. Development within a storage zone will not increase the computed water surface elevation, but loss of overbank storage can influence the peak and travel time of an actual flood wave.

Aerial photographs of past floods, interviews with people in the field, and other sources of historical information are generally helpful in determining effective-flow areas and the direction of flow.

A major difficulty with regard to modeling is simulating the flow distribution when the flow is high enough to flood the overbank areas. Since the flow area and conveyance are usually much larger for the overbanks than for the channel, the distribution of flow on the basis of conveyance places a large portion of the discharge in the overbanks. At the same time, it reduces the proportional amount of flow in the channel. Thus, the computed flow and velocity in the channel under these conditions can be far lower than at the bank-full stage.

The solution to this flow distribution problem is to reduce the overbank conveyance in some way, usually arbitrarily. A fixed lateral distance is sometimes assumed when the cross sections are defined. Sometimes, average velocities in the incremental areas of the floodplain fringe are used as the basis for limiting conveyance. In this case, a typical approach is to eliminate areas with velocities less than 1 ft/s. A more reasonable approach, suggested for handling this problem, is to reduce the overbank conveyance until the flow in the channel produces a channel velocity comparable to that at the bank-full stage.

Flood overflow situations. The "flood overflow" discussion in the reference cited is limited to situations in which a portion of flow leaves the general path of the channel and proceeds on a separate course. The portion leaving the channel may or may not rejoin the main-channel flow downstream at some point. The overflow may result from a natural drainage divide, a structure, or a natural watercourse that is higher than its overbank areas (perched stream). In the latter case, the overflow may occur at several different locations, making it difficult to model.

In addressing overflow problems, there are three basic considerations: (1) identifying the problem, (2) determining how much water is lost due to overflow, and (3) determining what happens to the overflow.

Identifying overflow conditions. To identify overflow, one can examine the output from a water surface profile computation to see if computed water surface elevations are higher than the controlling ground elevations in the study reach. If the computed water surface elevations are higher, an overflow problem exists.

Determining the amount of overflow. The method of determining how much water is lost depends on overflow conditions, which are categorized according to three cases. Overflow in case 1 is "confined and limited"; in case 2 it is "general and fills the overbank"; and in case 3 it is "general and not confined."[4]

An example of case 1 conditions is overflow at a drainage divide where there are a limited number of cross sections whose boundary geometry is exceeded by the computed water surface elevations. In this case, the split-flow option, described in Chap. 15, might be used to model the overflow discharge with a weir-flow equation, a rating curve, or normal-depth calculations.

An example of case 2 conditions is extensive outflow associated with low-level levees. If the overflow fills the overbank area and parallels flow in the channel, having access into and out of the channel, it may not require special treatment in the model. If the outflow does not parallel flow in the main channel, the problem becomes a more difficult two-dimensional flow problem—the perched-stream situation—typical of case 3.

In the case 3 situation, the overflow along the channel can be extensive, and the flow moving away from the channel may travel in different flow paths. If the overflow location can be considered stable, the split-flow option can be used to estimate the overflow. The split-flow option is not suitable for divided-flow conditions with complex multiple flow paths. Graphic procedures are available for analyzing separate paths of multiple-path flow.[5,6]

Estimating the area flooded by overflow. Three cases are considered in estimating the area flooded by overflow. The overflow is either contained in a storage area, conveyed in a defined separate flow path, or conveyed in an uncontrolled path. Reference should be made to the HEC floodway determination training document[4] for information on these cases.

Alluvial streams. Standard floodway determination procedures using a rigid-bed simulation model such as HEC-2 are applicable to some relatively inactive alluvial streams. However, many alluvial streams exhibit considerable mobility and readjustment of channel boundaries during flood events. For these, alternative floodway computation procedures may be necessary. One of the first steps in determining whether traditional floodway analytic methods are applicable is an evaluation of the stability of the alluvial stream. General methods of analysis to evaluate stability, involving both qualitative and quantitative techniques, have been developed.[7] Analytic methods and simulation models under development can be expected to modify methods for determining floodways on alluvial streams in the future. In the meantime, suggestions and guidelines for using current techniques for floodway determination on alluvial streams are available.[1,4]

High-velocity streams. Stream velocities greater than 2 to 6 ft/s may be considered high. Velocities in the neighborhood of 2 ft/s have the potential to sweep cars off the road. Velocities greater than 6 ft/s can have an adverse impact on the stability of many common bed materials.

Subcritical flow. In floodway analysis, subcritical high-velocity flows present different problems than supercritical high-velocity flows. For subcritical flows, the problems tend to be the greatest when the flow is near critical depth. Conveyance-based floodway computations are impractical when the base profile for existing conditions is at or near critical depth. At high velocities, the redistribution of flow resulting from the elimination of conveyance in the floodplain fringe will have a disproportionately higher impact on the computed velocity head and energy losses. Changes in water surface elevation due to encroachments tend to be more uneven. Encroachments may cause the water surface elevation to decrease—and even reach critical depth, in the most extreme cases. When this happens, the next cross section upstream is likely to have an unusually high increase in water surface elevation. The encroachment may have to be reduced to correct thisproblem and the profile computed again. This process may have to be repeated many times to obtain a reasonable balance between the encroachments and the water surface profile.

Supercritical flow. Computing water surface profiles for supercritical flow is more difficult than for subcritical flow. Generally, more cross sections are required to model the higher velocities, and variations in loss coefficients have a greater impact. Adding encroachments tends to compound the modeling problems.

The water surface profile for the base flood (existing conditions) should be used to define the floodway when there is little or no overbank area. If there is overbank area, the floodway can be delineated by blocking the conveyance in the fringe that contains flow less than 3 ft deep with a velocity less than 3 ft/s. It is unlikely that this area has supercritical flow.

Developed floodway.   There are two basic approaches to modeling individual structures in the floodplain. One is to define the structure in the cross-section geometry; the other is to adjust the friction loss coefficients to simulate the resistance to flow caused by the structure.

Defining the structure by the first of these methods may require additional cross sections. As indicated in Fig. 14.19, four cross sections may be needed to model an individual structure or several structures in a row. Two cross sections define the downstream and upstream faces of the buildings, and two define the extremities of ineffective flow downstream and upstream. If the development is fairly complete and the buildings are approximately in a line, they might be defined as a solid zone that is obstructed for the length of a city block.

Adjusting the friction loss coefficients to account for the impact of structures on flow in the floodplain is recommended if development is

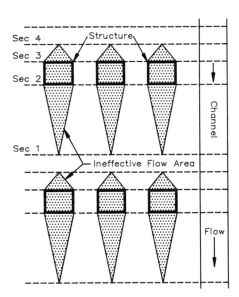

Figure 14.19 Modeling individual structures in the floodplain.

extensive. A procedure developed to estimate adjusted n values for urban areas[8] may be useful in this approach.

Channel Improvement

Introduction

The channel improvement option in HEC-2 modifies cross-section data to simulate trapezoidal channel excavation for analyzing modifications to existing streams. The trapezoidal excavation is defined by the location of the centerline, the elevation of the improved invert, a new channel reach length, a new n value, the left and right side slopes, and the bottom width. Up to five different bottom widths can be specified in program input for a single run. An illustration of the option's capabilities is shown in Fig. 14.20. Note that the improved section is obtained only by excavation; however, the old channel can be filled in completely prior to excavation. This provides for considerable latitude in specifying modified cross sections.

The CI line

The location and dimensions of the trapezoidal excavation and other input data for the channel improvement option are specified on CI

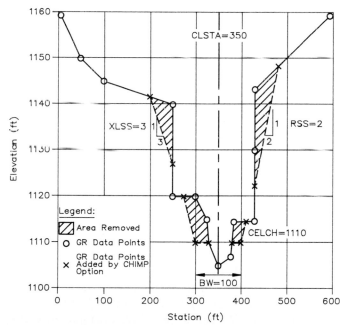

Figure 14.20   Illustration of channel improvements with HEC-2.

lines. This option is implemented by including a CI line immediately following the X1 line in the input file. The data to be included on the CI line is shown in Table 14.1.[9]

Centerline station. The station of the centerline of the trapezoidal channel excavation CLSTA is entered in field 1 of the CI line. The station must be consistent with the stationing of ground points indicated on GR lines. If -1 is entered in this field in lieu of a station number, the centerline will be located by the program midway between the right- and left-bank stations. If a zero is entered, the centerline station from the CI line for the previous cross section is repeated.

Channel invert elevation. The improved channel invert elevation is entered in field 2, unless a zero is entered to repeat the value specified for the previous cross section. Entering a -1 in this field sets the invert elevation equal to the minimum elevation of the natural cross section. Entering a number between 0.1 and 0.00001 instructs the program to compute an invert elevation, using the downstream minimum elevation, the slope, and the channel reach length.

Revised Manning's _n_ value and modified reach length. The variable XLCH.CNCH, located in field 3, has a dual function. The value for XLCH, on the left side of the decimal point, can be used to specify a new channel reach length. If the assigned the value is zero, the channel reach length specified on the X1 line will be used. The value for CNCH, on the right side of the decimal point, is the improved channel _n_ value. If this is set to zero, the previous _n_ value assigned to the channel will be used.

Side slopes of the trapezoidal excavation. The left and right side slopes of the excavation are entered in fields 4 and 5. If zeros are entered here, the side slopes from the CI line for the previous cross section are used. A vertical side slope can be approximated by using the value 0.01.

Bottom width of the trapezoidal excavation. Up to five different bottom widths for different trapezoidal excavations for a multiple profile run can be specified in fields 6 to 10. The program has the capability to compute 14 profiles incorporating these five different bottom widths.

The bottom-width variable BW has some additional functions besides specifying trapezoidal bottom width. The value 0.01 entered for this variable indicates to the program that the channel improvement option is to be turned off. As with some of the other options, once the

TABLE 14.1   Channel Improvement Input Data

Field	Variable	Value	Description
0	IA	CI	First two columns of line for line identification.
1	CLSTA	0	Value on previous cross section's CI line is used.
		+	Station of the centerline of trapezoidal channel excavation which is expressed in terms of the stations used in the existing cross section description (GR lines).
		− 1	CLSTA is determined by program halfway between bank stations.
2	CELCH	0	Value on previous cross section's CI line is used.
		+ or −	Elevation of channel invert (cannot be −1).
		− 1	Elevation of channel invert is equal to minimum elevation in cross section. (For pilot channel excavations, second and third CI lines, the channel invert elevation should be specified.)
		$0.1 \geq X \geq 0.00001$	Elevation of channel invert is based on CELCH (slope) × XLCH (channel reach length) + PELMN (downstream minimum elevation). X = value.
3	XLCH CNCH	0 or +	Value to the left of decimal point is channel reach length (XLCH). If 0, the channel reach length specified on X1 line will be used. Value to the right of decimal point is new channel n value (CNCH). If 0, previously specified n (CI or NC line) will be used.
4	XLSS	0	Value on previous cross section's CI line for left side slope of trapezoidal excavation is to be used. If not previously specified, the left side slope will be vertical.
		+	Left side slope of excavation expressed as number of horizontal units per one vertical unit.
5	RSS	0 or +	Same as XLSS, except for right side of trapezoid.

TABLE 14.1   Channel Improvement Input Data (Continued)

Field	Variable	Value	Description
6–10	BW	0	Value on previous cross section's CI line is used.
		0.01	End of channel improvement. If multiple CI lines are being used, then all the CI lines must have 0.01 to turn off the channel improvement. If not all the CI lines have a 0.01, then the lines that do not have a 0.01 will be used to do the channel improvement.
		+	Bottom width of channel. The field used for this profile is indicated by variable IBW on the J2 line.
		−	Same as +, but the existing channel will be filled to an elevation equal to the minimum bank elevation.

channel improvement option is implemented at a cross section, it continues to be implemented at succeeding cross sections until it is turned off. If multiple CI lines are being used, as when a pilot channel is defined, all the CI lines in the set must have a 0.01 to turn off the channel improvement option. If not all CI lines have a 0.01, the CI lines that don't will be used by the program to do the channel improvement. This feature is useful in varying channel improvements in a multiple profile run, as will be demonstrated in the example problem. Another function of the BW variable is to specify a fill operation. If a negative bottom width is specified, the program interprets this as an instruction to fill in the natural channel area—before making the specified trapezoidal excavation—up to an elevation equal to the lower of the two bank elevations.

The input variables on the CI line defining an improved cross section are diagramed in Fig. 14.21. The variable CHNIM, not discussed in connection with the CI line, is used for specifying a horizontal distance on each side of the channel to which the channel n value will be applied. This variable, which can be used to represent access roads, is entered in field 9 of the J2 line.

Multiple channel improvements

Up to three CI lines can be used at each cross section to define multiple improvements. The cross section in Fig. 14.22 shows an example of multiple cut-and-fill operations that can be specified with three CI lines ("cut" here means "excavation"). The channel improvements can

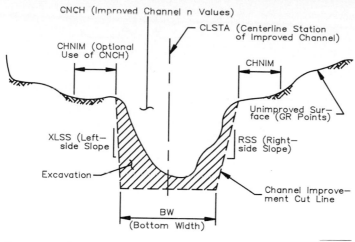

Figure 14.21 Diagram of channel improvement variables.

be accomplished in different sequences, as governed by the order of the CI lines. In one possible scenario, the first CI line could be used to fill the existing channel and cut the large channel on the right, the second to cut the pilot channel, and the third to cut the secondary channel on the left side.

The right- and left-bank stations are moved in the process and normally end up associated with the final cut. In this example, their final location would be on the last channel cut, located on the left side of the figure. If the cutting of the channels had been done in a different order, the bank stations could end up in different locations. However, it should be noted that bank stations are moved outward by the program, as in the case of widening an existing channel or a previous cut, but they are not moved inward. For example, if a pilot channel is cut after the main channel is cut, the bank stations are not moved to the banks of the pilot channel.

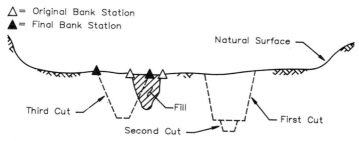

Figure 14.22 Example of multiple cut-and-fill operations specified with three CI lines of input.

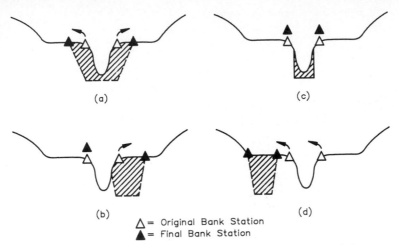

(a)   (c)

(b)   (d)

△ = Original Bank Station
▲ = Final Bank Station

Figure 14.23   Changes in bank stations associated with channel improvements.

Several changes in right- and left-bank designations are depicted in Fig. 14.23. What is probably the most typical improvement, widening a natural channel, is shown in Fig. 14.23a. Here, both bank stations are moved outward to the banks of the improved channel. In Fig. 14.23b the channel improvement is made on the right side of the natural channel, and only the right bank station, displaced by the excavation, is moved to a new location on the right bank of the improvement. In the case of a minor cut located within the natural channel, depicted in Fig. 14.23c, the bank stations of the natural channel are not moved. If the improved channel is completely outside of the existing channel, as shown in Fig. 14.23d, both bank stations are moved, as indicated.

It is possible to redefine bank designations to suit practically any situation with the use of a fictitious channel. This can be demonstrated in Fig. 14.24. A small fictitious channel is cut first, and then a pilot channel is cut inside of the natural channel. The bank stations are moved first from the natural channel to the fictitious channel and then from there to the pilot channel. In this way the right- and left-

△ = Original Bank Station
▲ = Final Bank Station

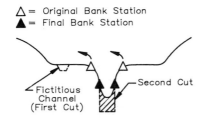

Fictitious
Channel
(First Cut)

Second Cut

Figure 14.24   Redefining bank stations by means of a fictitious channel improvement.

bank stations can be located on the pilot channel, which would not be possible under usual procedures.

The order in which the fill option is specified has significant consequences. When the fill option is invoked on a CI line, the program fills the area between the existing right- and left-bank stations. In the example shown in Fig. 14.22, if the fill option had been invoked on the third CI line instead of the first, the newly excavated main channel and pilot channel on the right side would have been filled instead of the natural channel.

Modification of channel reach length

The channel improvement routine includes an option to change the reach length. The change is accomplished by entering the new reach length in field 3 of the CI line. This option can be applied in channel realignments, such as a realignment to cut off meanders.

Example problem: channel improvement

In this example, three different channel improvements at a cross section are specified for a three-profile run. In Figs. 14.25 to 14.28, cross-section plots and corresponding data sets extracted from the input file are shown together.

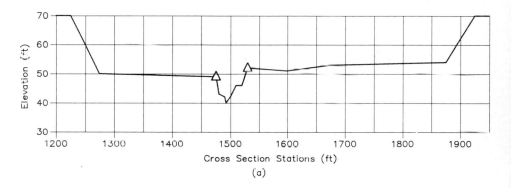

ID.....1.......2.......3.......4.......5.......6.......7.......8.......9......10

J2   1

CI	1500	200	.015	3	3	.01	200	.01	200
CI	1600	200	500.025	3	3	.01	.01	200	.01
CI	1500	195		1	1	.01	.01	.01	20

(b)

Figure 14.25 Existing conditions. (*a*) Cross section; (*b*) input file.

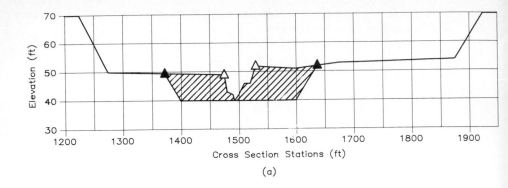

ID.....1.......2.......3.......4.......5.......6.......7.......8.......9......10

| J2 | 2 | | | | | | | 7 | | |

CI	1500	200	.025	3	3	.01	200	.01	200
CI	1600	200	500.025	3	3	.01	.01	200	.01
CI	1500	195				.01	.01	.01	20

(b)

Figure 14.26 Channel improvement centered on existing channel. (*a*) Cross section; (*b*) input file.

The CI line that applies to a particular profile is indicated in field 8 of the J2 line. For the first profile representing unimproved or existing conditions, shown in Fig. 14.25*b*, the J2 line is blank in field 8. With field 8 blank, the program defaults to field 6 on the CI line. A value of 0.01 in this field indicates that there is no channel improvement.

In Fig. 14.26*b*, the J2 line has a "7" in field 8, indicating to the program that the bottom width for the channel improvement should be read in field 7 of a CI line. The first CI line has bottom width of 200 entered in field 7, while the remaining two CI lines have the value 0.01 in field 7. Thus, the bottom width of 200 and the other values entered on the first CI line are used to determine the channel improvement for the first profile. To reiterate a point made earlier, the value 0.01 in this location instructs the program to disregard the data on the CI line for this profile.

In Fig. 14.27*b*, the J2 line has an "8" in field 8, indicating that field 8 on the CI lines should be read. In this case, the bottom width of 200 is read on the second CI line, and the other two CI lines have the value 0.01, indicating that these lines are inoperative. The data on the second CI line is for the same improved cross section, except that the

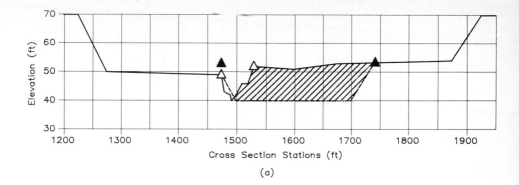

(a)

```
ID.....1.......2.......3.......4.......5.......6.......7.......8.......9......10

J2     3                                               8

CI  1500      200    .025      3     3    .01    200    .01    200
CI  1600      200 500.025      3     3    .01    .01    200    .01
CI  1500      195             1     1    .01    .01    .01    20
```

(b)

Figure 14.27 Channel improvement on right side of existing channel. (*a*) Cross section; (*b*) input file.

centerline station is 100 ft farther to the right at station 1600, and a reach length of 500 ft is indicated in field 3.

In Fig. 14.28*b*, the J2 line indicates that field 9 is to be read on the CI lines. Both the first and the third CI lines have bottom widths specified in field 9, indicating that two sets of channel improvements are to be employed. This is consistent with the main channel and pilot channel shown in the cross-section plot. The main channel as specified on the first CI line is excavated first; then the pilot channel specified on the third CI line is excavated.

Volume of excavation

Approximate volumes of excavation attributed to channel improvements are computed by the program. An example of cross-section output showing channel improvement values is shown in Fig. 14.29. The second line under the cross-section number is for excavation data. This data includes area of the excavation in the cross section AEX, the volume of excavation in the reach VEXR, and the total volume of excavation VEXT. Since the volumes are rounded to the nearest thousand cubic yards, there could be significant rounding errors in small-volume excavations.

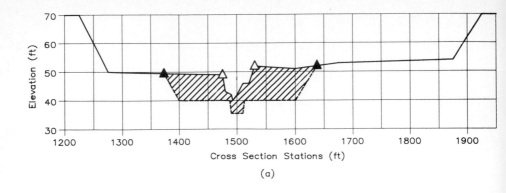

Figure 14.28 Channel improvement with pilot channel. (*a*) Cross section; (*b*) input file.

Use of the channel improvement (CHIMP) option with other options

Some of the other options can be used effectively with the channel improvement option; others cannot. The sediment option ELSED, specified on the X3 line in field 2, can be used for filling a cross section prior to cutting the improvement with the CHIMP option. The encroachment options can be used with the CHIMP option, but some cautions need to be observed. Encroachment methods 1 and 2 are executed prior to the channel improvement, while encroachment methods 3 through 6 are executed after. Since method 1 can, in effect, create vertical walls 100,000 ft high, the side walls of the channel improvement may intersect the encroachment and have its endpoints undefined, as shown in Fig. 14.30. This problem can be averted by limiting the encroachment heights with the variables ELENCL and ELENCR on an X3 line. Doing a channel improvement through a

```
*SECNO 1.820
CHIMP CLSTA=  18299.00 CELCH=     148.79 BW=      10.00 STCHL=  18150.00 STCHR=  18448.00
EXCAVATION DATA
AEX=     16.3SQ-FT    VEXR=      .8K*CU-YD    VEXT=      .8K*CU-YD
```

Figure 14.29 Example of channel improvement output showing volume of excavation.

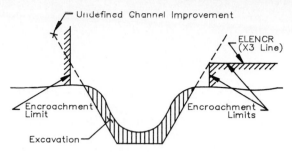

Figure 14.30 Restricting encroachments so that they are compatible with channel improvement cut lines.

bridge is not recommended, particularly if the normal-bridge method is being used. The channel improvement may eliminate some of the ground points that coincide with BT points. Using NH lines with channel improvements is also not recommended. The CHIMP routine has its own NH simulation capability.

Workshop Problem 14.1: Floodway Determination on North Buffalo Creek

Purpose

The purpose of this workshop problem is to provide an understanding of floodway determination with the method 4 and method 1 encroachment options in HEC-2.

Problem statement

The designated floodway for North Buffalo Creek is to be determined on the basis of the 1 percent chance flood of 8000 ft³/s. Maximum limits of encroachment are to be established according to the following guidelines:

1. Water surface elevations should not be increased by more than 1 ft.
2. Channel velocity increases should not be so large as to cause significant damage.
3. The top width should not be permitted to vary to the extent that the floodway boundary is impractical for implementation.

The watercourse reach to be analyzed is shown in Fig. P14.1. The partially completed input file for HEC-2 shown in Fig. P14.2 is to be

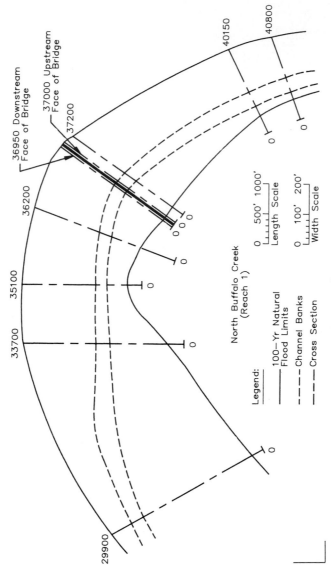

Figure P14.1 Plan of North Buffalo Creek floodway reach.

36950 Downstream
Face of Bridge

37000 Upstream
Face of Bridge

37200

40150

40800

36200

35100

33700

29900

North Buffalo Creek
(Reach 1)

Legend:
———— 100-Yr Natural
Flood Limits
––––– Channel Banks
— ·· — Cross Section

0   500' 1000'
Length Scale

0   100'   200'
Width Scale

```
ID.....1.......2,......3.......4.......5.......6.......7.......8.......9......10

T1      NORTH BUFFALO CREEK FLOODWAY DETERMINATION
T2      100-YR FLOOD
T3      PROFILE 1 - EXISTING CONDITIONS
NC    .12     .12     .055     .1      .3
QT      4    8000    8000    8000    8000
X1  29900     15     460     508                                   1
GR    712           699.2     42    695.2     71    691.2    150    691    400
GR  689.2    460    681.2    470    681.2    495    689.2    508    691.2    512
GR  693.2    530    695.2    563    701.2    595    709     612    712    630
NH      5    .12      75     .1     245    .055    285     .1     457    .12
NH    590
X1  33700     16     245     285    2600    3300    3000      1
GR    713           705.2     35    699.2     75    697.2     90    695.2    130
GR  695.2    240    693.2    245     685     250     685     280    693.2    285
GR  695.2    292    695.2    410    697.2    457    699.2    500    705.2    560
GR    713    590
NH      6    .12      48     .1     115     .04     150     .06     160     .1
NH    360    .12     410
X1  35100     11     115     150    1100    1000    1000      1
GR    714           706.7     36    705.2     48    699.2     92    689.2    115
GR    686    116     686     136    693.2    150    695.2    160    695.2    360
GR  714.2    410
NC    .12     .12     .055     .1      .3
X1  36200      9     230     270     700    1100     850
GR    /15            705     100     691     230     689     235     687     250
GR    689    265     691     270     705     500     715     600
X1  36950     14     187     238     700     950     800
X3     10                                                  704     704
GR    715           701.2     61    699.2     87    697.2    187    688    198
GR    688    228    697.2    238    697.2    290    699.2    310    701.2    370
GR  705.2    445    707.2    465    709.2    500    715.2    512
SB     .9    1.5     2.9              27     1.5     680    1.09     688    688
X1  37000                              50      50      50
X2             1     704    706.2
X3     10                                                 706.2    706.2
BT      6            715              38    706.4            310    706.2
BT    465    710            505    711.5            512    715.2
X1  37200                             200     200     200
NC                             .1      .3
X1  40150     10      95     145    2300    2800    2500
GR  720.2           719.2     22    709.2     70    699.2     95     693    105
GR    693    135    699.2    145    701.2    150    701.2    220    719.2    390
X1  40800                             650     700     650      1      3
EJ


ER
```

Figure P14.2 Partially completed input file for North Buffalo Creek HEC-2 model.

used as the basis for making several floodway determination runs. Data is to be added and revisions made as necessary to complete the runs required.

A starting water surface elevation of 698.3 ft should be used for the existing conditions, and a starting water surface elevation of 699.3 ft should be used with encroachments. A printout of the input file and summary printouts, including predefined tables 110 and 200, should be specified in input. No interpolated cross sections should be used, and no printer plots need to be requested.

Tasks

1. Compute profiles for existing conditions and for trial floodway encroachments using the method 4 encroachment option. Use more than one target change in water surface elevation with method 4. Also compute profiles with and without bridge encroachments.

2. After analyzing the results of the first run relative to the guidelines, adjust the encroachment stations obtained with method 4 if necessary, and recompute the profile using method 1 encroachments. Make a tracing of the map shown in Fig. P14.1, and enlarge it to a more convenient working size, if desired. Sketch in the final boundaries of the designated floodway.

3. Prepare a brief written report covering the following questions:
   a. Does the profile for the 1 percent chance flood computed for existing conditions appear to present serious problems in terms of excessive velocities, floodway widths, and depths?
   b. Do the floodway boundaries determined by method 4 violate any of the guidelines? Describe any significant violations.
   c. Were any adjustments made prior to running method 1? If so, discuss them.
   d. How close are the computed water surface elevations in the final run to the target increase of 1 ft?

Workshop Problem 14.2: Red Fox River Channel Improvement

Purpose

The purpose of this workshop problem is to provide an understanding of the channel improvement option in HEC-2 and how to use it.

Problem statement

A plan view of the Red Fox River channel improvement is shown in Fig. P14.3, with the improved channel marked at 100-ft intervals. A

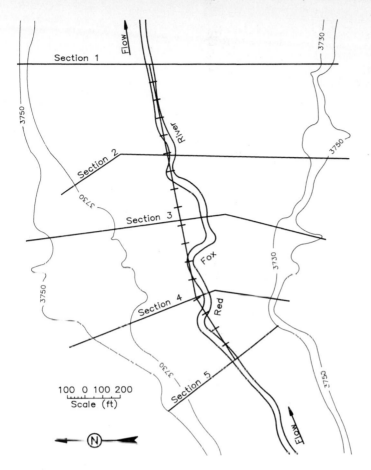

Figure P14.3 Plan of Red Fox River channel improvement.

data input file for this channel improvement model, complete except for missing channel improvement data, is shown in Fig. P14.4.

Tasks

1. Complete the CI lines in the input file as required to obtain four profiles based on the following conditions:
   a. Use the existing channel (not improved).
   b. Using the existing channel alignment, put in a channel improvement with a bottom width of 200 ft, side slopes of 3:1, a channel n value of 0.025, and bottom elevations based on a bottom slope of 0.006:1. No improvement is to be made at cross section 1. At cross section 5, the bottom width is to be changed to

```
ID.....1.......2.......3.......4.......5.......6.......7.......8.......9......10

T1   CHANNEL IMPROVEMENT EXERCISE
T2   RED FOX RIVER
T3   1000 CFS - NATURAL CONDITIONS
J1                                                       1000      3.8
J2                          -1                              0
NC                                      .1       .3
NH      5      .1     415     .05     650     .03     710     .05    1020      .1
NH   1635
X1      1      11     650     710
GR     25      20      18     110      17     415      14     650       1     675
GR      0     690       1     710      13     710      14    1020      14    1590
GR     25    1635
NC    .08      .1     .03
X1      2      10     575     640     560     510     520
CI     -1
CI    557
CI
GR     25      30      20     110      20     200      17     415      10     575
GR      4     580       4     615      18     640      18    1195      25    1250
NC     .1     .05     .03
X1      3      10     370     600     400     330     420
CI     -1
CI    350
CI
GR     25      40      22     260    18.7     370      15     420     7.1     500
GR    7.5     530    17.3     560      20     600      22     850      25     875
NC    .08     .07    .036
X1      4       8     330     460     460     400     450
CI     -1
CI    325
CI
GR     26      30      24     130      23     330     9.5     370      10     400
GR     22     460      22     610      26     700
NC    .06     .06    .036
X1      5      11     335     440     500     300     400
CI     -1
CI     -1
GR     30      30      26      65      24     235      24     335    12.1     375
GR   11.2     390      26     440      24     485      24     615      26     665
GR     30     700
EJ
T1   CHANNEL IMPROVEMENT EXERCISE
T2   RED FOX RIVER
T3   1000 CFS - IMPROVED CHANNEL WITH NO PILOT CHANNEL
J1                                                       1000      3.8
J2      2               -1                                  7
T1   CHANNEL IMPROVEMENT EXERCISE
T2   RED FOX RIVER
T3   1000 CFS - IMPROVED CHANNEL WITH PILOT CHANNEL (OLD CHANNEL FILLED)
J1                                                       1000      3.8
J2      3               -1                                  8
```

```
T1   CHANNEL IMPROVEMENT EXERCISE
T2   RED FOX RIVER
T3   1000 CFS - IMPROVED CHANNEL WITH PILOT CHANNEL (OLD CHANNEL NOT FILLED)
J1                                                         1000    3.8
J2   15                -1                                   9



ER
```

Figure P14.4 Partially completed input file for Red Fox River channel improvement model.

100 ft and the bottom is to be placed at the same elevation as that of the existing channel.

c. On the basis of the new channel alignment shown in Fig. P14.3, put in a channel improvement with a bottom width of 200 ft, side slopes of 3:1, a channel n value of 0.025, and bottom elevations based on the shortened channel length and a bottom slope of 0.00696:1. Fill the existing channel by the use of the CI line. Place a pilot channel with a bottom width of 25 ft, side slopes of 2:1, and a depth of 1 ft at cross sections 2, 3, and 4. Remember to alter the channel reach lengths where necessary. No improvement is to be made at cross section 1. At cross section 5, change the bottom width to 100 ft; place the improved channel bottom at the same elevation as that of the existing channel; do not put in a pilot channel; and do not fill in the existing channel.

d. Use the same channel improvements as in task c. The exception is that the existing channel is not to be filled in.

2. After completing the input file, execute the program and obtain cross-section plots to verify the improvements.

References

1. *Flood Insurance Study Guidelines and Specifications for Study Contractors*, FEMA 37, Federal Emergency Management Agency, Washington, D.C., September 1985.
2. *Floodway Design Considerations*, U.S. Army Corps of Engineers, South Atlantic Division, Atlanta, Ga., September 1978.

3. Hydrologic Engineering Center, *Floodway Determination Using Computer Program HEC-2*, Training Document 5, U.S. Army Corps of Engineers, Davis, Calif., May 1974.
4. Hydrologic Engineering Center, *Floodway Determination Using Computer Program HEC-2*, Training Document 5, rev. ed., U.S. Army Corps of Engineers, Davis, Calif., January 1988.
5. Ven T. Chow, *Open Channel Hydraulics*, McGraw-Hill, New York, 1959.
6. Hydrologic Engineering Center, *Application of the HEC-2 Split Flow Option*, Training Document 18, U.S. Army Corps of Engineers, Davis, Calif., April 1982.
7. Simons, Li & Associates, Inc., *Engineering Analysis of Fluvial Systems*, Fort Collins, Colo., 1982.
8. H. R. Hejl, "A Method for Adjusting Values of Manning's Roughness Coefficient for Flooded Urban Areas," *Journal of Research*, U.S. Geological Survey, vol. 5, no. 5, September–October 1977, pp. 541–545.
9. Hydrologic Engineering Center, *HEC-2 Water Surface Profiles, Users Manual*, U.S. Army Corps of Engineers, Davis, Calif., September 1982.

Supercritical Flow, Split Flow, and Other HEC-2 Options

Supercritical-Stream Analysis

The nature of supercritical flow

Upstream control. Flow is considered supercritical when inertial forces become dominant, as indicated by a Froude number greater than unity. The velocity of flow is high and may be described as "rapid" or "shooting." Upstream control is associated with this state of flow, because the effects of changes in flow conditions are transmitted downstream. Small gravity waves developed by disturbances or obstacles in the channel are not propagated upstream, because wave celerity is less than the velocity of flow in the channel. Channel expansions can cause a hydraulic jump, which is unstable in both location and height.[1]

Steep slopes. High-velocity supercritical flows are associated with relatively steep slopes not normally found in natural channels. Even in high-gradient mountain streams, the flow is generally subcritical due to the combined effects of channel and cross-sectional variations, bank roughness, and other factors. The effects of these factors increase with increasing discharge, creating extreme turbulence, high energy loss, and increased resistance to flow. During floods, energy is also required to transport eroded material. Thus, flow conditions may approach, but generally do not exceed, critical flow, except in very localized areas in the channel.[2]

Steep-slope profiles. For a given discharge and channel condition, normal-depth and critical-depth lines are used to delineate three zones for classifying flow profiles.[3] For supercritical flows, zone 1 is above critical depth, zone 2 is between normal depth and critical depth, and zone 3 is below normal depth.

S profiles.   An S1 profile begins with a hydraulic jump at the upstream end and becomes tangent to a horizontal pool level at the downstream end. An example is the profile of flow behind a dam or constricted bridge in a steep channel (Fig. 15.1*a*). The control is from downstream, and the profile is computed with a backwater analysis.

An S2 profile is a drawdown curve. It usually serves as a short transition between a hydraulic drop and uniform flow. One example is the profile formed on the downstream side of an enlargement of a channel cross section (Fig. 15.1*b*), and another is the profile on the downstream side of a change in channel slope from steep to steeper (Fig. 15.1*c*). The control is upstream, so the profile is determined with a downwater computation.

The S3 profile also is a transitional type, formed between an issuing supercritical flow and the normal-depth line to which the profile is tangent. An example is the profile on the steep-slope side of a change

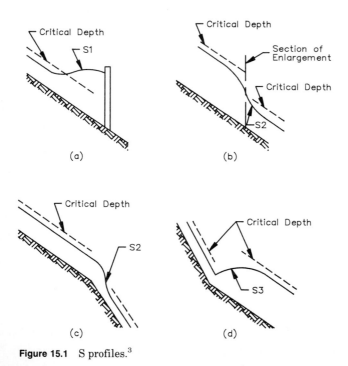

Figure 15.1   S profiles.[3]

in channel slope from steep to less steep (Fig. 15.1d). It is analyzed in a downwater computation.

Supercritical-flow analysis with HEC-2

Program limitations. Basic assumptions implicit in the analytic expressions of HEC-2 are discussed in Chap. 10. These assumptions are recapped here to remind the reader of their applicability in the analysis of supercritical-flow regimes. These assumptions, which also could be viewed as limitations, are as follows: steady flow, gradually varied flow, one-dimensional flow with correction for horizontal velocity distribution, constant friction slope between two cross sections, rigid boundaries, and small channel slope.

All of these limitations apply to supercritical flow as well as subcritical flow. The last two are particularly relevant to supercritical flow because of the relatively high velocities and steep slopes generally associated with this flow regime. High-velocity flows in natural channels usually are not consistent with the assumption of rigid-boundary conditions. Bed materials move and channel geometry changes with high velocity. Supercritical flows also are associated with steep slopes, which in some cases exceed the limitation of a 1:10 slope recommended for HEC-2 applications. On the other hand, if the limitations are reasonably complied with, the program can be used to compute supercritical-flow profiles. Frequently, the program is applied to troublesome segments of a subcritical reach where the profile goes through critical depth.

Input requirements. A supercritical profile requires that the cross-section data be arranged in input so that the computations can proceed from an upstream location to a downstream one. With one or two exceptions, the data input file is the same as one for a subcritical run in reverse order. Data files developed for a subcritical run often are converted for a supercritical computation. When this is done, particular attention must be given to repeated cross sections and special-bridge data.

A cross section created by repetition in a subcritical data file becomes the cross section which is repeated in a supercritical file. Making this conversion requires a revision of data for these cross sections. Similarly, cross sections associated with the modification factors PSXECR and PSXECE on X1 lines must be revised.

Special-bridge data input from a subcritical model which is converted into a supercritical model must be reordered. The X2 and BT lines associated with the upstream cross section in a subcritical data

ID.....1.......2......3......4.......5......6.......7.......8......9......10

ID	1	2	3	4	5	6	7	8	9	10
T1	MINERS CREEK AT ELDORADO FLATS									
T2	WATER SURFACE PROFILE									
T3	SUBCRITICAL									
J1					.001			3000	110	
J2	-1	0	-1	0	0	0	-1			
J3	38	43	1	2	26	58	62			
J6	1									
NC	.06	.06	.035	.1	.3					
X1	101	8	310	370						
GR	120	0	115	10	110	310	100	320	100	360
GR	110	370	115	670	120	680				
X1	102	0	0	0	2000	2000	2000	0	2	
NC	.05	.05	.015							
X1	201	8	110	170	10	10	10			
GR	122	0	117	10	112	110	102	120	102	160
GR	112	170	117	270	122	280				
X1	202	0	0	0	100	100	100			
X1	203				100	100	100			
X1	204				500	500	500			
X1	205				500	500	500			
X1	206				500	500	500			
X1	207				100	100	100			
NC	.06	.06	.035	.1	.3					
X1	301	8	310	370	1	1	1			
GR	140		135	10	130	310	120	320	120	360
GR	130	370	135	670	140	680				
X1	302				100	100	100		0.01	
X1	303				100	100	100		0.01	
X1	304				1000	1000	1000		.1	
X1	305				1000	1000	1000		.1	
X1	306				1000	1000	1000		.1	
EJ										

ER

Figure 15.2   Data input file for Miners Creek subcritical model.

file must be changed so that they are associated with the downstream
cross section in a supercritical file.

Three examples are presented to explain how HEC-2 is used in analyzing a supercritical-flow regime.

Miners Creek subcritical-profile example.   This subcritical-profile analysis is of a stream with a steep reach between two mild reaches. The
data input file is shown in Fig. 15.2. A profile plot from this run is
shown in Fig. 15.3. Observe that the profile is plotted at critical depth
for cross sections 204 to 207 in the steep part of the river.

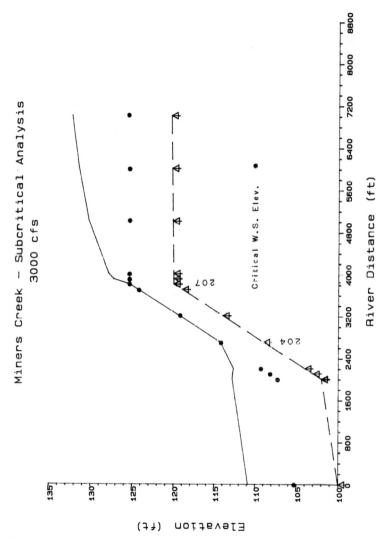

Figure 15.3 Profile plot from Miners Creek subcritical run.

```
ID.....1.......2.......3.......4.......5.......6.......7.......8.......9......10

T1 MINERS CREEK AT ELDORADO FLATS
T2 WATER SURFACE PROFILE
T3 SUPERCRITICAL
J1                          1      -1                      3000     110
J2    -1                   -1
J6     1
NC    .06     .06    .035     .1      .3
X1    306       8     310     370    1000    1000    1000       0     .32
GR    140       0     135      10     130     310     120     320     120     360
GR    130     370     135     670     140     680
X1    305       0       0       0    1000    1000    1000       0     -.1
X1    304       0       0       0    1000    1000    1000       0     -.1
X1    303       0       0       0     100     100     100       0     -.1
X1    302       0       0       0     100     100     100       0    -.01
X1    301       0       0       0       1       1       1             -.01
NC    .05     .05    .015
X1    207       8     110     170     100     100     100       0      18
GR    122       0     117      10     112     110     102     120     102     160
GR    112     170     117     270     122     280
X1    206       0       0       0     500     500     500       0      -1
X1    205       0       0       0     500     500     500       0      -5
X1    204       0       0       0     500     500     500       0      -5
X1    203       0       0       0     100     100     100       0      -5
X1    202       0       0       0     100     100     100       0      -1
X1    201                              10      10      10              -1
NC    .06     .06    .035     .1      .3
X1    102       8     310     370    2000    2000    2000       0       2
GR    120       0     115      10     110     310     100     320     100     360
GR    110     370     115     670     120     680
X1    101                                                              -2
EJ
```

ER

Figure 15.4 Data input file for Miners Creek supercritical model.

Miners Creek supercritical-profile example. The data input file from the preceding subcritical example, converted for a supercritical run, is shown in Fig. 15.4.

Compare the two data files, and observe the following changes that were made in the conversion:

1. The value in field 4 of the J1 line was changed from 0 to 1 to change the analysis from subcritical to supercritical.

2. The value ".001" in field 5 of the J1 line was changed to a " -1 " to specify the use of a starting water surface elevation equal to critical depth.

3. Cross-section data was reordered so that the upstream cross sec-

tion, no. 306, which is the last cross section in the subcritical run, is the first cross section in the supercritical data set.

4. The X1 lines (specifically, fields 2 to 4 and field 9) were modified to account for repeated GR data. Observe that it is not necessary to change reach lengths (fields 5 to 7 of the X1 lines).

Examine the profile plot from this supercritical run, shown in Fig. 15.5. Observe that the profile is plotted at supercritical depth in the steep reach. By combining the output from this run with the output from the previous subcritical run, one can obtain a complete profile for the three reaches. That is, the profile is complete except for the location of the hydraulic jump where the flow changes from subcritical to supercritical.

Supercritical special-bridge model example. This example is taken from an application of the special-bridge method in a supercritical-flow analysis. An excerpt from the output, shown in Fig. 15.6, reveals a serious problem in the computed water surface profile at the bridge. A combination of weir flow and low flow is indicated, which is physically impossible at this bridge. The bridge cross section has no low overbank area that could sustain weir flow in combination with low flow. Cross section and profile plots in Fig. 15.7a and b show this abnormal situation for flows of 3000 ft^3/s and 5000 ft^3/s at cross section 202.

This problem results from the effect of a high-velocity head on the operation of the decision logic in the program for determining weir flow. The final test for weir flow compares the total energy head (in this case, for low flow) with the elevation of the top of road ELTRD. The low-flow energy head, which is abnormally high because of the high-velocity-head component, is higher than the ELTRD, and weir flow is computed.

An arbitrary adjustment to the top-of-road and low-chord elevations in the data input will correct this problem. The results of raising these elevations 20 ft are shown in Fig. 15.8a and b. With the elevations of the top of road and low chord raised, the program determines that no weir flow exists and computes a low-flow solution. The results appear much more reasonable but should be carefully verified.

Split-Flow Analysis

Introduction

Split flow occurs when a portion of the flow leaves the main stream, and follows a separate path. Some split flows eventually return down-

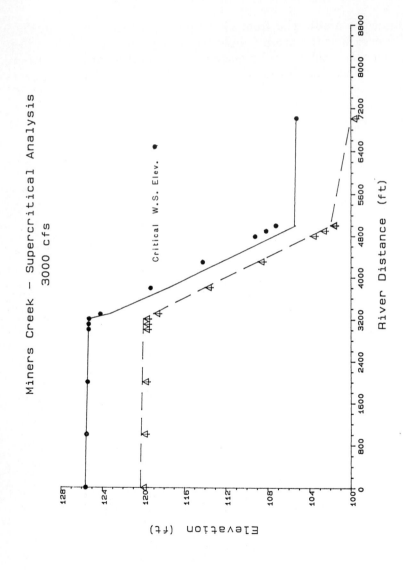

Figure 15.5 Profile plot from Miners Creek supercritical run.

```
*SECNO 202.000
6840 FLOW IS BY WEIR AND LOW FLOW
6870 D.S. ENERGY OF  104.08 HIGHER THAN COMPUTED ENERGY OF  103.88

3301 HV CHANGED MORE THAN HVINS

3420 BRIDGE W.S.=    95.53 BRIDGE VELOCITY=    23.20      CALCULATED CHANNEL AREA=    187.
```

EGPRS	EGLWC	H3	QWEIR	QLOW	BAREA	TRAPEZOID AREA	ELLC	ELTRD
99.72	103.88	.00	594.	4452.	380.	380.	100.60	102.00

202.00	12.76	102.71	.00	.00	104.08	1.38	.00	.00	99.95
5000.	66.	4868.	66.	78.	510.	78.	6.	1.	99.95
.01	.85	9.54	.85	.050	.015	.050	.000	89.95	63.79
.000534	25.	25.	25.	3	0	2	.00	152.42	216.21

Figure 15.6 Example of output from a supercritical model at a bridge location.

stream; others do not. The most common types of split flow result from the following conditions:

Islands or high ground

Overtopping of a levee

Overtopping of a watershed divide

Diversion structures

Program capabilities

The HEC-2 split-flow option is capable of directly analyzing all the conditions listed above except the bifurcation of flow at islands. Even this condition can be solved directly if a rating curve for one of the channels is available at the point where the flow divides; then the split flow can be treated as a diversion. A graphic interpolation procedure also can be used.[4]

The program has the following computational capabilities:

1. It can solve up to 100 split flows simultaneously.
2. It can compute up to 15 profiles in one run.
3. It can perform analyses with three different methods: weir flow, normal depth, and rating curve.
4. It has the option to return all or part of the split flow downstream.
5. It has the option to use either water surface elevation or energy elevation to compute the split flow.

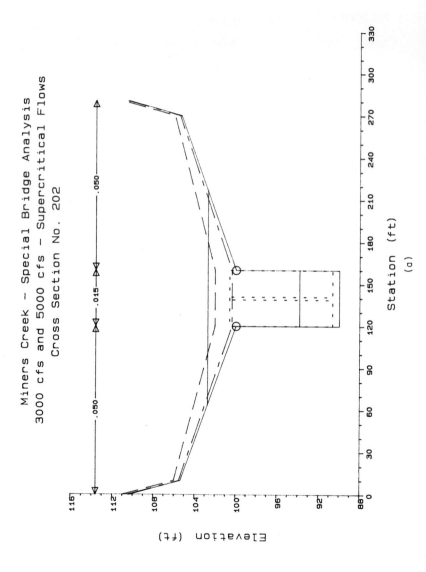

Miners Creek – Special Bridge Analysis
3000 cfs and 5000 cfs – Supercritical Flows
Cross Section No. 202

(a)

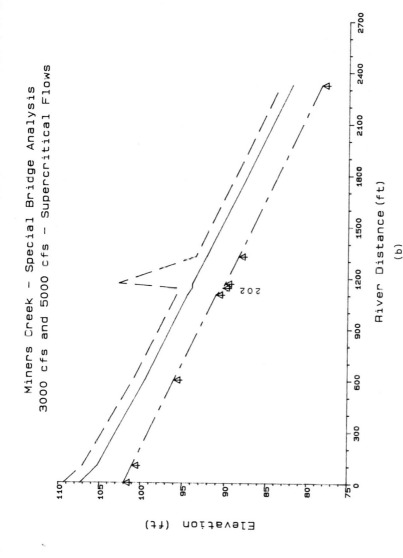

Figure 15.7 Cross-section and profile plots of supercritical flow at the same bridge location. (a) Cross section; (b) profile.

Miners Creek – Special Bridge Analysis
3000 cfs and 5000 cfs – Supercritical Flows
Cross Section No. 202

(a)

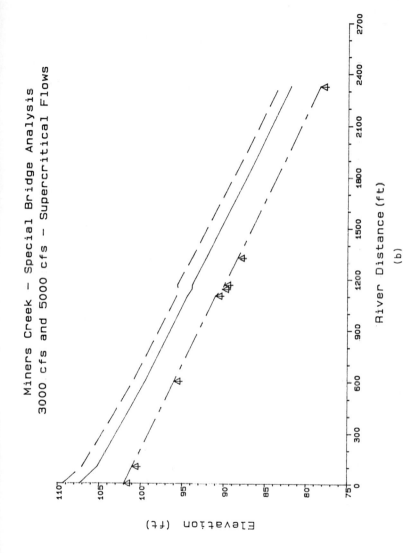

Figure 15.8 Cross-section and profile plots at the same bridge location after adjustment of top-of-road and low-chord elevations. (*a*) Cross section; (*b*) profile.

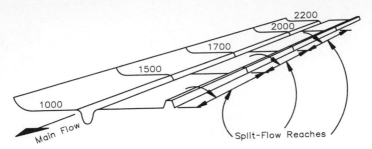

Figure 15.9   Illustration of split-flow reaches.

In addition, the split-flow option is compatible with other HEC-2 options, except for encroachment methods 3 to 6.

Computational procedure

The location of each split-flow reach is defined by the numbers of upstream and downstream cross sections, as shown in Fig. 15.9. The program uses a method of trial and error to compute the water surface profile in the main stream while reducing the discharge to account for the departure of the split flow. In the first iteration, the water surface profile in the main stream is computed under the assumption that no split flow occurs or that a specified amount occurs as indicated in input.

Water surface elevations or energy elevations, based on this initial profile, are used to compute the split flow from each of the designated reaches. These computed split flows are compared with the assumed value, which in the first iteration may be zero, and if the difference is greater than 2 percent, the program begins a new iteration. The water surface profile in the main channel is computed again with a new set of assumed split flows; the results of this computation are used as a basis for obtaining another set of computed split flows; the computed and assumed values are compared; and the iterations continue until the difference criterion of 2 percent is met. However, if the criterion is not met in 20 iterations, the procedure is terminated.

Program limitations

In addition to the general limitations in HEC-2, which apply, there are assumptions in the split-flow option. They are as follows:

1. Split flows can be estimated with reasonable accuracy by means of the standard weir equation, normal-depth computations, or a rating curve of outflow vs. elevation.

2. Submergence of weir flow is considered insignificant.

3. Weir flow varies linearly along the length of the weir on the basis of upstream and downstream water surface or energy elevations.

4. Normal-depth conveyance varies linearly along the length of the split-flow cross section on the basis of upstream and downstream water surface elevations.

5. The direction of the split flow is perpendicular to the direction of the main flow.

6. Split-flow boundaries are fixed; they do not erode or change with time.

7. Normal-depth slope must be a subcritical slope.

Input requirements

Split-flow input is specified on a set of data lines located at the beginning of an HEC-2 data input file. It consists of a split-flow title line, which executes the split-flow option; an optional set of job lines used to control the processing of data and output; a set of reach lines containing data required for the method of analysis being used (weir, normal depth, or diversion); and an end line.

The split-flow data input varies from standard HEC-2 format in that a set of split-flow data lines is always preceded by a title line. The order of the lines within each set must follow precisely a prescribed pattern. The sets of lines for various reaches do not have to be in a specific order, but it is recommended that they be arranged in a downstream sequence to make the file more readable.

Other lines may be added to the HEC-2 data input file at other locations to facilitate the use of the split-flow option. These include lines for starting backwater computations with a rating curve, for using rating curve elevations at any cross section in lieu of computed water surface elevations, and for specifying an initial split-flow discharge for each split-flow reach.

Modeling procedures

Numbering of cross sections. Cross sections must be numbered in ascending order from downstream to upstream locations. This is the order that the program follows in computing water surface elevations and split flows at designated reaches. It is recommended that cross sections be numbered according to distance (feet, meters, miles, etc.) up the main channel and that split-flow reaches be started and ended at numbered cross sections. However, if the downstream and upstream ends of split-flow reaches do not match with numbers of cross sections, the program will interpolate linearly in determining water surface or energy elevations.

Reach lengths.   The calculation of split flow in a reach is similar to integration by the trapezoidal rule, which approximates segments of a curve. The greater the number of segments (or the shorter each segment is), the more accurate the approximation. Similarly, the shorter the split-flow reaches, the more accurate the computation of split flow.

In an analysis of the effect of reach length on accuracy,[4] a 1000-ft reach was divided four different ways for the computation of split flow. The alternatives analyzed included a single reach of 1000 ft, two reaches of 500 ft, four reaches of 250 ft, and ten reaches of 100 ft. The results of this analysis, shown in Fig. 15.10, reveal how the computation of the water surface profile and the spatial distribution of split flow can be distorted by a long reach length.

Hydrologic considerations.   In water surface profile computations it is assumed that the peak discharge from a hydrograph can be analyzed as steady flow. This is discussed in Chap. 10. In split-flow analysis, this assumption can cause problems, and its effect should be considered. The split-flow option reduces discharge in the main stream by amounts of split flow leaving the stream. This reduced flow continues unless or until all or part of the split flow is returned downstream, where the discharge is again adjusted accordingly. The effects of split-flow reductions on downstream hydrograph peaks may be significant, depending on the timing of tributary and local inflows.

An example[4] illustrates this problem. A split-flow reach above the confluence of a tributary is shown in Fig. 15.11a. Hydrographs at the confluence, without split flow, are shown in Fig. 15.11b. If flows above 3000 ft^3/s are permanently lost due to split flow, the modified hydrographs in Fig. 15.11c result. A comparison of peak discharges on the lower main stem, with and without split flow, reveals a difference of 500 ft^3/s. The HEC-2 split-flow option by itself would reduce the flow in the main stem by 1000 ft^3/s.

The effect of a split-flow reduction on a downstream hydrograph peak also depends on flood routing. Flood routing tends to reduce the magnitude of peak discharge through a routing reach. It would also tend to reduce the effect of upstream changes in peak discharge due to split-flow reductions. A procedure to account for the hydrologic aspects of split flow is as follows:[4]

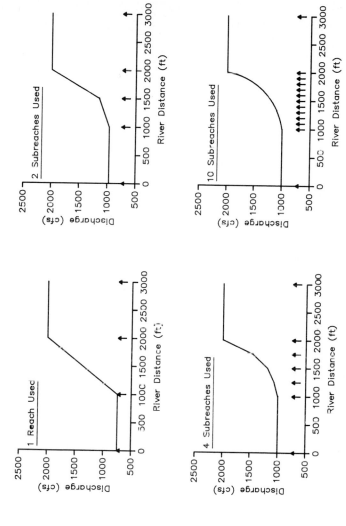

Figure 15.10 Approximation curves of split flow for a 1000-ft reach based on four different divisions of the reach into subreaches for computation.

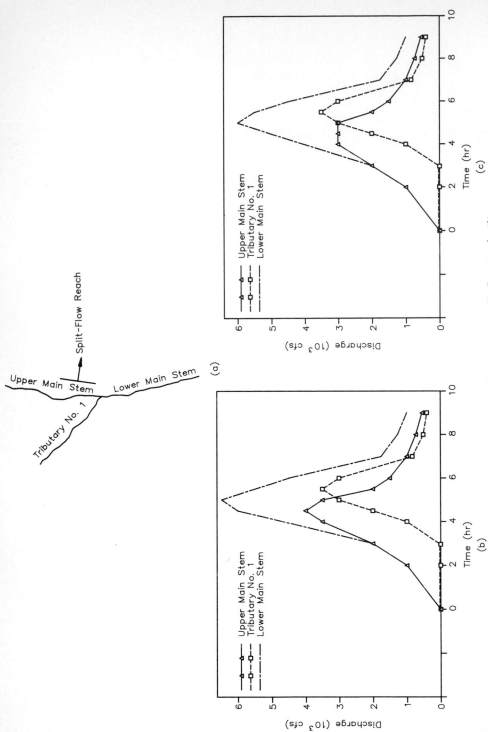

Figure 15.11 Effects of split flow on downstream hydrographs. (*a*) Plan showing split-flow reach; (*b*) hydrographs at the confluence without split flow; (*c*) hydrographs at the confluence with split flow.

1. Compute split flows through the entire study reach on the basis of an initial peak discharge from a runoff hydrograph.

2. Adjust the hydrologic model (HEC-1) at the split-flow reaches to reflect the lost flows found in step 1, and compute new hydrographs.

3. Analyze the effects of the split-flow losses on the peak discharge downstream from the split-flow reaches. If the peak discharge is reduced by the total amount of the losses found in step 1, no further analysis is required. If only part of the split-flow loss is reflected by the reduction in peak discharge, the other part should be returned to the main stream. This can be accomplished by modifying the discharge data in HEC-2 to reflect the return flows downstream.

4. Repeat this procedure, using the new peak discharge found in step 2, and the modified HEC-2 data file in step 3, until the split-flow loss balances with the reduction in peak discharge downstream.

This procedure is suitable for split-flow applications that involve a maximum of three or four split-flow reaches. More complex conditions should be analyzed by means of unsteady-flow techniques.[5]

Example problem: Red Fox River split flow

A plan view of a levee system on the Red Fox River is shown in Fig. 15.12. Profiles of the river bed, levees, and overflow weir are shown in Fig. 15.13. The starting water surface elevation is based on a normal-depth calculation using a slope of 0.005 ft/ft. The weir coefficients for the levee and the overflow weir are 3.4 and 2.7, respectively. The coefficient for the overflow weir is relatively low because of anticipated submergence caused by tailwater in the floodway.

The split-flow data, which is located at the beginning of an HEC-2 input file, is shown in Fig 15.14. Split flow is computed in this example in three reaches: between cross sections 4 and 3, 3 and 2, and 2 and 1. Following the split-flow title line SF, the first set of lines is for the reach 4–3. The TW line in the set is used to identify the reach. The WS line includes the number of coordinates used to specify the levee geometry, the location of downstream and upstream limits of the reach, and the cross-section number where the split flow returns. In this example, a " – 1" indicates that the split flow does not return. The WS line has the station and elevation coordinates for the weir. Sets of data lines for the reaches 3–2 and 2–1 are the same as the set just described.

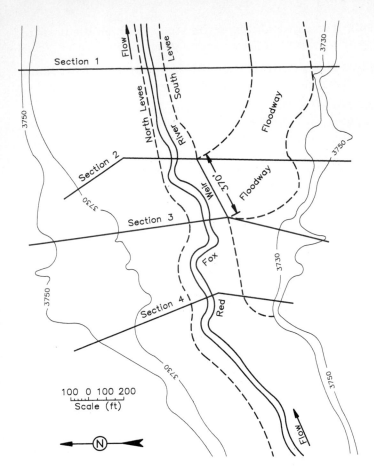

Figure 15.12 Plan of split-flow reach on the Red Fox River.

The split-flow output for this example is shown in Fig. 15.15. The split-flow-output variables are defined as follows:

ASQ	Assumed split flow
QCOMP	Computed split flow
ERRAC	Percentage of error between assumed and computed values
TASQ	Total assumed split flow
TCQ	Total computed split flow
TABER	Percentage of error between totals
NITER	Number of iterations the program completed
DSWS	Computed downstream water surface elevation
USWS	Computed upstream water surface elevation
DSSNO	Downstream cross section where split flow begins
USSNO	Upstream cross section where split flow ends

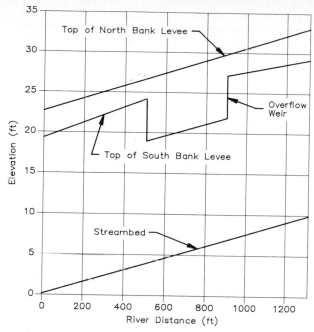

Figure 15.13 Profiles of the bed, levees, and overflow weir for the Red Fox River.

Analysis of Ice-Covered Streams

Introduction

The ice-covered-flow option in HEC-2 can be used to determine water surface profiles for streams with a stationary or floating ice cover. Different ice thicknesses can be specified for the channel and the right and left overbanks. A composite Manning's n is determined by the program with the Belokon-Sabaneev equation. This option also determines the potential for the formation of ice jams through the application of Pariset's ice stability function.

SF	SPLIT-FLOW DATA				
TW	SOUTH BANK LEVEE BETWEEN SECTIONS 3 AND 4				
WS	2	3	4	-1	3.4
WC	0	26	400	28	
TW	FLOODWAY OVERFLOW SECTION				
WS	2	2	3	-1	2.7
WC	0	19	370	21	
TW	SOUTH BANK LEVEE BETWEEN SECTIONS 1 AND 2				
WS	2	1	2	-1	3.4
WC	0	19	500	24	

Figure 15.14 Split-flow input data for Red Fox River example.

TW   SOUTH BANK LEVEE BETWEEN SECTIONS 3 AND 4

ASQ	QCOMP	ERRAC	TASQ	TCQ	TABER	NITER	DSWS	USWS	DSSNO	USSNO
.00	.00	.00	.00	.00	.00	9	24.025	25.070	3.000	4.000

TW   FLOODWAY OVERFLOW SECTION

ASQ	QCOMP	ERRAC	TASQ	TCQ	TABER	NITER	DSWS	USWS	DSSNO	USSNO
5878.83	5860.35	.31	5878.83	5860.35	.31	9	22.478	24.025	2.000	3.000

TW   SOUTH BANK LEVEE BETWEEN SECTIONS 1 AND 2

ASQ	QCOMP	ERRAC	TASQ	TCQ	TABER	NITER	DSWS	USWS	DSSNO	USSNO
34.18	30.31	12.00	5913.01	5890.66	.38	9	19.372	22.478	1.000	2.000

Figure 15.15   Sample output for the split-flow example.

A basic assumption in water surface profile analysis of ice-covered streams is that the ice cover floats. The ice cover on wide streams tends to crack from thermal expansion and contraction and from water-level fluctuations. Thus, it is free to rise and fall with changing flows. On small streams, the cover is more likely to be held in place by rocks and trees and to have sufficient strength to bridge the flow. A stationary ice cover like this is more complicated to analyze, because open-channel conditions may occur beneath the ice during low flow, and a combination of pressure flow and open-channel conditions (above the ice) may occur during high flow.

Water surface profiles are computed for ice-covered streams by means of standard step calculations, with allowances made for the flow area blocked by ice, the increased wetted perimeter, and the difference in roughness between the ice and the streambed. The position of the floating ice relative to the free water surface (piezometric head) is determined by the specific gravity of the ice, typically about 0.92. Computation of the backwater profile requires an estimated ice-cover thickness, an adjustment of the wetted perimeter and cross-sectional area of flow, and a composite roughness coefficient.

Hydraulic parameters

The hydraulic parameters of ice-covered flow are shown in Fig. 15.16. The parameters shown in this figure are defined as follows:

A Open-flow area under the ice
W_p Wetted perimeter of the channel
B Wetted perimeter of the ice
n_b Manning's n for the streambed
n_i Manning's n for the ice

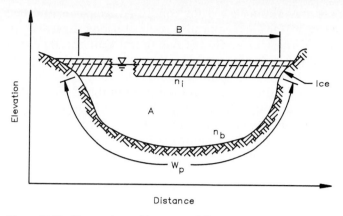

Figure 15.16   Parameters of ice-covered flow.

The open-flow area under the ice is determined by

$$A = A_t - 0.92hB$$

where A_t is the total area under the free water surface and h is the thickness of the ice cover.

The hydraulic radius is determined by

$$R = \frac{A}{W_p + B}$$

For wide ice-covered channels, the wetted perimeter is almost double the wetted perimeter of the same channel without ice.

Ice-cover thickness.   The change in the hydraulic radius and the reduction in cross-sectional area attributed to ice cover can significantly affect water surface elevations. Studies have indicated that the water surface profile for ice-covered flow may attain a uniform slope upstream at a depth 30 percent greater than the uniform-flow depth for equivalent open-channel flow.

For streams which produce frazil-ice deposits underneath the ice cover, such deposits must be included in reductions of flow area. Frazil ice can reduce cross-sectional area by 65 percent or more. A rule of thumb for determining when the total thickness of the ice cover is significant enough to warrant reducing the area of flow is as follows: If the total ice-cover thickness is greater than 5 percent of the average water depth, the reduction should be made.[6] Regardless of how small the ice thickness may be, the effect on the hydraulic radius is significant, and the change must be incorporated.

Ice-cover roughness. A Manning's n for combined ice-cover and bed roughness is used in water surface profile computations. Research in Sweden, Canada, and the United States has found that the Belokon-Sabaneev formula [Eq. (15.1)] for a composite n gives satisfactory results. Another formula for determining composite roughness is presented in the Corps of Engineers ice engineering manual.[6] Values for ice-cover roughness that might be used in determining a composite value range from about 0.01 to 0.05. Probably the most accurate method of determining composite roughness is to compute the value from uniform-flow equations based on field measurements of flow during the winter. However, it is important to note that ice roughness may change with time, especially during melting conditions, when ripple patterns may develop on the underneath surface.

$$n_c = \left(\frac{n_i^{3/2} + n_b^{3/2}}{2} \right)^{2/3} \tag{15.1}$$

where n_c = composite n value
$\quad\ n_b$ = Manning's n value for streambed
$\quad\ n_i$ = Manning's n value for ice

The effect of ice-cover roughness on computed water surface elevations is not as significant as that of changes in flow area and wetted perimeter related to ice-cover thickness. The results of changing composite roughness from 0.04 to 0.045, other conditions being equal, are shown in Table 15.1. A difference of only 0.13 ft in 4000 ft is indicated. By contrast, for the same distance, an increase in ice thickness of half an inch increases the computed water surface elevation 0.46 ft.[6]

TABLE 15.1   Effects of Variation in Ice-Cover Conditions on the Computed Water Surface Profile

Upstream distance, ft	Bed elevation, ft	Open water $(n_b=0.04)$	$n_c = 0.04$ $n_i = 0.04$ $h = 0.543*$	$n_c = 0.045$ $n_i = 0.05$ $h = 0.543*$	$n_c = 0.04$ $n_i = 0.04$ $h = 0.76*$	$n_c = 0.04$ $n_i = 0.04$ $h = 1.08$
			Ice-cover conditions			
0	558.0	562.82	562.80	562.80	562.80	562.80
1000	559.0	562.82	562.89	562.91	562.91	562.94
2000	560.0	562.83	563.10	563.17	563.17	563.28
2500	560.5	562.87	563.30	563.39	563.39	563.56
3000	561.0	562.98	563.57	563.69	563.70	563.93
3500	561.5	563.18	563.93	564.06	564.10	564.37
4000	562.0	563.50	564.37	564.50	564.55	564.83
4500	563.0	564.40	565.03	565.16	565.23	565.52
5000	564.0	565.12	565.98	566.09	566.18	566.48

*In feet.

Ice jams

Formation. An ice jam, which is an accumulation of ice in a stream, reduces the cross-sectional area available to carry flow and increases the water surface elevation. Ice jams tend to reoccur at locations where conditions exist both for generating an adequate supply of ice and for obstructing its downstream transport. They usually form at natural or constructed obstructions and at abrupt changes in bottom slope, channel alignment, and cross-sectional shape. A common feature at many sites is the existence of a long backwater condition extending upstream. A jam can form through the transport and deposition of upstream ice with no consolidation of the ice jam. However, if external forces exceed the internal strength of the mass, consolidation will occur. Since ice jams typically occur in the same locations, historical records of floods should be reviewed to determine high-incidence locations for use in a water surface profile study.

Classification. Several classifications of ice jams have been identified, including freeze-up, breakup, moving, stationary, floating, and grounded. The first two classes characterize the evolutionary process involved. "Freeze-up" jams are caused by the accumulation of frazil ice, which eventually forms a continuous ice cover. "Breakup" jams are caused by a rapid rise in stage and flow resulting from rainfall or snowmelt. The rise in stage releases large amounts of ice cover, which are transported downstream to form the ice jam. Because of the large volumes of ice and the high discharges that may be involved, this type of jam is predominant in flood events caused by ice jams.

The other classifications characterize the physical state of ice jams. "Moving" jams cause increases in stage, but their effect is minor compared to that of "stationary" jams, which may be either "floating" or "grounded." Both floating and grounded types can cause significant backwater effects. The floating type, typical of deep rivers, is not grounded on the bottom, so significant flow can pass underneath the ice. The grounded type, typical of shallow confined streams, is at least partially grounded on the bottom and may divert most of the flow to the overbanks.

Roughness coefficients for ice jams range from 0.05 to 0.10. Composite roughness n_c near the downstream end of a jam may be as high as 0.13 but may decrease to a much lower value of 0.04 near the upstream edge. The high value at the downstream end, where blockage of the channel is greater, can be attributed more to drag than to roughness.

Volume and thickness. Not all the ice existing in a stream network is transported downstream when the ice cover breaks up. Some remains along the shore or melts in place. An approximate volume of ice in a jam can be estimated from the river mileage of the contributing network, the ice-cover thickness, and the average top width of the river. After this total volume has been calculated, the portion or percentage reaching the site must be estimated. Ten percent has been suggested as a rough approximation. An equation incorporating an ice loss coefficient also may be used to determine the ice jam volume.[6]

Empirical relationships have been used for determining average jam depths and jam lengths.[6] Studies have found that the jam length does not exceed 10 percent of the upstream river length that contributes ice. It is possible for the ratio of ice jam thickness to river depth in a shallow stream to reach 0.7 near the toe of the jam, and the jam may even ground and completely block the channel.

Effect on water levels. Ice jams can cause large increases in water levels. Studies have shown that an ice jam can cause a 10-year flood event to have a water level equivalent to that of an ice-free 100-year event.[6]

Ice stability analysis. A stable ice jam is defined as a stationary consolidated accumulation of ice. An unstable jam will tend to consolidate and move, frequently causing greater blockage. Stability criteria, utilized in HEC-2, that are suitable for analyzing cohesionless ice in deep, wide channels were developed by Pariset.[7] Spring "breakup" ice has negligible cohesion and can be analyzed by these criteria. The Pariset stability function employs a dimensionless ice stability indicator to analyze the ratio of ice thickness T to upstream open-water depth H. A graph of this is shown in Fig. 15.17. The stability function has the form

$$X = \frac{Q^2}{C^2 BH} \tag{15.2}$$

where Q = water discharge, ft^3/s
   C = Chezy's coefficient
   B = stream width, ft
   H = upstream water depth, ft
   X = ice stability indicator

Example problem: ice stability analysis

This example problem is presented to demonstrate the application of the Pariset criteria through an analysis of the stability of an ice jam in a rectangular channel. The upstream distances and corresponding

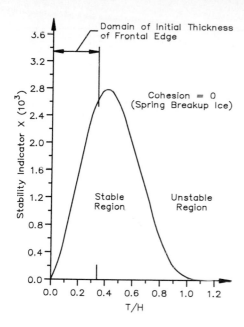

Figure 15.17 Ice stability indication curve.

bed elevations of the cross sections are taken from Table 15.1. The ice jam has a top width of 175 ft and a uniform thickness h of 3.27 ft. The composite roughness n_c is 0.04.

The backwater profile is computed with HEC-2, under the assumption that a floating ice jam of constant thickness exists between cross sections. The analysis is made for three discharges of 1000 ft³/s, 2000 ft³/s, and 4000 ft³/s, and corresponding downstream (starting) water surface elevations of 567 ft, 567.2 ft, and 568 feet.

The ice stability analysis is described with reference to Table 15.2. The computed water surface elevations are shown in column (2). In terms of the stability analysis, these are considered upstream water surface elevations. The upstream depths are found by subtracting the bed elevations and are shown in column (5). The ratios of thickness to upstream depth are shown in column (6), and the dimensionless stability index is shown in column (7). The state of ice jam stability is determined by the position of points located by coordinates, h/H and X, on the stability function diagram (Fig. 15.17). If a point is inside of the stable domain defined by the bell-shaped curve, stability is indicated. Vertical intercepts on the curve corresponding to X values are shown in column (8) for comparison. The results are shown in column (9): the ice jam is stable for a distance upstream of 4500 ft for a flow of 1000 ft³/s; it is stable for a distance upstream of 4000 ft for a flow of 2000 ft³/s; and it is completely unstable for a flow of 4000 ft³/s. An unstable reach has to be adjusted in the computations by distributing the

TABLE 15.2   Ice Stability Computations

Upstream distance, 10^3 ft (1)	Water surface elevation, ft (2)	H_u, ft (3)	V, ft/s (4)	H, ft (5)	h/H (6)	X, 10^{-3} ft (7)	Pariset curve, 10^{-3} ft (8)	Stable cover (9)
			$Q = 1000$ ft^3/s		$h = 3.27$ ft			
0.0	567							
1.0	567.2	5.2	1.12	8.2	0.40	0.70	2.77	Yes
2.0	567.5	4.5	1.28	7.5	0.44	1.02	2.69	Yes
3.0	568.0	4.0	1.44	7.0	0.47	1.39	2.01	Yes
4.0	568.7	3.7	1.55	6.7	0.49	1.68	2.54	Yes
4.5	569.2	3.2	1.78	6.2	0.53	2.39	2.52	Yes
5.0	569.9	2.9	1.94	5.9	0.55	2.93	2.24	No
			$Q = 2000$ ft^3/s		$h = 3.27$ ft			
0.0	567.2							
1.0	567.6	5.83	2.05	8.6	0.38	2.17	2.77	Yes
2.0	568.5	5.57	2.10	8.6	0.39	2.35	2.78	Yes
3.0	569.4	5.41	2.13	8.4	0.39	2.46	2.78	Yes
4.0	570.3	5.33	2.15	8.3	0.39	2.53	2.78	Yes
4.5	570.9	4.86	2.35	7.9	0.47	3.24	2.78	No
5.0	571.6	4.56	2.50	7.6	0.43	3.84	2.78	No
			$Q = 4000$ ft^3/s		$h = 3.27$ ft			
0.0	568							
1.0	568.9	6.9	3.34	9.9	0.33	4.74	2.74	No
2.0	570.4	7.4	3.10	10.4	0.32	3.76	2.67	No
3.0	571.6	7.6	3.00	10.6	0.31	3.37	2.64	No
4.0	572.8	7.8	2.92	10.8	0.30	3.17	2.67	No
4.5	573.4	7.4	3.10	10.4	0.30	3.76	2.62	No
5.0	574.1	7.1	3.24	10.1	0.32	4.29	2.62	No

ice downstream to the stable reaches; then the analysis described above is repeated.

Water surface profile analysis of ice-covered streams with HEC-2

Computational procedures. Floating ice cover is treated by the program as if it were a floating bridge. Techniques used to account for flow areas blocked by a bridge deck and the additional wetted perimeter caused by a submerged low chord are employed in ice-cover analysis. The low chord of the "floating bridge" is determined for each trial in the standard step computation by subtracting the product of the specific gravity and thickness of the ice from the trial water surface elevation. A diagram showing the variables used by the program is in

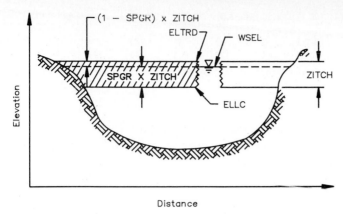

Figure 15.18   Ice-covered-flow variables in HEC-2.

Fig. 15.18. The top and bottom elevations of the ice cover are computed as follows:

$$ELTRD = WSEL + (1 - SPGR)*ZITCH$$

$$ELLC = WSEL - (SPGR*ZITCH)$$

where ELTRD = top elevation of the ice cover
      ELLC = bottom elevation of the ice cover
      WSEL = trial water surface elevation
      SPGR = specific gravity of ice
      ZITCH = thickness of ice in the channel

The Belokon-Sabaneev formula [Eq. (15.1)] is used to determine a composite n value. The calculation of a composite n is compatible with the full range of n-value options in the program (NC, NH, and NV) as well as the optional n-value adjustment factor, FN. Ice n values and cover thicknesses can be modified independently of corresponding channel values with the adjustment factor FZ entered on the IC line.

Program Input.   Ice-cover parameters are specified on the ice-cover data line—the IC line. The data input includes the thickness of ice in the channel and in each of the overbanks, the n value for the underside of the ice, the specific gravity of the ice, and the adjustment factor FZ. Fields 6 to 10 on the IC line function like the same fields on the channel improvement line—the CI line. The placement of FZ values in these fields, in coordination with values for IBW in field 8 of the J2 lines, can be used to select varying sets of conditions for multiple profile runs.

```
4478 FLOATING ICE COVER, ICE THICKNESS LOB= 3.3   CH= 3.3   ROB= 3.3

    SPGR    ICE N    B-S N     C       B       H      T/H     X*K
    .920    .0400    .0399    43.    175.     8.1     .40     .697
```

Figure 15.19 Example of HEC-2 output for ice-covered-flow computations.

The IC line is located in the data input file just in front of the X1 line for the cross section where the ice-cover computations are to begin. Ice-cover computations continue with this set of data through succeeding cross sections until another IC line is encountered, which changes parameters or terminates the ice cover.

Program output. At locations where ice-cover computations are made, the program prints out a note indicating ice cover, followed by 11 ice-related variables, as shown in Fig. 15.19. The variables are defined as follows:

LOB	Left-overbank ice thickness, ft
CH	Channel ice thickness, ft
ROB	Right-overbank ice thickness, ft
SPGR	Specific gravity of ice
ICE N	ZIN, Manning's n value for ice
B-S N	Composite n value
C	Chezy's coefficient
B	Wetted perimeter, ft
H	Depth of flow, ft
T/H	Ratio of ice thickness to depth
X*K	Ice stability variable times 1000

Other ice-related variables available for summary table output include the MAX STABLE X*K, the maximum value on the stability curve corresponding to X*K; ELTRD, the elevation of the top of ice; ELLC, the elevation of the bottom of ice; and K*WTN, composite n value. The state of ice stability at a cross section can be determined by comparing values for X*K and MAX STABLE X*K.

Other Options in HEC-2

Tributary stream profile computations

In subcritical-flow regimes, profiles for tributaries can be computed together with profiles for the main channel in a single run. Data sets are arranged to first compute the profile for the main channel from

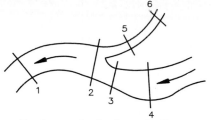

Figure 15.20 Cross-section numbering for single-tributary simulation.

Numbers refer to Cross Sections

the downstream end to the upstream end. Data for the tributary stream, whose starting water surface elevation is determined when the main-channel profile is calculated, follows the data for the main channel.

Refer to Fig. 15.20 for a single-tributary example. Cross sections 1 through 4 are entered in the data set first to compute the profile for the main channel. Following cross section 4, the data for cross section 2 is repeated so that cross section 2 serves as the starting cross section for the tributary. The cross-section identification number "2" is changed to " − 2" to signal to the program that it is to begin the tributary computation. When the program encounters the negative section number, it searches its memory for the corresponding positive cross-section number. It then takes the previously computed water surface elevation for cross section 2 and begins profile computations for the tributary, proceeding in this example from cross section 2 to 5 to 6. Since the tributary computation starts in the main stream, the discharge downstream from cross section 2 is reinstated in input at cross section −2.

In some cases, it may be desirable to compute a profile for a stream system with a second-order tributary. This can be done in a single run if the data for the first tributary with a tributary is treated as a portion of the main channel and the main channel upstream from the junction with the tributary is treated as a tributary. This arrangement is illustrated in Fig. 15.21. Cross sections 1 through 8 are in the main stream, 11 through 16 are in the first-order tributary, and 21 and 22 are in the second-order tributary. The arrangement of cross-section data for an analysis of this system in a single run is as follows: 1, 3, 4, 11, 12, 13, 14, 15, 16, −4, 5, 6, 7, 8, −14, 21, 22.

The method of computing tributary profiles in a single run with the main stream should not be used with encroachment methods 3 through 6.

Determination of Manning's *n*

Manning's *n* can be computed by HEC-2 if the discharge and the water surface elevation, as indicated by high-water marks, are specified for

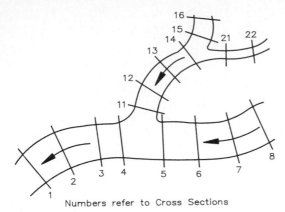

Numbers refer to Cross Sections

Figure 15.21 Cross-section numbering for a second-order-tributary model.

each cross section. Estimates of the channel and right- and left-overbank n values are specified by the user for the first cross section. The program computes a ratio between channel and overbank n values based on the first cross section and uses this relationship for all subsequent cross sections unless altered by another set of specified n values.

The elevations of high-water marks for all cross sections are entered in input as known water surface elevations. These are treated by the program as computed elevations in a reverse computation to find the n values. Since the average friction slope equation [Eq. (10.16)] is utilized in this process, it may be appropriate to verify the validity of the computed n values if another friction loss equation is being used for computing the water surface profile. This can be done by comparing the water surface elevations computed with the new n values with the high-water-mark elevations.

Because of the sensitivity of the n-value computations to slight errors in high-water-mark elevations, a length-weighted n value is computed for each cross section. When an adverse slope of the energy grade line is encountered, as might result from erroneous high-water marks, the n-value computations are restarted with n values from the previous cross section. However, the weighted-n computations continue without restarting and provide a frame of reference for evaluating the reasonableness of computed n values for all cross sections.

A trial-and-error method of determining n values is often preferred because of uncertainty associated with high-water-mark data. In this approach, an assumed set of n values and a specified discharge are used to compute a water surface profile, which is compared with the high-water-mark profile. The process can be repeated until a suitable fit is obtained.

Computation of storage-outflow data

Output from an HEC-2 multiple profile run contains data that can be used as input for the HEC-1 flood hydrograph program for doing stream routing with the modified Puls method. A user-defined summary output table can be designed to tabulate volume and discharge data for a routing reach. Travel time through the reach also can be included. An excerpt from such a table for a five-profile run (five discharges) is shown in Fig. 15.22. Only the data tabulated for the last cross section in the reach is shown in the excerpt.

SECNO	CWSEL	CRIWS	VOL	Q	TIME	OLOSS	QLOB	QCH	QROB
3000.000	4542.94	.00	5.74	390.00	.17	.02	.00	390.00	.00
3000.000	4544.71	4542.77	12.31	570.00	.20	.01	125.68	444.32	.00
3000.000	4544.94	.00	13.32	590.00	.20	.01	149.78	440.22	.00
3000.000	4546.79	4543.58	23.44	855.00	.21	.01	335.92	507.64	11.44
3000.000	4548.34	.00	32.59	1790.00	.18	.01	755.14	954.73	80.14

Figure 15.22 Storage-outflow data in summary output.

If the reach contains ineffective-flow areas, such as those normally located next to bridges, the storage data in the summary printout should be adjusted accordingly. This is an important point to remember, because a normal HEC-2 run is likely to have a number of ineffective-flow areas that should be included as storage.

Buildings located in the floodplain constitute a source of storage that may or may not be included in a normal HEC-2 run, depending on how the buildings are modeled. Four approaches to modeling buildings in the floodplain are depicted in Fig 15.23. In the first approach, shown in Fig. 15.23b, storage is eliminated by blocking out the buildings with ground points. In Fig. 15.23c, an overall roughness factor for the floodplain is subjectively increased to account for the buildings. The storage is unaffected. In Fig. 15.23d, two steps are used: first, the buildings are blocked out with extra ground points, and the water surface elevation is computed; then the extra ground points are removed, and the computed water surface elevation is introduced for storage computation. In Fig. 15.23e, which shows the preferred approach, the building widths are assigned high n values. This approach has the ad-

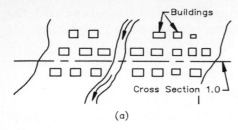

(a)

Plan View

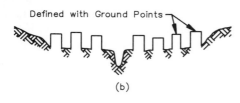

(b)

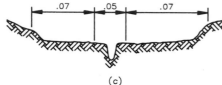

(c)

2—Stage Procedure

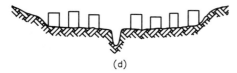

(d)

High n Values

(e)

Cross Section 1.0

Figure 15.23 Alternatives for modeling buildings in the floodplain. (*a*) Plan; (*b*) block out; (*c*) represent with composite *n* values; (*d*) block out, and then remove for storage computations; (*e*) represent with high *n* values.

vantage of effectively blocking conveyance at building locations without removing the storage capacity of the buildings.

Since water surface profile analysis with HEC-2 requires discharge data frequently provided by a flood hydrograph program such as HEC-1, coordinated application of the programs may be needed. One approach is to first run HEC-2 with several different discharges to obtain the storage-discharge and travel time data needed for an HEC-1 run. This data is then used in HEC-1 to compute flood hydrographs. Finally, HEC-2 can be run again, using discharge data from the HEC-1 output to compute the water surface profiles.

Workshop Problem 15.1: Yuba River Split-Flow Analysis

Purpose

The purpose of this workshop is to provide an understanding of the split-flow option in HEC-2 and how to apply it.

Problem statement

Marysville, California, is situated at the confluence of the Feather and Yuba rivers. The city is surrounded by a high ring levee 3 to 6 ft higher than levees protecting nearby communities. A flood insurance study for the City of Marysville indicates that a 500-year flood on the Yuba River may overtop the levee protecting Linda, an unincorporated community south of Marysville. A map of the Yuba River and its levee system in the vicinity of Marysville and Linda is shown in Fig. P15.1.

An input file for an HEC-2 model of this reach of the Yuba River is shown in Fig. P15.2. The cross sections used in the model are shown on the map. Concurrent 500-year-flood hydrographs for the Feather and Yuba rivers at a location immediately upstream from their confluence are shown in Fig. P15.3. A rating curve for the Feather River just below the mouth of the Yuba is shown in Fig. P15.4. Stations and elevations for the top of the Linda levee are given in Table P15.1.

Task

Complete the input file for the Yuba River model so that it will compute split flow at the Linda levee. Run the model, and determine the maximum discharge that will overtop the levee.

Figure P15.1 Map of the Yuba River and its levees in the vicinity of Marysville and Linda, California.

```
C
C      4
C    .492 WESTERN PACIFIC RAILROAD BRIDGE
C    .631 E STREET/HIGHWAY 70 BRIDGE
C   1.042 SOUTHERN PACIFIC RAILROAD BRIDGE
C   1.681 SIMPSON LANE BRIDGE
T1       ADVANCED HEC-2 SPLIT-FLOW PROBLEM
T2       LINDA LEVEE OVERFLOW PROBLEM - 500-YR FLOOD
T3       YUBA RIVER (START BASED ON CONFLUENCE CONTROL)
J1     0     2                6                      75
JR     0    73  50000   75.3 100000   77.2 150000   78.8 200000   80.3
JR245000   81.6
J2     1          -1
J3    38    43     1     3    10    39    63     2
NC    .1    .1   .04    .1    .3
QT     2 245000 280000
X1   .27    14  2440  2730  1800  2100  2100
X2         80.6
GR    81     0    81    50    50   240    50   400    60   500
GR    61  1600    70  2180    60  2320    60  2400    50  2440
GR    40  2530    40  2710    50  2730    65  2770
X1   .49    14  2140  2650  1200  1250  1200
GR    80     0    60    50    55   650    62   900    60  1140
GR    60  2020    50  2140    40  2260    40  2540    50  2580
GR    55  2650    60  2750    75  2790    78  2800
X1  .491    59  2100  2750    10    10    10
GR    80     0    60    50    52    73    77  73.1    77  86.9
GR    52    87    52   233    77 233.1    77 246.9    52   247
GR    60   393    77 393.1    77 406.9    60   407    60   480
GR    60   525    77 525.1    77 534.9    60   535    60   630
GR    78 630.1    78 639.9    60   640    60   740    78 740.1
GR    78 749.9    60   750    60   850    78 850.1    78 859.9
GR    60   860    60   960    78 960.1    78 969.9    60   970
GR    60  1070    78 1070.1   78 1079.9   60  1080    60  1120
GR    78  1130    78  2000    60  2100    45  2150    45  2192
GR    73 2192.1   73 2207.9   45  2208    45  2392    73 2392.1
GR    73 2407.9   45  2408    45  2592    73 2592.1   73 2606.9
GR    45  2607    55  2660    60  2750    78  2800
X1  .492                      20    20    20
X2          81
BT     8          80    80    50    80  77.1   407    79  77,1
BT   630    79  78.1  2000    78    78  2000    78  73.1  2750
BT    78  73.1  2800    78    78
X1  .493    14  2140  2650    10    10    10
GR    80     0    60    50    55   650    62   900    60  1140
GR    60  2020    50  2140    40  2260    40  2540    50  2580
GR    55  2650    60  2750    75  2790    78  2800
X1   .63    11  1800  2200   300   760   760
GR    81     0    60    80    50   200    60   500    50  1750
GR    40  1800    38  1900    38  2150    40  2200    50  2220
GR    86  2360
X1  .631    20  1800  2170    10    10    10
X2          81
GR    81     0    60    80    50   200    60   500    90 500.1
GR    90 519.9    50   520    53  1300    95 1300.1   95 1319.9
```

GR	53	1320	50	1720	40	1800	38	2000	99	2000.1
GR	99	2009.9	38	2010	40	2170	50	2200	86	2360
X1	.632				40	40	40			
X1	.633	11	1800	2200	10	10	10			
GR	81	0	60	80	50	200	60	500	50	1750
GR	40	1800	38	1900	38	2150	40	2200	50	2220
GR	86	2360								
X1	.80	13	1550	1710	550	920	920			
X2		81								
GR	81	0	60	80	55	320	60	600	50	1050
GR	50	1340	45	1370	40	1520	38	1550	38	1670
GR	40	1710	50	1810	86.5	1930				
X1	.96				700	850	850	1.24		
X2		81								
X1	1.04	12	3100	3230	1200	200	200			
GR	81.5	0	75	30	60	600	65	1900	50	2370
GR	50	3000	40	3100	35	3125	35	3200	50	3230
GR	75	3300	86.5	3430						
X1	1.041	50	3100	3230	10	10	10			
GR	81.5	0	81.5	450	60	600	60	841	80	841.1
GR	80	869.9	60	870	60	1100	80	1100.1	80	1129.9
GR	60	1130	60	1400	80	1400.1	80	1429.9	60	1430
GR	60	1650	80	1650.1	80	1679.9	60	1680	65	1875
GR	65	2070	80	2070.1	80	2089.9	65	2090	60	2280
GR	80	2280.1	80	2299.9	60	2300	60	2490	80	2490.1
GR	80	2509.9	60	2510	60	2675	55	2870	76.5	2870.1
GR	76.5	2889.9	53	2890	50	3000	76.5	3000.1	76.5	3019.9
GR	50	3020	40	3100	35	3120	76.5	3120.1	76.5	3129.9
GR	35	3130	35	3200	50	3230	75	3300	86.5	3430
X1	1.042				20	20	20			
X2		81.5								
BT	6	0	81.5	81.5	450	81.5	81.5	450	81.5	80
BT	2675	81.5	80	2675	81.5	76.5	3430	81.5	76.5	
X1	1.043	12	3100	3230	10	10	10			
GR	81.5	0	75	30	60	600	65	1900	50	2370
GR	50	3000	40	3100	35	3125	35	3200	50	3230
GR	75	3300	86.5	3430						
X1	1.3	9	4040	4440	1600	1300	1370			
X2		81								
GR	80.0	0	65.0	2900	62.0	3930	67.0	4040	32.0	4120
GR	36.0	4270	57.0	4440	59.0	5600	86.0	5700		
X1	1.58	10	5320	5780	1600	1500	1500			
X2		81								
GR	80.5	0	70.0	800	65.5	3300	65.0	5260	69.0	5320
GR	25.0	5430	67.0	5780	54.0	6800	61.0	7300	85.0	7400
X1	1.68	12	5320	5730	530	530	530			
GR	81		70	880	65.5	3300	65	5260	69	5320
GR	30	5420	30	5600	69	5730	65	5830	55	6800
GR	62	7300	85.5	7400						
SB	1.25	1.5	1.8		180	8	15000-	2.2	30	30
X1	1.681				35	35	35			
X2		81	1	83	55					
BT	12	0	81	91	880	70	70	3300	65.5	65.5
BT	3900	65.5	65.5	4600	67	65	5320	84.5	81.5	5520
BT	86	83	5730	84.5	81.5	5830	82	65	6800	55

BT	55	7300	62	62	7400	85.5	85.5			
X1	1.682				50	50	50			
X1	1.96	12	7040	7760	435	935	765			
X2		81.5								
GR	81.6		70	400	65	3200	68	4500	67	5900
GR	71	7040	32	7380	58	7540	62	7700	74	7760
GR	64	8160	86	8300						
X1	2.27	10	7420	7840	1500	1900	1800			
X2		82								
GR	81.9		69	2000	67.5	4500	71.5	7420	37.5	7600
GR	41	7750	70.5	7840	69.5	8120	64	8260	85	8400
X1	2.72	12	6560	7340	1800	2000	2000			
X2		82.5								
GR	82		70	1300	67.5	4600	71.5	6560	49	7140
GR	28	7220	73	7340	71	8360	82	8420	64	8460
GR	64.5	8500	87	8700						
X1	3	15	7340	8400	1930	1750	1800			
X2		83.5								
GR	82		70	600	73.5	4900	73	7340	39.5	7400
GR	39.5	7510	48	7630	69.5	7680	55.5	7830	55.5	8150
GR	75.5	8400	68	8820	78	9200	78	9660	87	9700
X1	3.38	15	7920	8460	2400	1500	2000			
X2		84.4								
GR	81.2		75	2000	70	3300	75	4800	74.5	5100
GR	76	5610	63.5	5920	76	6400	75.5	7580	79	7920
GR	34.5	8180	76	8460	67	9690	87.5	9780	99	10500
EJ										

ER

Figure P15.2 Input file, without split-flow data, for a Yuba River HEC-2 model.

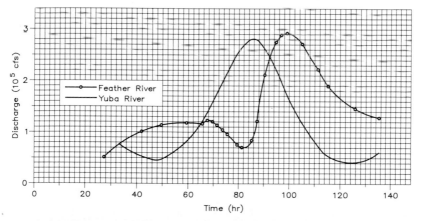

Figure P15.3 Concurrent 500-year-flood hydrographs for the Yuba and Feather rivers at a location immediately above their confluence.

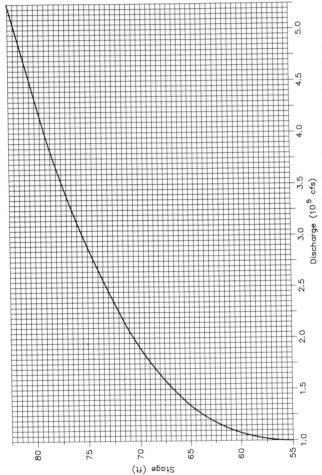

Figure P15.4 Rating curve for the Feather River immediately below the mouth of the Yuba River.

TABLE P15.1   Stations and Elevations, Top of Linda Levee

River mile	Levee station	Levee elevation
0.000	0	80.4
0.270	650	80.6
0.492	1400	81.0
0.631	1500	81.0
0.800	1650	81.0
0.960	2350	81.0
1.042	4600	81.5
1.300	6400	81.0
1.580	9100	81.0
1.681	9900	81.0
1.930	10900	81.5
2.270	12300	82.0
	12800	82.0
2.720	13000	82.5
3.000	15000	83.5
3.380	17500	84.4

References

1. Simons, Li & Associates, Inc., *Engineering Analysis of Fluvial Systems,* Fort Collins, Colo., 1982.
2. Robert D. Jarrett, "Hydraulics of High-Gradient Streams," *Journal of Hydraulic Engineering,* ASCE, vol. 110, No. 11, November 1984.
3. Ven T. Chow, *Open Channel Hydraulics,* McGraw-Hill, New York, 1959.
4. Hydrologic Engineering Center, *Application of the HEC-2 Split Flow Option,* Training Document 18, U.S. Army Corps of Engineers, Davis, Calif., April 1982.
5. D. L. Fread, "National Weather Service Operational Dynamic Wave Model," *Verification of Mathematical and Physical Models in Hydraulic Engineering,* Proceedings of the 26th Annual Hydraulics Division, Special Conference, ASCE, College Park, Md., 1978.
6. *Ice Engineering,* EM 1110-2-1612, U.S. Army Corps of Engineers, Office of the Chief of Engineers, October 1982.
7. Ernest Pariset, Rene Hausser, and Andre Gagnon, "Formation of Ice Covers and Ice Jams in Rivers," *Journal of the Hydraulics Division,* ASCE, vol. 92, no. HY6, November 1966.

Conversion Factors

Length:
1 mm = 0.0394 in
1 m = 3.28 ft
1 km = 0.621 mi

Area:
1 m^2 = 10.764 ft^2
1 km^2 = 0.386 m^2
1 ha = 2.471 acres

Volume:
1 m^3 = 35.31 ft^3
     = 1.308 yd^3
     = 0.81×10^{-3} acre-feet
1 L = 0.264 gal

Weight:
1 kg = 2.205 lb

Velocity:
1 m/s   = 3.281 ft/s
1 km/h = 0.621 mi/h

Pressure:
1 kg/cm^3 = 14.22 lb/in^2
          = 0.968 atm
          = 98.07 kPa

Flow:
$1 \text{ m}^3/\text{s}$ = 35.31 ft^3/s

SOURCE: Adapted from Jack J. Fritz, *Small and Mini Hydropower Systems*, McGraw-Hill Book Company, New York, 1984.

Index

Alluvial streams in floodway analysis, 443
Annual series versus partial-duration series, 186–187
Arithmetic mean:
  defined, 180
  of gaged data, 20

Balanced design storm, 218–220
Base flow, 4, 60–62
Basin hydrology simulation, 3–6
Basin models for rivers (see River basin models)
Basin rainfall (see Rainfall analysis)
Basin subdivision, 136–139
Basin urbanization (see Urbanizing basin analysis)
Belokon-Sabaneev formula, 486
Breaks in records, 184
Bridge encroachments, 432–433
Bridge-flow analysis:
  combined flow, 386–387, 404–406
  externally computed bridge losses, 412–413
  introduction to, 383–384
  low flow, 385–386, 397–400
  method selection for, 413–419
  nature of flow, 384–385
  normal-bridge method:
    cross section location, 388–390
    example (culvert analysis), 390–396
    input data validity, 371–372
    output for, 397
    overview of, 383–384, 413
    selection of, 413–419
    structure definition, 387–388, 396–397
    workshop, 421–422

Bridge-flow analysis (Cont.):
  pressure flow, 386–387, 400–402
  special-bridge method:
    combined flow, 404–406
    cross section location, 406–407
    example, 407–410
    input data validity, 372–373
    low-flow classification, 397–400
    output for, 411–412
    overview of, 384, 397, 413
    pressure flow, 400–402
    selection of, 413–419
    supercritical-flow analysis example, 469–475
    weir flow, 402–403
    workshop, 419–421
  weir flow, 386–387, 402–403
Building modeling in HEC-2, 370–371, 444–445, 495–497

Calibration concept in HEC-1, 117, 139–140
Catchment roughness factors, 260
Central limit theorem, 182
Channel cross section selection in HEC-1, 263–264
Channel improvement analysis in HEC-2:
  channel improvement (CI) line, 445–448
  example, 451–454
  introduction to, 445
  modification of channel reach length, 451
  multiple channel improvements, 448–451
  use of other options, 454–455
  volume of excavation, 453, 454
  workshop, 458–461

Channel improvement (CI) line in HEC-2, 445–448
Channel routing, 76–79, 136, 141–142
CHIMP option in HEC-2 (*see* Channel improvement analysis in HEC-2)
CI (channel improvement) line in HEC-2, 445–448
Clark synthetic unit hydrograph method, 49–56, 62–66, 121, 123, 251–252
Class A, B, and C low flow, 385–386, 398–400
Class frequency, 177
CNs (curve numbers), runoff, 30–37, 62–66, 121, 123, 250
Coefficient of variation, 181
Collector channels in HEC-1, 260–262, 265–266
Combined flow at bridges, 386–387, 404–406
Combining, hydrograph, 136, 141
Composite roughness parameter, 332–334
Computed water surface elevation in HEC-2, 362
Computer simulation (*see* Simulation programs)
Confidence limits, 199–200, 203
Continuous probability distribution, 176–177
Continuous simulation model for precipitation runoff, 7–8
Contraction coefficient, 334, 369
Conversion factors for units, 505
Conveyance values in HEC-2, 363
Coriolis coefficient, 288–289
Correlation techniques (*see* Regression analyses)
Critical depth, 288–292, 319, 463–465
Cross sections in HEC-2, 321, 335–339, 370–371, 388–390, 406–407
Culvert analysis, 390–396, 400–402, 417
Cumulative frequency distributions, 178–179
Curve numbers (CNs), runoff, 30–37, 62–66, 121, 123, 250

Dam analysis, 97, 417
Data storage system (HECDSS), 96
Depth-area option of HEC-1:
  computational procedure, 226–227
  concept, 225–226
  example (Never Dry River basin), 228–233

Depth-area option of HEC-1 (*Cont.*):
  overview of, 98–99
  workshop, 244–248
Design storms:
  balanced, 218–220
  data for:
    description of, 216–219
    sources of, 215–216
    discharge-frequency curve development using, 234–235
    introduction to, 215
  PMS (*see* Probable maximum storm)
  procedures for developing:
    Davis, California example, 222–226
    general, 219–220
    Texas panhandle example, 220–223
  workshop, 244–248
  (*See also* Depth-area option of HEC-1)
Deterministic models versus stochastic models, 173–174
Direct runoff, 4
Discharge for water surface profile computations, 322, 363
Discharge-frequency curve development using design storms, 234–235
Discharge frequency determinations, 11–13
Discharge hydrographs, 4–6
Discrete probability distribution, 176
Diversion simulation in HEC-1, 97, 142–143
Dynamic-wave equations, 71, 89–90

Effective-flow areas in HEC-2, 319–320, 388–389, 391–394
Encroachment analysis (*see* Floodway determination in HEC-2)
Energy principles in water surface profile computations (*see* Water surface profile computations, energy principles)
Energy slope in HEC-2, 363
English units:
  conversion to metric units, 505
  in HEC-1, 95–96
  in HEC-2, 318
Equivalent roughness parameter, 331–332
Exceedance frequency distributions, 179–180
Expansion coefficient, 334, 369
Exponential loss function, 29–30, 121, 123

Flood characteristics, 1
Flood damage determination, 10–11
Flood Flow Frequency Analysis Program
  (see HECWRC Flood Flow Frequency
  Analysis Program)
Flood Hydrograph Package (see HEC-1
  Flood Hydrograph Package)
Flood overflow situations, 441–443
Flood risk, 173
Flood routing analysis:
  concept of, 69
  HEC-1 parameter optimization option
    and, 127
  methods for, 70–72, 89–90
  purposes of, 69–70
  workshop, 90–92
  (See also Modified Puls method;
    Muskingum method; Routing)
Flood volume frequency analysis,
  205
Floodplain hydrologic and hydraulic
  (H&H) analysis:
  application of, 10–13
  basin hydrology simulation, 3–6
  flood hydrograph analysis (see HEC-1
    Flood Hydrograph Package)
  frequency analysis (see Frequency
    analysis; HECWRC Flood Flow
    Frequency Analysis Program)
  introduction to, 1–3
  simulation programs for (see Simula-
    tion programs)
  water surface profile analysis (see
    HEC-2 Water Surface Profiles
    Program; Water surface profile
    computations)
Floodway determination in HEC-2:
  encroachment methods:
    bridge encroachments, 432–433
    method 1, 427–429
    method 2, 431
    method 3, 432
    method 4, 429–430
    method 5, 430–431
    method 6, 431
    overview of, 427
  example (Cow Creek), 435–440
  existing condition analysis, 426–427
  guidelines for, 423–424
  input data validity, 373–374
  introduction to, 423
  modeling procedures, 424–425
  modulating flow boundaries,
    425–426

Floodway determination in HEC-2
  (Cont.):
  problems encountered in:
    alluvial streams, 443
    developed floodway, 444–445
    flood overflow situations, 441–
      443
    high-velocity streams, 443–444
    low-gradient streams, 441
    overview of, 440–441
  workshop, 455–458
Frequency analysis:
  data required for:
    data adjustments, 184–186
    sources of, 182–184
  flood volume, 205
  fundamentals of:
    frequency distributions, 177–183
    probability theory, 175–177
    sampling theory, 174–175
  HECWRC (see HECWRC Flood Flow
    Frequency Analysis Program)
  introduction to, 6–8, 173–174
  Log Pearson Type III distribution for
    (see Log Pearson Type III distribu-
    tion)
  regional (see Regional frequency
    analysis)
  STATS, 205
  streamflow (see Streamflow frequency
    analysis)
  of ungaged basins:
    breakdown of techniques, 205
    field reconnaissance at ungaged
      locations, 211
    precipitation-runoff simulation,
      210–211
    selection of a technique, 207, 210
  in urbanizing areas:
    effects of urbanization, 270
    overview of methods, 270–271
    period-of-record rainfall-runoff
      simulation, 273
    reconstructed historical streamflows,
      271–273
    regression relationships, 273–275
    statistical analysis of historical
      streamflows, 271
  workshop, 211–213
Frequency distributions, 177–183
Friction loss in a channel reach,
  299–302, 320
  (See also Manning's roughness
    coefficient)

Froude number, 287–288, 463

Gaussian distribution, 181–183
Gradually varied flow, rapidly varied
    flow versus, 286–287
Green-Ampt loss function, 26, 27
Group A, B, C, and D soils, 31–32

H&H analysis (*see* Floodplain hydrologic
    and hydraulic analysis)
Hazen plotting positions, 188–189
HEC (Hydrologic Engineering Center)
    programs (*see* Simulation programs)
HEC-1 Flood Hydrograph Package:
    base flow simulation in, 60–62
    calibration concept in, 117, 139–140
    capability of, 96–97
    computational sequence in, 144
    dam safety analysis in, 97
    data storage system (HECDSS) for,
        96
    depth-area option (*see* Depth-area
        option of HEC-1)
    design storm generation with, 215, 225
        (*See also* Depth-area option of
        HEC-1)
    diversion simulation in, 97, 142–143
    HEC-2 storage-outflow data input to,
        70–71, 79–80, 495–497
    hydrograph generation approaches in:
        kinematic-wave method (*see*
            Kinematic-wave rainfall-runoff
            analysis, in HEC-1)
        overview of, 45
        unit hydrograph methods (*see*
            Synthetic unit hydrograph
            methods)
    hydrologic routing methods in:
        modified Puls (*see* Modified Puls
            method)
        Muskingum (*see* Muskingum
            method)
        overview of, 72
    hyetographs in, 24, 45, 47, 49
    inputs:
        example, 106–109
        input file format, 100–101
        input file organization, 101–105
        model parameters, 100
        options, 105–106
        for parameter estimation, 121–123
        units, 95–96
    introduction to, 14, 95
    limitations of, 268–270

HEC-1 Flood Hydrograph Package
        (*Cont.*):
    loss functions in (*see* Loss functions)
    multiplan and multiflood options in
        (*see* Multiplan-multiflood analysis
        in HEC-1)
    outputs:
        example, 111–116
        general, 109
        input file listing, 109, 110
        intermediate simulation results, 109,
            111
        for parameter estimation, 122–126
        summary output, 111
    parameter estimation:
        guidelines for, 125, 127
        inputs for, 121–123
        introduction, 97–98, 116–117
        objective function computation,
            119–120
        optimization methodology, 117–
            118
        outputs for, 122–126
        for river basin models, 139–140
        routing parameters, 127
        univariate gradient search proce-
            dure, 120–121
        workshop, 130–133
    pumping option in, 97
    rainfall data overview, 24
    river basin component definition in,
        99, 100
    river basin modeling with (*see* River
        basin models)
    simulation process in, 99–100
    snowfall and snowmelt simulation in,
        97
    stack principle in, 144
    storage-outflow analysis, 70–71, 79–80,
        495–497
    stream-network diagrams in, 102, 109,
        110
    units in, 95–96
    workshops:
        Clark unit hydrograph, 62–66
        depth-area option, 244–248
        multiplan-multiflood analysis,
            279–281
        multiple-subbasin model, 166–170
        parameter estimation, 130–133
        regionalization of unit hydrograph
            parameters, 164–166
        SCS loss rate, 62–66
        single-basin model, 127–129

HEC-2 Water Surface Profiles Program:
  assumptions in, 9, 284
  bridge analysis with (*see* Bridge-flow
    analysis)
  building modeling in, 370–371,
    444–445, 495–497
  capabilities of, 317–318
  channel improvement option in (*see*
    Channel improvement analysis in
    HEC-2)
  conservation of energy in, 9–10,
    297–298
  critical depth in, 288–292, 319,
    463–465
  cross sections in, 321, 335–339,
    370–371, 388–390, 406–407
  data required for:
    contraction coefficient, 334
    cross sections, 335–339
    discharge, 322
    expansion coefficient, 334
    flow regime, 322
    overview of, 321–322
    reach lengths, 339–340
    roughness coefficient, 324, 327–334
    starting water surface elevation,
      322–326
    validity of, 340–343
  effective-flow areas in, 319–320,
    388–389, 391–394
  encroachment analysis in (*see*
    Floodway determination in HEC-2)
  floodway determination in (*see*
    Floodway determination in HEC-2)
  friction loss computations in, 320
  HEC-1 use of storage-outflow data
    generated by, 70–71, 79–80,
    495–497
  ice-covered flow option in (*see* Ice-
    covered stream analysis in
    HEC-2)
  inputs:
    error checking of, 351–352
    example, 345–351
    input file format and organization,
      344–345
    sensitivity analyses for, 375–376
    units, 318
    validity of, 340–343
  interpolated cross sections in, 321
  introduction to, 317–318
  Manning's roughness coefficient
    computation in, 324, 331,
    493–495

HEC-2 Water Surface Profiles Program
    (*Cont.*):
  multiple profiles in, 318
  output analysis:
    changes in key variables, 362
    computed water surface elevation,
      362
    conveyance, 363
    discharge, 363
    energy slope, 363
    example (Bison Creek), 363–368
    importance of, 355
    overview of, 362
    top width, 362–363
    velocity, 363
  outputs:
    comments, 360
    control options for, 356
    cross-section data, 356–359
    flow distribution, 360–361
    labels, 356
    plots, 359–362
    program banner, 355–356
    special notes, 358
    standard, 355
    tables, 359–361
    traces, 361
  overview of, 14–15
  pressure distribution in a channel
    cross section, 295–296
  project review:
    importance of, 355
    introduction to, 366, 368
    questions for, 369–374
    sensitivity of data, 375–376
    steps in, 369
    verification of model, 374–375
  rapidly varied flow versus gradually
    varied flow in, 286–287
  sensitivity analyses, 375–376
  split-flow analysis in (*see* Split-flow
    analysis in HEC-2)
  standard step method of (*see* Standard
    step method)
  steady flow versus unsteady flow in,
    284–285
  storage-outflow analysis, 70–71, 79–80,
    495–497
  subcritical flow versus supercritical
    flow in, 287–288, 291–292
  supercritical flow analysis in (*see*
    Supercritical flow, HEC-2 analysis
    of)
  theoretical basis of, 9–10, 284

HEC-2 Water Surface Profiles Program
(*Cont.*):
tributary stream profile computations
in, 492–494
uniform flow versus varied flow in,
285–287
units in, 318
velocity distribution in a channel cross
section, 292–295
verification of model, 374–375
workshops:
channel improvement, 458–461
floodway determination, 455–458
input file construction and execution,
352–353
normal-bridge method, 421–422
output data evaluation, 376–381
special-bridge method, 419–421
split-flow analysis, 497–503
standard step method hand calculations, 310–315
(*See also* Water surface profiles
computations)
HECDSS (data storage system), 96
HECWRC Flood Flow Frequency
Analysis Program:
capabilities of, 203–204
example problem (Fishkill Creek):
inputs, 204–205
outputs, 205–209
input data, 204
overview of, 15, 203
workshop, 211–213
HMR52 Probable Maximum Storm
program (Eastern United States):
example (watershed above Jones
Reservoir), 240–246
hyetographs in, 237–244
overview of, 16
program description, 237, 239
Holtan loss function, 37–39, 121,
123
Homogeneity of frequency analysis data,
187–188
Hydraulic radius, 295, 485
Hydraulic routing, 70–71, 89–90
Hydrograph combining, 136, 141
Hydrographs:
discharge, 4–6
IUH, 49–56
unit (*see* Unit hydrograph analysis)
(*See also* HEC-1 Flood Hydrograph
Package)

Hydrologic Engineering Center (HEC)
programs (*see* Simulation programs)
Hydrologic and hydraulic (H&H)
analysis (*see* Floodplain hydrologic
and hydraulic analysis)
Hydrologic routing, 70–72, 89–90, 127
(*See also* Modified Puls method;
Muskingum method)
Hyetographs:
in HEC-1, 24, 45, 47, 49
in HMR52, 237–244

Ice-covered stream analysis in HEC-2:
Belokon-Sabaneev formula, 486
computational procedures, 490–491
hydraulic parameters, 484–485
ice-cover roughness, 486
ice-cover thickness, 485
ice jams, 487–490
introduction to, 483–484
Pariset's ice stability function, 488–490
program input, 491–492
program output, 492
Incomplete records, 184
Infiltration, 25
(*See also* Loss functions)
Initial loss rates, 26–29, 123
Instantaneous unit hydrograph (IUH),
49–56
Interpolated cross sections in HEC-2, 321
Isohyetal method, 21–22
Isopluvial maps, 216, 217
IUH (instantaneous unit hydrograph),
49–56

Kinematic-wave approximation, 71
Kinematic-wave equations, 255, 257
Kinematic-wave rainfall-runoff analysis:
concept of, 254–256
in HEC-1:
accuracy of finite-difference solution,
268–270
channel cross section selection,
263–264
collector channels, 260–262, 265
computational increment selection,
268–270
example (Waller Creek basin),
264–268
main channels, 261, 263, 266
overland-flow planes, 258–260, 265
overview of, 257–259
kinematic-wave equations, 255, 257

Kinematic-wave rainfall-runoff analysis
    (*Cont.*):
  limitations of, 255

Levees:
  confined flow within, 319–320
  effect on routing, 78
  HEC-1 pumping option for, 97
  split-flow analysis workshop (Linda
      Levee), 497–503
Linear regression, 159–160
Log Pearson Type III distribution:
  adjustment for expected probability,
      199–202
  confidence limits, 199–200, 203
  outliers, 197–200
  parameters of, 193–195
  skew coefficients, 193–198
  (*See also* HECWRC Flood Flow
      Frequency Analysis Program)
Lognormal distribution, 182
Looped storage-outflow relationships,
    82–83
Loss functions:
  initial and uniform, 26–29, 121,
      123
  nonuniform:
    exponential, 29–30, 121, 123
    Green-Ampt, 26, 27
    Holtan, 37–39, 121, 123
    SCS, 30–37, 62–66, 121, 123,
        250
  parameter adjustment with frequency
      curves, 234
  regionalization of parameters (*see*
      Regional frequency analysis)
  urbanization effects on, 249–250
Low flow at bridges, 385–386, 397–400
Low-gradient streams in floodway
    analysis, 441

Main channels in HEC-1, 261, 263, 266
Manning's equation, 255, 257, 294, 295
Manning's roughness coefficient:
  in channel improvement analysis, 446
  composite, 332–334
  equations for, 255, 257, 294, 295
  equivalent for, 331–332
  HEC-2 computation of, 324, 331,
      493–495
  for ice cover, 486
  project review of values, 369
  values for, 324, 327–331, 486

Mean (arithmetic):
  defined, 180
  of gaged data, 20
Meandering stream models in HEC-2,
    337–338
Median plotting positions, 188–191
Metric units:
  conversion to English units, 505
  in HEC-1, 95–96
  in HEC-2, 318
Mixed populations, 185
MLRP (*see* Multiple Linear Regression
    Program)
Modified Puls method:
  applicability of, 89–90
  channel storage routing, 76–79
  HEC-1 inputs, 142
  number of routing steps, 82
  reservoir storage routing, 72–76
  storage-outflow relationships, 79–82
  for urbanizing basins, 254
  workshop, 91–92
Momentum balance, 398
Mountains and basin subdivision, 137
Multiplan-multiflood analysis in HEC-1:
  example, 278–279
  multiflood ratios, 275–276
  multiplan analysis, 277–278
  overview of, 98
  workshop, 279–281
Multiple Linear Regression Program
    (MLRP):
  capabilities of, 160
  example regression analysis using,
      161–162
  overview of, 15
  workshop, 164–166
Multiple profiles in HEC-2, 318
Multiple regression, 160
Muskingum method:
  example of, 88–89
  HEC-1 inputs, 141–142
  HEC-1 parameter optimization option
      and, 127
  looped storage-outflow relationship,
      82–83
  number of subreaches, 87–88
  parameters for gaged reaches, 84–86
  parameters for ungaged reaches,
      86–87
  routing equation, 83–84
  for urbanizing basins, 253–254
  workshop, 90–91

National Oceanic and Atmospheric
    Administration (NOAA) design
    storm information, 215–216
National Weather Service (NWS):
  design storm information, 215–216
  frequency analysis data, 182, 184
NOAA (National Oceanic and Atmo-
    spheric Administration) design storm
    information, 215–216
Nonuniform loss rates (*see* Loss func-
    tions, nonuniform)
Normal-bridge method (*see* Bridge-flow
    analysis, normal bridge method)
Normal distribution, 181–183
NWS (*see* National Weather Service)

Objective function computation in
    HEC-1, 119–120
Orifice equation, 400–402
Outliers, 185, 197–200
Overland-flow planes in HEC-1, 258–260,
    265

Paired data versus time-series data,
    174
Parameter estimation in HEC-1 (*see*
    HEC-1 Flood Hydrograph Package,
    parameter estimation)
Pariset's ice stability function, 488–490
Partial-duration series, annual series
    versus, 186–187
Peak discharges, statistical analysis of,
    6
Pier shape coefficients, 399
Plotting positions, 188–192
PMF (probable maximum flood), 235,
    240–246
PMP (probable maximum precipitation),
    236–238
PMS (*see* Probable maximum storm)
Precipitation depth-area relationship (*see*
    Depth-area option of HEC-1)
Precipitation runoff:
  analysis of, 6–8
  overview of, 3–4
  simulation of, 210–211
  (*See also* Rainfall runoff)
Pressure distribution in a channel cross
    section, 295–296
Pressure flow at bridges, 386–387,
    400–402
Prism storage, 83–84
Probabilistic models, deterministic
    models versus, 173–174

Probability density function, 176–177
Probability theory, 175–177
Probable maximum flood (PMF), 235,
    240–246
Probable maximum precipitation (PMP),
    236–238
Probable maximum storm (PMS):
  concept of, 235–236
  HMR52 analysis of (*see* HMR52
      Probable Maximum Storm
      Program)
  probable maximum precipitation
      (PMP), 236–238
Project review for HEC-2 analysis (*see*
    HEC-2 Water Surface Profiles
    Program, project review)
Pumping option in HEC-1, 97

Rahway River Basin workshops (*see*
    under Workshop problems)
Rainfall analysis:
  overview of, 19
  spatial distribution of rainfall, 19–
      22
  temporal distribution of rainfall,
      22–24
  workshop, 39–44
Rainfall loss analysis:
  definition of loss, 24–25
  infiltration, 25
  loss functions for (*see* Loss functions)
  parameter adjustment with frequency
      curves, 234
  workshop, 39–44
  (*See also* Rainfall analysis)
Rainfall runoff:
  curve numbers (CNs) for, 30–37,
      62–66, 121, 123, 250
  modeling:
    kinematic-wave method for (*see*
        Kinematic-wave rainfall-runoff
        analysis)
    UH approach for (*see* Unit
        hydrograph analysis)
  overview of, 3–4
  parameter estimation (*see* HEC-1 Flood
      Hydrograph Package, parameter
      estimation)
  (*See also* Precipitation runoff)
Rapidly varied flow versus gradually
    varied flow, 286–287
Reach lengths in HEC-2, 339–340
Red Fox River workshops (*see under*
    Workshop problems)

Red River basin model (*see* River basin
    models, HEC-1 example)
Regional frequency analysis:
  basic steps in, 153, 158
  introduction to, 6, 7, 153
  for ungaged basins, 207, 210
  workshop, 164–166
Regression analyses:
  acceptance of regression analysis, 161
  evaluation of residuals, 162–164
  example regression analysis, 161–162
  linear regression, 159–160
  MLRP (*see* Multiple Linear Regression
    Program)
  multiple regression, 160
  overview of, 158–159
  for urbanizing basins, 249–253,
    273–275
  (*See also* Regional frequency analysis)
Reservoir routing, 72–76, 136, 141–142
Residuals maps, 162–164
River basin components in HEC-1, 99,
  100
River basin models:
  concepts of:
    basin characteristics and model
      constraints, 136–137
    basin subdivision, 136–139
    introduction, 135
    model components, 135–136
    parameter estimation, 139–140
    study scope and objectives, 138–139
  HEC-1 computational sequence, 144
  HEC-1 example (Red River basin):
    basin layout, 144, 145
    input file, 144–153
    model schematic, 144, 145
    outputs, 154–157
  HEC-1 inputs:
    diversion, 142–143
    hydrograph combining, 141
    hydrograph comparison, 143–144
    introduction, 140
    routing, 141–142
    subbasin runoff, 140–141
  regionalization of unit hydrograph and
    loss function parameters (*see*
    Regional frequency analysis)
  workshops (Rahway River Basin):
    multiple-subbasin model, 166–170
    regionalization of unit hydrograph
      parameters, 164–166
    single-basin model, 127–129
River storage routing, 76–79

Roughness coefficients:
  composite, 332–334
  equivalent, 331–332
  Manning's (*see* Manning's roughness
    coefficient)
Roughness factors, catchment, 260
Routing:
  channel, 76–79, 136, 141–142
  reservoir, 72–76, 136, 141–142
  urbanization effects on, 253–254
  (*See also* Flood routing analysis;
    Modified Puls method; Muskingum
    method)
Runoff:
  curve numbers (CNs) for, 30–37,
    62–66, 121, 123, 250
  direct, 4
  overview of, 3–4
  precipitation (*see* Precipitation runoff)
  rainfall (*see* Rainfall runoff)
  subbasin, 135, 140–141
  urbanization effects on, 249–253

S profiles, 464–465
Saint-Venant equations, 71
Sampling theory, 174–175
SCS (*see* Soil Conservation Service)
Sediment deposition modeling in HEC-2,
  319
Short records, 185–186
Simulation models for precipitation
  runoff, 7–8
Simulation programs:
  advances in, 2
  data storage system (HECDSS) for,
    96
  HEC-1 (*see* HEC-1 Flood Hydrograph
    Package)
  HEC-2 (*see* HEC-2 Water Surface
    Profiles Program)
  HECWRC (*see* HECWRC Flood Flow
    Frequency Analysis Program)
  HMR52 (*see* HMR52 Probable Maxi-
    mum Storm program)
  MLRP (*see* Multiple Linear Regression
    Program)
  overview of, 13
  personal computer versions of, 13–14,
    96
  procurement of, 14
  STATS, 16, 205
Single-event simulation model for
  precipitation runoff, 8
Skew coefficients, 181, 193–198

Snowfall and snowmelt simulation in HEC-1, 97

Snyder synthetic unit hydrograph method, 56–57, 251

Soil Conservation Service (SCS):
  loss function, 30–37, 62–66, 121, 123, 250
  synthetic unit hydrograph method, 57–60

Soil Groups A, B, C, and D, 31–32

Spatial distribution of rainfall, 19–22

Special-bridge method (*see* Bridge-flow analysis, special-bridge method)

Specific energy, 288–290

Split-flow analysis in HEC-2:
  computational procedure, 476
  example (Red Fox River), 481–483
  input requirements, 477
  introduction to, 469, 471
  modeling procedures:
    cross section numbering, 477–478
    hydrologic considerations, 478–481
    reach lengths, 478, 479
  program capabilities, 471, 476
  program limitations, 476–477
  workshop, 497–503

Stack principle in HEC-1, 144

Standard deviation, 180–181

Standard step method:
  computational procedure, 301–304
  example hand calculations, 306–310
  overview of, 9–10, 14, 283–284
  tabulation of computations, 304–306
  workshop, 310–315

Starting water surface elevation for water surface profile computations, 322–326

Statistical analysis of peak discharges, 6

Statistical Analysis of Time Series Data (STATS) program, 16, 205

Steady flow versus unsteady flow, 284–285

Stochastic models, deterministic models versus, 173–174

Storage changes and frequency analysis, 187–188

Storage-outflow analysis:
  channel, 76–79, 136, 141–142
  modified Puls method for (*see* Modified Puls method)
  Muskingum method for (*see* Muskingum method)
  reservoir, 72–76, 136, 141–142

Storage-outflow analysis (*Cont.*):
  using HEC-1 and HEC-2, 70–71, 79–80, 495–497

Storm type and frequency analysis, 188

Straddle-stagger method in HEC-1, 127

Stream-network diagrams in HEC-1, 102, 109, 110

Streamflow frequency analysis:
  analysis techniques:
    analytical, 189, 193
    comparison of, 203
    graphic, 188–192
    (*See also* Log Pearson Type III distribution)
  annual series versus partial-duration series, 186–187
  factors affecting homogeneity of data, 187–188
  workshop (Rahway River Basin), 211–213

Subbasin runoff, 135, 140–141

Subcritical flow:
  in floodway analysis, 443
  Miners Creek subcritical-profile example, 466–467
  versus supercritical flow, 287–288, 291–292

Subdivision of a basin, 136–139

Supercritical flow:
  in floodway analysis, 443–444
  HEC-2 analysis of:
    input requirements, 465–466
    Miners Creek subcritical-profile example, 466–467
    Miners Creek supercritical-profile example, 468–470
    program limitations, 465
    special bridge model example, 469–475
  nature of, 463–465
  subcritical flow versus, 287–288, 291–292

Synthetic unit hydrograph (UH) methods:
  Clark, 49–56, 62–66, 121, 123, 251–252
  overview of, 47, 49
  SCS, 57–60
  Snyder, 56–57, 251

Tatum method, 127

Temporal distribution of rainfall, 22–24

Thiessen polygon method, 20–22, 39–44

Time-series data:
  paired data versus, 174
  statistical analysis of (*see* Statistical
      Analysis of Time Series Data
      program)
Top width in HEC-2, 362–363
Tributary stream profile computations in
    HEC-2, 492–494

UH analysis (*see* Unit hydrograph
    analysis)
Undulating flow boundaries, 425–426
Ungaged basin frequency analysis (*see*
    Frequency analysis, of ungaged
    basins)
Uniform flow versus varied flow, 285–287
Uniform loss rates, 26–29, 121, 123
Unit hydrograph (UH) analysis:
  application of, 47, 49
  assumptions for, 46
  computational time interval for, 46
  derivation of, 46–48
  linearity principle for, 45–46
  overview of, 45
  regionalization of parameters (*see*
      Regional frequency analysis)
  synthetic methods for (*see* Synthetic
      unit hydrograph methods)
  time invariance principle for, 45–46
  urbanization effects on, 249–253
U.S. Army Corps of Engineers programs
    (*see* Simulation programs)
U.S. Geological Survey (USGS) frequency
    analysis data, 182, 184
U.S. Water Resources Council (WRC)
    categories for ungaged-basin
    analysis, 205
Units:
  conversion factors for, 505
  in HEC-1, 95–96
  in HEC-2, 318
Univariate gradient search procedure in
    HEC-1, 120–121
Unsteady flow:
  equations for, 71, 89–90
  steady flow versus, 284–285
Urbanizing basin analysis:
  frequency curve development in (*see*
      Frequency analysis, in urbanizing
      areas)
  kinematic-wave method for (*see*
      Kinematic-wave rainfall-runoff
      modeling)

Urbanizing basin analysis (*Cont.*):
  multiplan-multiflood (*see* Multiplan-
      multiflood analysis in HEC-1)
  routing effects, 253–254
  runoff effects, 249–253
USGS (U.S. Geological Survey) frequency
    analysis data, 182, 184

Variance, 180–181
Varied flow, uniform flow versus, 285–287
Velocity changes in HEC-2, 363
Velocity distribution coefficient, 288–290,
    293–295

Water Resources Council (WRC)
    categories for ungaged-basin
    analysis, 205
Water surface elevation:
  computed, in HEC-2, 362
  starting, for water surface profile
      computations, 322–326
Water surface profile computations:
  for bridges (*see* Bridge-flow analysis)
  data requirements for (*see* HEC-2
      Water Surface Profiles Program,
      data required for)
  energy principles:
    energy equation, 297–298
    energy losses, 298–299
    total energy head for a cross section,
        296–297
  friction loss in a channel reach,
      299–302, 320
  overview of, 9–10, 283
  standard step method (*see* Standard
      step method)
  workshop, 310–315
Water Surface Profiles Program (*see*
    HEC-2 Water Surface Profiles
    Program)
Wedge storage, 83–84
Weibull plotting positions, 188–189
Weir analysis, 417
Weir flow at bridges, 386–387, 402–403
Workshop problems:
  channel improvement (Red Fox River),
      458–461
  Clark synthetic unit hydrograph
      (Rahway River Basin), 62–66
  depth-area option of HEC-1 (Rahway
      River Basin), 244–248
  design storm development (Rahway
      River Basin), 244–248

Workshop problems (*Cont.*):
   flood routing (Rahway River Basin),
      90–92
   floodway determination (North Buffalo
      Creek), 455–458
   frequency analysis (Rahway River
      Basin), 211–213
   input file for HEC-2 (Red Fox River),
      352–353
   modified Puls method (Rahway River
      Basin), 91–92
   multiplan-multiflood analysis (Rahway
      River Basin), 279–281
   multiple-subbasin model (Rahway
      River Basin), 166–170
   Muskingum method (Rahway River
      Basin), 90–91
   normal-bridge method for HEC-2,
      421–422
   output data evaluation for HEC-2
      (Baker Creek, Easy Creek, Rocky
      Creek, and Trail Creek), 376–381
   parameter estimation (Rahway River
      Basin), 130–133

Workshop problems (*Cont.*):
   rainfall analysis (Rahway River
      Basin), 39–44
   rainfall loss analysis (Rahway River
      Basin), 39–44
   regionalization of unit hydrograph
      parameters (Rahway River Basin),
      164–166
   SCS loss function (Rahway River
      Basin), 62–66
   single-basin model (Rahway River
      Basin), 127–129
   special-bridge method for HEC-2,
      419–421
   split-flow analysis (Yuba River),
      497–503
   standard step hand calculations (Red
      Fox River), 310–315
WRC (Water Resources Council)
   categories for ungaged-basin
   analysis, 205

Yarnell equation, 398–399

Zero-flow years, 184–185

ABOUT THE AUTHOR

Daniel H. Hoggan is a professor in the Department of Civil and Environmental Engineering and the Utah Water Research Laboratory at Utah State University. His more than 30 years' experience as a civil engineer specializing in water resource planning and allocation includes a position as a hydraulic engineer (visiting scholar) with the U.S. Army Corps of Engineers' Hydrologic Engineering Center from 1985 to 1988, during which time he worked on the development and application of real-time water control software as well as on technical support for HEC-1 and HEC-2 programs. A registered civil engineer in Utah and California, he is a member of the American Society of Civil Engineers, the American Water Resources Association, and the American Society for Engineering Education.